THE FOOD FOREST SOLUTION: STRATEGIC PERMACULTURE DESIGNS FOR EVERY U.S. ZONE VOLUME TWO

BY

AUDRA HALDANE

Disclaimer

While every effort has been made to ensure the accuracy and success of the food forest plans and plant information presented in this book, the author assumes no responsibility for the specific outcomes of any planting or gardening project. It is important to note that the growth, survival, and thriving of plants are influenced by a variety of factors, including local climate conditions, soil health, water availability, and individual care practices. Readers are encouraged to conduct their own research on local growing conditions. Additionally, the edibility and suitability of certain plants may vary depending on individual health conditions, allergies, or sensitivities. Before consuming any plants mentioned in this book, please consult healthcare professionals to ensure their safety for you and your family. The author is not responsible for any adverse effects resulting from the consumption of plants without proper research and consultation. By following the guidelines and planting recommendations provided in this book, readers accept full responsibility for the care and success of their food forest, and the author assumes no liability for any issues that may arise. While the author has used their best efforts in preparing this book, they make no representations or warranties with respect to the accuracy or completeness of the contents of this book and specifically disclaim any implied warranties of merchantability or fitness for a particular purpose. No warranty may be created or extended by sales representatives or written sales materials. The advice and strategies contained herein may not be suitable for your situation. You should consult with a professional when appropriate. The author shall not be liable for any loss of profit or any other commercial damages, including but not limited to special, incidental, consequential, personal, or other damages.

BOOKS IN THIS SERIES

The Food Forest Solution: Strategic Permaculture Designs for Every U.S. Zone Volume One

The Food Forest Solution: Strategic Permaculture Designs for Every U.S. ZoneVolume Two

The Food Forest Solution: Strategic Permaculture Designs for Every U.S. Zone Collection

The Food Forest Solution: Strategic Permaculture Designs for Every Canadian Zone

The Food Forest Solution: Strategic Permaculture Designs For Every European Zone

The Homestead Blueprint: Introductory Self-Sufficiency With Homesteading

Homesteading Mastery: Advanced Skills and Techniques on the Modern Homestead

Homesteading for Total Self-Sufficiency

The Urban and Suburban Homestead

Homesteading and Natural Building Techniques

Finding Your Homestead: A Comprehensive Guide to Purchasing Land in the U.S., Canada, and Europe

CONTENTS

INTRODUCTION

Envision a woodland teeming with fruit, herbs, vegetables, nuts, and healing plants—an ecosystem that requires minimal care, flows in step with nature, and renews itself season after season. Now, picture designing such a landscape in your own yard, customized for your unique climate with a plan that has been thoroughly investigated and expertly devised.

Welcome to *The Food Forest Solution*, Volume Two, the pioneering resource offering exhaustive, zone-tailored food forest designs for every USDA Plant Hardiness Zone in the United States. This volume is more than a manual—it's a call to reimagine our methods of cultivating food and engaging with our surroundings. Grounded in the principles of permaculture, where nature itself is the ultimate architect, this Almanac presents 10 distinct food forest plans for each USDA zone—from the icy realms of Zone 1 to the sultry territories of Zones 12 & 13. Every plan is meticulously developed to mirror the natural cycles of the ecosystem, ensuring each plant collaborates harmoniously with its neighbors—just as nature intends.

A Vision Rooted in Permaculture

At the heart of this book is the concept of permaculture, a design system that mimics the natural ecosystems around us. When we build food forests, we're not just growing food; we're cultivating a regenerative system that supports life on multiple levels—plants, animals, soil, and people alike. This approach allows us to work with nature, not against it, by designing food systems that become more productive and self-sustaining over time.

Permaculture isn't new, but what makes *The Food Forest Solution* groundbreaking is the level of specificity it provides. Each of the seven traditional layers of the food forest—canopy, understory, shrubs, herbaceous plants, groundcover, root crops, and climbers—are represented in every plan. There is a bonus Fungal layer included in some of the zones. Each plant is carefully selected for its hardiness, productivity, and ability to thrive in a particular growing zone. But more than that, these designs are dynamic. They consider the unique needs of each zone: How much sunlight? What kind of soil? How long is the growing season? And most importantly, how can we create a food forest that will endure for generations?

A First Of Its Kind

No other publication has achieved what this Almanac does. It stands as the inaugural collection of food forest plans organized by USDA Plant Hardiness Zones, delivering a detailed, zone-specific approach to permaculture design. For the first time, readers can reference a resource that provides precise, actionable food forest blueprints tailored to their climate. Whether you reside in the frost-bitten expanses of Zone 1 or the tropical embrace of Zone 13, you will discover a guide that resonates with your environmental reality—making this volume an indispensable asset for homesteaders, gardeners, permaculturists, and anyone intrigued by sustainable, regenerative cultivation.

While many guides offer broad gardening advice or general overviews of permaculture, *The Food Forest Solution* raises the bar. Each food forest plan has been rigorously researched to ensure it is not only suitable for your zone but optimized for success. These are not one-size-fits-all solutions; they are finely tuned blueprints addressing the unique challenges and opportunities of every climate. Here, you will find expert insights distilled into a format that is both practical and inspiring.

Though deeply anchored in scholarly research and permaculture principles, this book is also crafted for delight. After all, gardening should be an enjoyable pursuit! *The Food Forest Solution* beckons you to discover the diverse and exhilarating world of food forests. Every plan bears a distinctive, memorable title that sparks your imagination—transforming the process of designing and constructing your own food forest into an engaging and creative adventure.

The language throughout is both accessible and authoritative, ensuring that beginners and seasoned practitioners alike can confidently apply the guidance. No advanced degree in horticulture or permaculture is required—only a passion for nature, an eagerness to learn, and a commitment to cultivating food in harmony with the earth.

Inside this book you'll find:

Comprehensive food forest plans for every USDA Plant Hardiness Zone—from Zone 1 to Zones 12 & 13. Each design spans the full spectrum of permaculture layers, featuring thoughtful companion planting and ecological interconnections.

Step-by-step instructions to bring these designs to life, including detailed plant profiles, companion planting benefits, and strategies for building resilient ecosystems.

A wide array of edible and medicinal species, each selected for its ability to prosper in its designated zone. From fruit and nut trees to shrubs, herbs, vegetables, groundcovers, vines, and root crops—the entire spectrum of your food forest is addressed.

Inspiring design ideas to transform any space into a self-sustaining food forest. Whether you have a modest urban plot or extensive acreage, these plans can be tailored to your specific conditions and objectives.Creative design inspiration for transforming your backyard into a self-sustaining food forest. Whether you have a small urban plot or acres of land, these designs can be adapted to your specific space and goals.

We live in an era where our bond with the planet is more crucial than ever. The way we cultivate our food is a vital component of that bond. Industrial agriculture has disrupted ecosystems, exhausted soils, and accelerated climate change. In contrast, permaculture—and specifically food forests—offers an alternative: a method to produce food that rejuvenates the land, supports biodiversity, and enhances resilience in the face of global change.

The Food Forest Solution Volume Two is not merely a book about growing food. It is a roadmap to creating abundance, fostering community, and deepening our connection with the natural world. It serves as both a practical tool and a wellspring of inspiration. Whether you are an experienced permaculturist or just embarking on your journey towards sustainable living, this volume will be your steadfast companion as you design and cultivate your own food forest.

So, let's begin. Your zone-specific food forest is ready to be designed, and the possibilities are as varied and plentiful as nature itself. With *The Food Forest Solution* Volume Two, you possess the knowledge to build a thriving, regenerative ecosystem—one that will nourish your body, enrich your spirit, and restore the land for future generations. Let's delve into a brief history before we embark on this transformative journey.

FOOD FOREST HISTORY

The term **"food forest"** is rooted in the **permaculture** movement, and it describes a sustainable, self-sufficient agricultural system designed to mimic the structure and function of natural forest ecosystems.

The idea of a food forest is based on the observation that forests are productive, resilient, and require minimal maintenance once established. A food forest is designed to replicate this, but with a focus on plants that provide food, medicine, and other useful products for humans.

Origins of the Term

The term "food forest" was popularized in the late 20th century, but the concept has much older roots, tracing back to ancient human agroforestry systems. Some of the earliest known examples of managed, food-producing forests are in parts of Africa, Asia, and Central America, where indigenous people have cultivated forested landscapes for thousands of years.

However, Robert Hart, a British pioneer of agroforestry, is often credited with coining the term "food forest" or at least bringing it into widespread use in the 1970s and 1980s. Hart developed the concept while experimenting with small-scale permaculture principles in his own garden in Shropshire, England.

He envisioned a low-maintenance, sustainable, and productive garden that mimicked natural woodland ecosystems by using perennial plants in layered designs, much like a forest. His food forest approach focused on using trees, shrubs, and ground-level plants that would provide food and other resources while functioning together as a resilient ecosystem.

Robert A. de J. Hart (1913–2000) was a British farmer and early permaculture advocate who is often recognized as one of the pioneers of the food forest concept. Hart initially managed a commercial fruit farm in Shropshire, but his interest in more sustainable forms of agriculture led him to experiment with agroforestry principles. Inspired by both traditional woodland systems and his brother's experiences of caring for disabled individuals, Hart began to create a garden that would minimize work while maximizing productivity and ecological benefits.

In the 1960s, Hart started his first "forest garden," which featured seven layers of plant life: large fruit and nut trees, smaller trees and shrubs, herbs, ground covers, root crops, and climbing plants. His work attracted attention in the 1970s and became a key inspiration for the permaculture movement, influencing a generation of permaculture designers. Hart's focus was on the holistic benefits of food forests, particularly their ability to support biodiversity, enhance resilience, and reduce the need for inputs like fertilizer and pesticides.

Relationship Between Food Forests and Permaculture

The concept of a food forest is closely tied to permaculture, a design philosophy that seeks to create sustainable, regenerative human habitats by mimicking natural ecosystems. The word "permaculture" itself was coined in the 1970s by Bill Mollison and David Holmgren. It is a combination of "permanent" and "agriculture" (or later, "culture"), indicating a long-term, sustainable approach to agriculture and society.

Permaculture is based on principles of ecology and systems thinking. Its goal is to create self-sustaining systems that require minimal external inputs and maintain themselves over time. A key aspect of permaculture design is the integration of different elements in ways that support each other, much like how plants and animals in an ecosystem interact. The idea of a food forest fits perfectly into permaculture, as it involves layering plants to replicate natural forests while providing food, medicine, and other useful resources to humans.

While permaculture encompasses a wide range of practices, including water management, soil care, and energy efficiency, the food forest is one of its most well-known and visually striking concepts. The layered approach of food forests—often structured around the seven layers of trees, shrubs, ground covers, climbers, and so on—mirrors the way natural forests grow in layers, with each layer supporting the others.

Key Figures

Robert Hart (1913–2000): His experiments with forest gardening and his work promoting agroforestry laid much of the groundwork for modern food forest design.

Bill Mollison (1928–2016): An Australian ecologist and co-founder of the permaculture movement. His work, particularly his 1978 book "Permaculture One," written with David Holmgren, popularized the permaculture philosophy and brought global attention to sustainable agriculture techniques like food forests.

David Holmgren (b. 1955): Co-originator of permaculture, Holmgren's ongoing work continues to influence sustainable design and agriculture, with food forests being a key aspect of permaculture projects worldwide.

Permaculture And Principles

Permaculture is often defined as a design philosophy for creating sustainable human habitats that mimic the patterns and relationships found in natural ecosystems. It was initially developed as an agricultural system but has since evolved to encompass much more, including housing, energy systems, community building, and more.

Some of the core permaculture principles include:

Observe and interact: Learn from nature by observing natural patterns before making design decisions.

Catch and store energy: Use renewable resources like sunlight and water efficiently.

Obtain a yield: Ensure that systems produce food or other resources to meet human needs.

Apply self-regulation and accept feedback: Adjust systems based on their performance.

Use and value renewable resources: Make the most of nature's abundance.

Design from patterns to details: Use larger ecosystem patterns to guide the smaller design decisions.

Integrate rather than segregate: Elements in a system should support each other.

Use small and slow solutions: Smaller, slower systems are often more sustainable and resilient over time.

The concept of a food forest embodies many of these principles, particularly in its focus on mimicking natural ecosystems, valuing biodiversity, and minimizing external inputs.

Through permaculture, food forests have become an integral part of the global movement toward sustainable, regenerative agriculture.

Now let's quickly look at Plant Hardiness Zones.

PLANT HARDINESS ZONES

A Plant Hardiness Zone refers to a geographically defined area where specific categories of plants are most likely to thrive, based on climatic conditions. This system is a vital tool for gardeners, farmers, and foresters, helping them to select plants that are most suitable for the local climate.

The USDA Plant Hardiness Zone Map is the standard by which growers can determine which plants are most likely to survive winter conditions in their region. It divides the United States and other countries into zones based on average annual minimum winter temperatures. The system helps predict which plants will be resilient enough to survive harsh winters and thrive in their respective zones.

History of Plant Hardiness Zones

The concept of Plant Hardiness Zones dates back to the early 20th century, but the USDA (United States Department of Agriculture) introduced the first formal version of the Plant Hardiness Zone Map in 1960. This map was created to assist gardeners in selecting appropriate plant species for their region. Initially based on the average annual minimum temperature, the zones provided a simple, effective way to gauge whether a plant would survive a region's winter.

Since its inception, the USDA has updated the map several times to reflect shifts in climate patterns. For example, the most recent version (released in 2012) accounts for factors like increased urbanization, warming trends, and elevation, providing a more nuanced and accurate representation of growing conditions. With the help of modern technology, this system has become even more precise, allowing growers to adjust planting strategies based on real-time weather trends.

Today, Plant Hardiness Zones are recognized across the globe, although different countries might adapt them to their regional climates. In the United States, this system has become a cornerstone of agriculture and gardening, influencing everything from backyard landscaping to large-scale farming operations.

Breakdown of Plant Hardiness Zones in the United States

Each zone is divided into two sub-zones (a and b) with slight temperature variations. These zones are based on the average annual minimum winter temperature over a period of 30 years, reflecting long-term trends rather than year-to-year variability.

ZONE 1

Temperature Range: -60°F to -50°F (-51°C to -45°C)

Regions: Parts of Alaska

Zone 1a: -60°F to -55°F (-51°C to -48°C)

Zone 1b: -55°F to -50°F (-48°C to -45°C)

Zone 1 represents the coldest areas of the United States, including interior Alaska and extremely northern latitudes. Plants in this zone must be highly resilient to freezing temperatures.

ZONE 2

Temperature Range: -50°F to -40°F (-45°C to -40°C)

Regions: Northern parts of Alaska, parts of the

Yukon, northern Canada

Zone 2a: -50°F to -45°F (-45°C to -43°C)

Zone 2b: -45°F to -40°F (-43°C to -40°C)

Zone 2 includes the harshest climates in the U.S. outside of Zone 1, where only the hardiest plant species can survive, like cold-tolerant perennials and conifers.

ZONE 3

Temperature Range: -40°F to -30°F (-40°C to -34°C)

Regions: Northern Minnesota, Montana, North Dakota, parts of Alaska

Zone 3a: -40°F to -35°F (-40°C to -37°C)

Zone 3b: -35°F to -30°F (-37°C to -34°C)

Zone 3 stretches across northern latitudes of the U.S. and areas with high altitudes. Plants must be adapted to short growing seasons and freezing winter temperatures.

ZONE 4

Temperature Range: -30°F to -20°F (-34°C to -29°C)

Regions: Northern parts of New York, Wisconsin, Michigan, South Dakota, Montana

Zone 4a: -30°F to -25°F (-34°C to -32°C)

Zone 4b: -25°F to -20°F (-32°C to -29°C)

In Zone 4, winter temperatures are still extreme, but a wider variety of plants can thrive here, including certain fruit trees, cold-hardy vegetables, and ornamental shrubs.

ZONE 5

Temperature Range: -20°F to -10°F (-29°C to -23°C)

Regions: Central U.S., including parts of Illinois, Ohio, Pennsylvania, Iowa

Zone 5a: -20°F to -15°F (-29°C to -26°C)

Zone 5b: -15°F to -10°F (-26°C to -23°C)

Zone 5 covers a broad area of the Midwest and Northeast. It allows for more diversity in plants, including many types of perennials, vegetables, and cold-tolerant trees.

ZONE 6

Temperature Range: -10°F to 0°F (-23°C to -18°C)

Regions: Mid-Atlantic, parts of New Jersey, Kentucky, Missouri, Kansas

Zone 6a: -10°F to -5°F (-23°C to -21°C)

Zone 6b: -5°F to 0°F (-21°C to -18°C)

With a milder winter climate, Zone 6 supports a wide variety of deciduous trees, herbs, and cool-season vegetables.

ZONE 7

Temperature Range: 0°F to 10°F (-18°C to -12°C)

Regions: Southern U.S., including parts of Virginia, North Carolina, Arkansas, Oklahoma, and parts of Texas

Zone 7a: 0°F to 5°F (-18°C to -15°C)

Zone 7b: 5°F to 10°F (-15°C to -12°C)

Zone 7 offers a moderate climate, allowing a range of perennials, fruit trees, and subtropical plants to grow.

ZONE 8

Temperature Range: 10°F to 20°F (-12°C to -7°C)
Regions: Southeastern U.S., including Georgia, Alabama, South Carolina, northern Florida, Texas, and parts of Oregon and Washington
Zone 8a: 10°F to 15°F (-12°C to -9°C)
Zone 8b: 15°F to 20°F (-9°C to -7°C)

Zone 8's relatively warm winter makes it ideal for many fruits, nuts, and herbs, and it features long growing seasons.

ZONE 9

Temperature Range: 20°F to 30°F (-7°C to -1°C)
Regions: Southern Texas, southern Florida, parts of Louisiana and California
Zone 9a: 20°F to 25°F (-7°C to -4°C)
Zone 9b: 25°F to 30°F (-4°C to -1°C)

In Zone 9, the climate is warm year-round, making it suitable for a range of subtropical and tropical plants, including citrus and avocado trees.

ZONE 10

Temperature Range: 30°F to 40°F (-1°C to 4°C)
Regions: Southern Florida, parts of southern California, and southern Texas
Zone 10a: 30°F to 35°F (-1°C to 2°C)
Zone 10b: 35°F to 40°F (2°C to 4°C)

Zone 10 experiences little frost, allowing for year-round growth of tropical fruits and many warm-weather vegetables.

ZONE 11

Temperature Range: 40°F to 50°F (4°C to 10°C)
Regions: Southernmost tip of Florida and Hawaii
Zone 11a: 40°F to 45°F (4°C to 7°C)
Zone 11b: 45°F to 50°F (7°C to 10°C)

Zone 11 has no frost, supporting a wide variety of tropical plants, including bananas, pineapples, and palms.

ZONES 12 & 13

Temperature Range: 50°F and above (10°C and above)
Regions: Limited to tropical regions, such as Hawaii and U.S. territories
Zone 12a: 50°F to 55°F (10°C to 13°C)
Zone 12b: 55°F to 60°F (13°C to 16°C)
Zone 13a: 60°F to 65°F (16°C to 18°C)
Zone 13b: 65°F to 70°F (18°C to 21°C)

Zones 12 & 13 are rare in the United States and are primarily found in tropical climates, supporting an extensive range of tropical fruits, flowers, and plants.

Now let's go into the layers of the food forest that you will encounter in the food forest design plans to follow.

FOOD FOREST LAYERS

The layers of a food forest are modeled after the natural structure of forest ecosystems, where plants grow at different heights and serve various ecological functions. Each layer plays a vital role in maximizing space, enhancing biodiversity, and fostering resilience within the ecosystem.

By designing with these layers, food forests optimize vertical space, improve productivity, and promote sustainability and self-sufficiency.

Here is a detailed look at each layer and how they work together in a synergistic way.

Canopy Layer (Tall Trees)

Also called: Overstory, Upper Layer

What it is: The canopy layer consists of the tallest trees in the food forest, typically fruit or nut-bearing trees. These trees form the uppermost layer and provide shade and shelter to the layers below.

Examples:

Fruit trees: apple, pear, cherry, mulberry

Nut trees: walnut, pecan, chestnut

Other trees: oak (for acorns), maple (for sap)

Purpose:

Provide food through fruits and nuts.

Offer shade to lower layers, regulating temperature and moisture.

Create a microclimate that protects sensitive plants from wind and extreme weather.

Support wildlife and beneficial insects.

Synergy: The canopy protects and nurtures the layers below, while its leaf fall contributes organic matter to the soil, fostering a healthy ecosystem.

Understory Layer (Small Trees and Large Shrubs)

Also called: Lower Trees, Sub-canopy

What it is: This layer is made up of smaller, shade-tolerant trees or large shrubs that thrive in the dappled sunlight beneath the canopy.

Examples:

Small fruit trees: dwarf apple, dwarf pear, plum, peach

Nut trees: hazelnut

Berry bushes: elderberry, gooseberry, serviceberry

Purpose:

Produce fruits, berries, and nuts.

Provide habitat and shelter for birds, insects, and other wildlife.

Add diversity to the food forest, increasing resilience.

Synergy: These plants benefit from the partial shade provided by the canopy and, in turn, protect the layers below by further regulating light and temperature.

Shrub Layer (Shrubs and Bushes)

Also called: Bush Layer

What it is: The shrub layer includes smaller bushes and shrubs, many of which are highly productive in terms of fruits and berries. These plants are intermediate in height and can thrive in either the shade of taller layers or in full sun, depending on the design.

Examples:

Berry bushes: blueberry, currant, raspberry, blackberry

Medicinal and aromatic shrubs: lavender, rosemary, sage

Other shrubs: sea buckthorn, autumn olive (often nitrogen-fixers)

Purpose:

Provide fruits, berries, and medicinal herbs.

Fix nitrogen in the soil (e.g., certain species like sea buckthorn and autumn olive), enriching the soil for other plants.

Attract pollinators and beneficial insects.

Synergy: The shrub layer enhances soil fertility and improves pollination for the other layers, while also serving as a buffer for more delicate herbaceous plants below.

Herbaceous Layer (Herb Layer)

Also called: Herb Layer, Ground Layer

What it is: This layer consists of smaller, non-woody plants that grow close to the ground. These can be perennial herbs, vegetables, and wildflowers. Some may die back in the winter, but others persist year-round.

Examples:

Culinary herbs: basil, oregano, parsley, mint

Vegetables: kale, chard, lettuce, spinach, rhubarb

Medicinal plants: comfrey, echinacea, yarrow

Ground cover plants: clover, creeping thyme, chamomile

Purpose:

Provide herbs, vegetables, and medicinal plants for human use.

Improve soil fertility (e.g., clover fixes nitrogen, comfrey mines nutrients).

Attract pollinators and beneficial insects.

Synergy: Herbaceous plants cover the soil, reducing erosion, suppressing weeds, and helping retain moisture. They also attract beneficial insects and pollinators, which help increase the productivity of fruiting layers.

Groundcover Layer

Also called: Ground-Creeping Layer

What it is: The ground cover layer includes low-growing, spreading plants that form a living mulch across the soil, protecting it from erosion and maintaining moisture levels.

Examples:

Low-growing plants: strawberries, creeping thyme, clover, sweet woodruff

Cover crops: vetch, purslane, yarrow

Purpose:

Protect the soil by providing a living mulch.

Conserve moisture by shading the ground.

Suppress weeds and prevent soil erosion.

Some plants add nutrients to the soil (e.g., nitrogen-fixing clover).

Synergy: These plants work in tandem with the taller layers, benefiting from partial shade while protecting the soil and adding fertility.

Root Layer (Rhizosphere)

Also called: Underground Layer, Root Crop Layer

What it is: The root layer consists of plants that grow mainly underground, focusing their energy on developing tubers, roots, and bulbs. These plants are often harvested for their nutritious roots or serve other ecological functions such as improving soil structure.

Examples:

Root vegetables: carrots, potatoes, onions, garlic, beets, radishes

Perennial root crops: Jerusalem artichoke, horseradish, ginger, comfrey

Nitrogen fixers: legumes like lupines

Purpose:

Produce edible roots and tubers for human

consumption.

Improve soil structure by loosening compacted soils with deep roots.

Bring nutrients up from deeper soil layers, making them available to surface plants.

Synergy: The root layer helps aerate the soil and make it more fertile for other layers, while also providing a food source that doesn't compete for sunlight with above-ground plants.

Vertical Layer (Vines and Climbers)

Also called: Vine Layer, Climbing Layer

What it is: The vertical layer consists of climbing plants that use trees, shrubs, or other supports to reach sunlight. This layer maximizes vertical space without competing directly with other plants for soil space.

Examples:

Food-producing vines: grapes, kiwi, passionfruit, hops

Beans: pole beans, runner beans

Other climbers: hardy kiwifruit, cucumbers, climbing roses

Purpose:

Provide fruits, vegetables, or flowers for harvest.

Maximize vertical space without using up additional ground area.

Enhance biodiversity by adding another layer of plant life.

Synergy: Climbers benefit from the support of trees and shrubs while taking advantage of sunlight higher up, making them highly efficient in terms of space usage.

Fungal Layer

Also called: Mycelial Layer

What it is: The fungal layer consists of beneficial fungi, primarily mycorrhizal fungi, that live in the soil and form symbiotic relationships with the roots of plants. Fungi help decompose organic matter, recycle nutrients, and enhance soil health.

Examples:

Edible mushrooms: shiitake, oyster, lion's mane

Decomposing fungi: king stropharia (wine cap mushroom)

Mycorrhizal fungi: Glomus species (beneficial for tree roots)

Purpose:

Break down organic matter, turning it into nutrients accessible to plants.

Form symbiotic relationships with plant roots, increasing nutrient uptake.

Enhance soil health and biodiversity.

Synergy: The fungal layer works below the surface to decompose organic matter and facilitate nutrient exchange between plants and the soil. This critical layer supports the health of all plants in the food forest and adds a dynamic element to the system.

How the Layers Work Together Synergistically

The layers of the food forest create a dynamic and resilient ecosystem, each layer playing a specific role while interacting in synergy.

Resource Sharing: Plants in different layers optimize their use of vertical and horizontal space, ensuring that they don't compete directly for sunlight or nutrients. Deep-rooted plants access nutrients from below, while shallow-rooted plants benefit from surface resources.

Nutrient Cycling: Leaves, fruit, and organic matter fall from the canopy and shrub layers, decomposing with the help of the fungal layer and enriching the soil. Nitrogen-fixing plants enhance the soil naturally without synthetic inputs.

Biodiversity and Pest Control: A variety of plants attracts beneficial insects, pollinators, and wildlife that help manage pests. The diversity of plant species ensures ecological balance.

Water Management: Ground cover plants help retain moisture and prevent erosion. The canopy and shrub layers create a cooler, moister microclimate, reducing water needs.

By mimicking nature, a food forest creates a low-maintenance, highly productive system that supports biodiversity, enhances sustainability, and provides abundant resources for people and wildlife.

THE FOOD FOREST DESIGN PLANS

ZONE 1

Plant Hardiness Zone 1

Plant Hardiness Zone 1 includes some of the most frigid areas in North America, primarily situated in northern Alaska and a few elevated regions in the Rocky Mountains. This zone is known for its intense cold and brief growing periods, making it one of the hardest zones for plant cultivation.

Climate in Plant Hardiness Zone 1

Annual Low Temperature: Zone 1 can experience freezing lows between -60°F and -50°F (-51°C to -45°C). Growing Period: The growing window is notably short, generally spanning between 45 and 90 days. In certain areas, summer may consist of only a few weeks of temperatures above freezing. Daylight: In many sections of Zone 1, particularly at higher latitudes, the short growing period aligns with extended daylight hours, occasionally offering 24-hour sunlight during peak summer, known as the Midnight Sun. This extra sunlight can boost rapid plant growth, offsetting the brief season. Precipitation: Rainfall in Zone 1 can be inconsistent, and much of the moisture is trapped in ice and snow for most of the year. Though summer showers are sparse, they are critical to the short growing season. Coastal proximity can lead to more frequent precipitation, but these regions still endure severe cold.

Soil Characteristics of Plant Hardiness Zone 1

Permafrost: A large portion of Zone 1 sits atop

permafrost, which limits root penetration and disrupts water drainage. The soil remains frozen beneath the surface for most of the year, thawing only in the brief summer.

Poor Drainage: The cold and permafrost contribute to poorly drained soils, creating wet or marshy conditions in some places.

Low Nutrient Levels: The soils in Zone 1 are often nutrient-poor, making it tough for plants to grow unless managed properly, or with hardy native plants that can cope with low-nutrient environments. Acidic Soils: Many places in Zone 1 have acidic soils that favor plants such as blueberries, cranberries, and conifers.

Vegetation in Plant Hardiness Zone 1

Tundra: In the harshest parts of Zone 1, tundra vegetation dominates, featuring hardy, low-growing plants such as sedges, mosses, lichens, and dwarf shrubs. These plants are well-suited to surviving in windy, cold, and nutrient-poor environments.

Boreal Forest (Taiga): In the less extreme areas of Zone 1, boreal forests are common, mainly consisting of cold-tolerant conifers like spruce (Picea), pine (Pinus), and fir (Abies), along with some deciduous trees like birch (Betula) and aspen (Populus).

Alpine Zones: In higher elevation areas of Zone 1, alpine ecosystems exist, resembling tundra conditions and featuring cold-resistant perennials, grasses, and low shrubs.

Difficulties in Growing in Plant Hardiness Zone 1

Severe Cold: Winter temperatures frequently fall well below freezing, with the ground staying frozen for most of the year. Only the most resilient perennials and species adapted to the cold can thrive.

Zone 1

Brief Growing Season: With under three months for plant growth, many standard crops lack sufficient time to fully mature. Growers must concentrate on fast-maturing or perennial plants that can produce early or continue year after year without replanting.

Limited Plant Range: The cold, combined with the short growing season, restricts the variety of plants that can survive in Zone 1. Only species suited to low temperatures, nutrient-poor soils, and limited daylight can thrive, leading to reduced plant diversity compared to warmer zones.

Soil Challenges: The presence of permafrost, poor drainage, and low nutrient content make it hard for many conventional plants to grow. Perennials with shallow roots or those that do well in boggy, acidic soils are better suited to this environment.

Pests and Wildlife: In certain regions, insect pests can be prevalent during the short summer, and wildlife such as moose, caribou, and bears may compete for food sources. Cold-resistant birds and mammals depend on native plants, which may attract them to human-grown crops.

Planting Strategies for Zone 1

Cold-Hardy Perennials: Focus on cold-tolerant perennials that can withstand severe winter conditions and return annually without replanting. Native species adapted to Zone 1's conditions should take priority, such as conifers, dwarf shrubs, and hardy berries.

Creating Microclimates: Utilize methods like windbreaks, snow barriers, and tree lines to shield plants from harsh winds and create microenvironments with slightly warmer temperatures and more favorable growth conditions. Using raised beds and mounds with good drainage can reduce waterlogging, while planting on south-facing slopes can optimize exposure to sunlight and warmth.

Polyculture and Companion Planting: Polyculture mimics natural ecosystems, allowing plants to support one another. For instance, nitrogen-fixing plants like alders can enrich the soil for nearby species. Groundcovers and climbing plants can be used to maximize vertical space and prevent soil erosion.

Mulching and Soil Improvement: Mulching with organic materials like straw or wood chips helps insulate the soil, maintain moisture, and suppress weeds. It also enriches the soil as it decomposes. Consider incorporating compost or manure to enhance soil fertility, as Zone 1 soils tend to be nutrient-poor.

Greenhouses and Season-Extending Techniques:

In the coldest regions of Zone 1, greenhouses or cold frames can lengthen the growing season by shielding plants from frost and providing additional warmth. Use row covers or cloches to safeguard young plants from late frosts or unseasonably cold weather.

Successfully growing in Zone 1 demands selecting appropriate plants, implementing techniques to overcome environmental obstacles, and thoroughly understanding the local ecosystem. By focusing on resilient native plants, utilizing microclimates, and making the most of the short growing period, it is possible to create productive food forests in even the most inhospitable conditions.

Now, here are the 10 unique design plans for your Plant Hardiness Zone 1 food forest. Each one includes all seven traditional layers with two plant options for each layer. Plus, I've added a researched fungal layer specific to Zone 1 for each plan. Enjoy!

ZONE 1 PLAN ONE: CHILLY CANOPY QUEST

This plan is designed to brave the cold with a creative mix of layered edibles. It focuses on maximizing winter yield while reinforcing the ecosystem's resilience against frigid conditions.

1. CANOPY LAYER:

Amur Cork Tree (*Phellodendron amurense*): Though not typically grown for its fruit, the inner bark of the Amur Cork Tree has been used in traditional medicine. This hardy tree provides excellent canopy cover in cold climates and is valued for its ability to suppress weeds and improve soil conditions with its deep-rooted system.

Arctic Crabapple (*Malus baccata*): This crabapple produces small, tart fruits that are edible and can be used in jellies, ciders, or preserves. It is highly frost-resistant and thrives in extreme cold, offering pollination support to nearby apple varieties. Its flowers attract bees, making it a valuable plant for pollinator health.

2. UNDERSTORY LAYER (SMALL TREES):

Alpine Currant (*Ribes alpinum*): Producing small, edible red berries, the Alpine Currant is a hardy understory shrub that tolerates cold temperatures and poor soils. The berries are tart but can be used in jams and jellies, and the shrub provides shelter for wildlife.

Arctic Willow (*Salix purpurea*): This small willow produces young shoots that are edible and can be used as a vegetable in some cultures. It thrives in the cold climates of Zone 1 and provides both erosion control and wildlife habitat, while also offering medicinal properties.

3. SHRUB LAYER:

Arctic Juniper (*Juniperus communis var. saxatilis*):

This low-growing juniper produces edible berries that are used as a spice or in beverages like gin. It is highly cold-tolerant and adds evergreen interest to the shrub layer, providing habitat for wildlife and supporting soil health.

Arctic Raspberry (*Rubus arcticus*): This low-growing shrub produces small, sweet, and highly fragrant raspberries. It thrives in cold climates, making it ideal for the shrub layer in northern food forests. The berries are a delicious and nutrient-rich addition, supporting both human consumption and wildlife.

4. HERBACEOUS LAYER:

Alaska Violet (*Viola langsdorfii*): This cold-hardy violet produces edible leaves and flowers that can be used in salads or as a garnish. It thrives in moist, shaded areas, supporting soil health and attracting pollinators like bees and butterflies to the food forest.

Alpine Arnica (*Arnica angustifolia*): This herbaceous plant produces edible leaves and flowers that are used in traditional medicine. It thrives in cold, alpine environments and supports pollinators, making it a valuable part of the food forest ecosystem. Its bright yellow flowers add beauty to the landscape.

5. GROUNDCOVER LAYER:

Alpine Bearberry Willow (*Salix reticulata*): A creeping groundcover, Alpine Bearberry Willow produces edible young leaves that can be cooked and eaten. It thrives in cold, rocky environments, providing excellent erosion control and stabilizing soils in Zone 1. It supports wildlife by offering habitat and food for local fauna.

Alpine Forget-Me-Not (*Myosotis alpestris*): This charming groundcover is well-suited for cold climates, producing small, edible flowers that can be used as a garnish. It grows well in rocky, moist soils, stabilizing the ground and attracting pollinators with its vibrant blue blossoms.

6. ROOT LAYER:

Alaskan Wild Rhubarb (*Polygonum alaskanum*): The roots of Alaskan Wild Rhubarb are edible and were traditionally consumed by Indigenous communities in the far north. It thrives in cold, wet soils, providing both a valuable food source and a way to stabilize soil in flood-prone areas.

Alpine Bistort (*Bistorta vivipara*): The tuberous roots of Alpine Bistort are edible and can be consumed raw or cooked. This low-growing perennial is highly cold-tolerant, thriving in arctic and alpine environments. It also provides groundcover, prevents soil erosion, and supports pollinators with its small white or pink flowers.

7. VERTICAL LAYER:

Alaska Blueberry Vine (*Vaccinium alpinum*): This unique variety of blueberry grows as a climbing vine, making it suitable for vertical gardening in extremely cold climates. The small blue berries are edible and packed with antioxidants. It thrives in rocky or poor soils, providing valuable ground cover and habitat for pollinators.

Alaska Coral Pea (*Clianthus puniceus var. alaskanus*): This rare vine produces striking red flowers and edible pods. The plant is nitrogen-fixing, improving soil health, and can climb trellises or fences. It's cold-hardy and provides a splash of color and food in challenging climates.

8. FUNGAL LAYER:

Alaska Matsutake (*Tricholoma magnivelare*): This aromatic, edible mushroom is highly valued in Asian cuisine for its distinct flavor. It forms symbiotic relationships with pine and spruce trees, helping them absorb nutrients. It thrives in cold climates, and its presence signifies healthy forest ecosystems.

Angel Wings Mushroom (*Pleurocybella porrigens*): A cold-loving edible mushroom that grows on decaying conifer wood. It is highly valued for its mild taste and tender texture, often used in soups or stir-fries. It helps decompose organic matter, improving soil structure and nutrient availability for other forest plants.

Zone 1 Plan Two: Frostfire Fusion

Combining hardy, cold-tolerant plants in a layered approach, this design plan harnesses the spark of life in icy environments. It emphasizes nutritional yield and a self-sustaining food system even under extreme chill.

1. CANOPY LAYER:

Arctic Willow (*Salix arctica*): The young shoots of this cold-adapted willow are edible and can be used as a vegetable. It grows low to the ground but can still serve as a windbreak and helps stabilize the soil. Its branches are often used in basket weaving and other crafts.

Blue Honeysuckle (*Lonicera caerulea*): Though typically a shrub, this species can grow large in optimal conditions and produces edible berries that are tart and high in antioxidants. It thrives in extreme cold and grows well alongside other hardy species, adding to the productivity of the forest.

2. UNDERSTORY LAYER:

Bearberry (*Arctostaphylos uva-ursi*): Also known as Kinnikinnick, this plant produces small, edible red berries that can be used in teas or jellies. It is a creeping groundcover that tolerates cold temperatures and helps prevent soil erosion, making it a valuable addition to the food forest.

Black Crowberry (*Empetrum eamesii*): Producing edible black berries, the Black Crowberry is well-suited to cold, harsh environments. The berries are rich in antioxidants, and the plant itself provides ground cover and erosion control in the understory, helping stabilize the soil.

3. SHRUB LAYER:

Bearberry Willow (*Salix uva-ursi*): This low-growing shrub produces edible young shoots that can be consumed as a vegetable. It thrives in the cold and helps prevent soil erosion, making it a valuable part of the shrub layer in Zone 1 food forests.

Cloudberry (*Rubus chamaemorus*): Cloudberry is a rare northern plant that produces bright orange, edible fruits known for their high vitamin C content. It thrives in cold, boggy areas and adds a unique fruit to the food forest while supporting pollinators and other wildlife.

4. HERBACEOUS LAYER:

Alpine Bluebell (*Campanula rotundifolia*): Also known as Harebell, this hardy herb produces small, bell-shaped flowers that are edible and can be used in salads. It thrives in cold climates, adding beauty and diversity to the herbaceous layer while supporting pollinators like bees and butterflies.

Alpine Coltsfoot (*Tussilago alpina*): This hardy herb produces edible young leaves and flowers that can be cooked or used in herbal teas. It thrives in cold, rocky environments and provides early-season greens, supporting soil health and attracting pollinators to the food forest.

5. GROUNDCOVER LAYER:

Alpine Speedwell (*Veronica alpina*): This creeping groundcover has small, bright blue flowers and leaves that can be eaten in small amounts in salads. It thrives in cold, rocky environments and helps prevent erosion. Its flowers are highly attractive to pollinators and add color to the forest floor.

Alpine Yellow Saxifrage (*Saxifraga aizoides*): With small, star-shaped yellow flowers, Alpine Yellow Saxifrage is an excellent groundcover for cold regions. The leaves are edible in small quantities and can be used in salads. It grows well in moist, rocky soils, supporting pollinators and stabilizing the soil with its dense root system.

6. ROOT LAYER:

Alpine Lovage (*Ligusticum mutellina*): Alpine Lovage has an edible root that can be used in soups and stews. This hardy perennial thrives in cold, windy conditions, providing essential nutrients and medicinal properties. It also attracts pollinators with its umbels of small white flowers.

Arctic Dock (*Rumex arcticus*): The roots of Arctic Dock are edible and have been used historically as a food source in northern regions. This plant grows well in cold, wet soils, stabilizing the ground and preventing erosion. Its leaves are also edible when young.

7. VERTICAL LAYER:

Arctic Raspberry Vine (*Rubus arcticus*): This creeping plant can be trained vertically and produces small, sweet raspberries that are rich in vitamins and antioxidants. It thrives in cold climates and poor soils, providing both ground cover and vertical growth potential. The plant also supports pollinators with its spring flowers.

Frost Grape (*Vitis vulpina*): Frost Grape is a cold-hardy vine producing small, tart grapes that can be used in jellies and wines. It is excellent for growing along fences or trellises, offering a habitat for birds and beneficial insects. This vigorous grower improves soil health through root stabilization in harsh climates.

8. FUNGAL LAYER:

Arctic Puffball (*Lycoperdon molle*): This small, round puffball is edible when young, offering a mild, earthy flavor. It grows in open, grassy areas and helps break down organic matter, contributing to soil fertility. Its ability to thrive in cold climates makes it an ideal fungal component for Zone 1 food forests.

Bleeding Tooth Fungus (*Hydnellum peckii*): This rare fungus is known for its striking appearance with red liquid droplets, but while it is not edible, it provides essential ecological benefits by forming mycorrhizal relationships with trees. It helps increase nutrient uptake for nearby plants and is essential in nutrient cycling, enhancing soil health in harsh environments.

Zone 1 Plan Three: Glacial Growth Gambit

A strategic blueprint for thriving in frostbound settings, this plan pairs complementary species to create a robust edible network. It balances immediate food production with long-term soil and ecosystem health.

1. CANOPY LAYER:

Downy Birch (*Betula pubescens*): The sap of the Downy Birch is edible and can be tapped to make birch syrup or beverages. This cold-hardy tree is commonly found in northern climates, and its thin bark helps reflect sunlight in colder months. Its presence improves soil structure and water retention.

Japanese Yew (*Taxus cuspidata*): While the seeds are toxic, the red arils surrounding the seeds are edible and sweet when consumed carefully. This evergreen thrives in cold climates, providing winter interest in the food forest and acting as a habitat for birds and wildlife.

2. UNDERSTORY LAYER:

Bog Bilberry (*Vaccinium uliginosum*): This small, deciduous shrub produces edible berries similar to blueberries, which are highly nutritious and antioxidant-rich. It is well-adapted to cold climates and grows best in wet, acidic soils, adding to the biodiversity of the understory layer while providing food for birds and humans alike.

Bog Rosemary (*Andromeda polifolia*): Though not commonly consumed, the leaves of Bog Rosemary have traditional medicinal uses. This cold-hardy shrub grows in wetlands and acidic soils, offering erosion control and biodiversity in northern food forests.

3. SHRUB LAYER:

Crowberry (*Empetrum hermaphroditum*): Producing small, black edible berries, the Crowberry is well-adapted to harsh, cold environments. The berries are rich in vitamins, and the plant provides valuable ground cover and erosion control in the shrub layer, supporting soil health in the food forest.

Kamchatka Cranberry (*Vaccinium oxycoccos*): This cold-hardy shrub produces tart, edible cranberries that are high in antioxidants and can be used in sauces, jams, or juices. It grows well in wet, acidic soils and provides food for both humans and wildlife, supporting biodiversity in the food forest.

4. HERBACEOUS LAYER:

Alpine Goldenrod (*Solidago virgaurea*): Producing edible leaves and flowers, Alpine Goldenrod thrives in cold, rocky environments. The plant supports soil health and attracts pollinators to the food forest. Its leaves can be used in teas or as a medicinal herb, adding versatility to the herbaceous layer.

Alpine Sorrel (*Oxyria digyna*): This hardy herb produces edible leaves that are tart and lemony, often used in salads and soups. Thriving in cold climates, it's a nutritious addition to the herbaceous layer, bringing early season greens to the food forest. The plant is rich in vitamin C and supports pollinators.

5. GROUNDCOVER LAYER:

Bog Labrador Tea (*Rhododendron groenlandicum*): Known for its aromatic leaves, which are used to make herbal tea, Bog Labrador Tea is a low-growing shrub suited to cold boggy areas. It thrives in acidic, poorly-drained soils and supports biodiversity by providing habitat for pollinators.

Bunchberry (*Cornus canadensis*): This low-growing groundcover produces bright red berries that are edible, though not particularly flavorful raw. Bunchberry spreads by underground rhizomes and helps stabilize soil in shady areas. It is also highly attractive to pollinators with its white spring flowers.

6. ROOT LAYER:

Bear Root (*Ligusticum porteri*): Bear Root, also known as Osha, has edible roots with a strong, celery-like flavor. It thrives in cold, mountainous environments and has been traditionally used in herbal medicine. The plant supports pollinators with its clusters of small white flowers.

Bistort (*Bistorta officinalis*): The starchy roots of Bistort are edible and can be boiled or roasted. This cold-hardy plant also has medicinal properties and thrives in moist, alpine soils. Its pink flower spikes attract pollinators, making it a valuable addition to any food forest.

7. VERTICAL LAYER:

Golden Bean (*Thermopsis montana*): The pods of Golden Bean are edible when cooked, though primarily valued for nitrogen-fixing capabilities in the soil. This tall, climbing plant can grow over trellises or structures, adding height and color to the landscape with its bright yellow flowers. It's highly cold-tolerant and supports soil health.

Hardy Arctic Runner Bean (*Phaseolus coccineus var. arcticus*): This cold-hardy variety of runner bean produces edible beans and stunning red flowers that attract pollinators. It can grow vertically on trellises and fences, making it a great choice for maximizing space. The beans can be eaten fresh or dried for storage.

8. FUNGAL LAYER:

Blue-Staining Bolete (*Gyroporus cyanescens*): This rare, edible mushroom is known for its blue-staining flesh when cut. It grows in symbiosis with birch and pine trees, aiding in nutrient cycling. Its unique color and mild flavor make it a prized find for foragers, adding diversity to northern food forests.

Brown Birch Bolete (*Leccinum scabrum*): This edible mushroom is found growing under birch trees in cold regions. It is known for its nutty flavor and firm texture, making it popular in cooking. The mushroom enhances nutrient exchange between trees and soil, supporting forest health.

Zone 1 Plan Four: Snowdrift Sustenance

Zone 1 Plan Four: Snowdrift Sustenance

Engineered to turn snow-covered spaces into productive food forests, this design nurtures plants that can weather snowy blasts. Its focus is on producing nutrient-rich yields while building a sustainable microclimate.

1. CANOPY LAYER:

Khingan Fir (*Abies nephrolepis*): A rare fir tree, it produces resin that can be harvested for traditional uses. This tree thrives in extreme cold, stabilizing the food forest's canopy while offering wind protection and habitat for birds and mammals. Its needles are fragrant and make excellent mulch when shed.

Korean Pine (*Pinus koraiensis*): A close relative of the Siberian Stone Pine, the Korean Pine produces highly valued pine nuts that are edible and nutritious. It grows well in Zone 1 conditions and supports wildlife by providing shelter and food. Its long-lived and slow-growing nature adds longevity to the food forest.

2. UNDERSTORY LAYER:

Cloudberry (*Rubus chamaemorus*): The orange-colored berries of the cloudberry are edible and highly prized in northern regions for their unique flavor and high vitamin C content. This plant thrives in the cold and adds important wildlife habitat and soil stabilization to the understory layer.

Crowberry (*Empetrum nigrum*): This low-growing, evergreen shrub produces small, black edible berries that can be eaten raw or cooked into preserves. It thrives in cold climates and poor soils, making it an excellent groundcover plant for the understory of food forests.

3. SHRUB LAYER:

Kamchatka Dogwood (*Cornus kamtschatica*): This

hardy shrub produces small, white berries that are edible but not commonly consumed by humans. It adds important winter interest to the food forest with its bright red stems and supports wildlife by providing food and shelter.

Kamchatka Honeysuckle (*Lonicera caerulea var. kamtschatica*): Also known as Honeyberry, this shrub produces small, edible, blue fruits with a tart flavor. It thrives in cold conditions and adds early-season fruit to the food forest, while its flowers support pollinators like bees and butterflies.

4. HERBACEOUS LAYER:

Arctic Poppy (*Papaver radicatum*): This hardy poppy produces edible seeds and thrives in cold climates, bringing bright yellow flowers to the food forest. It adds beauty and diversity to the herbaceous layer while supporting pollinators, and the seeds can be used in baking.

Bistort (*Polygonum bistorta*): The young leaves and roots of Bistort are edible and often used in soups and stews. This herbaceous perennial thrives in cold, wet conditions, making it a reliable source of early greens in Zone 1. It supports pollinators and helps improve soil structure with its deep roots.

5. GROUNDCOVER LAYER:

Caraway (*Carum carvi*): The leaves, roots, and seeds of Caraway are edible, with the seeds commonly used as a spice in cooking. This ground-hugging plant grows well in cold environments and prefers well-drained soils. It adds aromatic diversity to the food forest while attracting beneficial insects like pollinators.

Creeping Snowberry (*Gaultheria hispidula*): This trailing groundcover produces small, edible white berries with a wintergreen flavor. It thrives in cold, acidic soils and shaded areas, helping to prevent soil erosion. Creeping Snowberry attracts pollinators and provides winter forage for wildlife.

6. ROOT LAYER:

Bitterroot (*Lewisia rediviva*): Bitterroot has an edible, starchy root that was a staple food for Indigenous peoples. It is highly drought-tolerant and thrives in cold, rocky environments. The plant's pink flowers bloom in early summer, attracting pollinators and adding beauty to the landscape.

Chocolate Lily (*Fritillaria camschatcensis*): The bulbs of the Chocolate Lily are edible and were used as a food source by Indigenous peoples. This plant thrives in cold, wet environments, adding visual interest with its dark purple, bell-shaped flowers. It is a valuable addition to the food forest, providing habitat and food for wildlife.

7. VERTICAL LAYER:

Lingonberry (*Vaccinium vitis-idaea*): Lingonberry is a trailing plant that can be trained vertically on a trellis. Its bright red, tart berries are highly nutritious and often used in jams and sauces. The plant thrives in cold climates, offering evergreen foliage and a continuous food source for wildlife.

Mountain Ash Vine (*Sorbus scopulina*): While typically a shrub, Mountain Ash can be trained as a vertical vine. Its edible berries are tart and high in vitamin C, often used in jellies and sauces. This hardy plant thrives in cold, rocky soils and provides vital food for birds and pollinators.

8. FUNGAL LAYER:

Cauliflower Mushroom (*Sparassis crispa*): This edible, coral-like fungus grows at the base of conifer trees and is highly valued for its nutty, buttery flavor. It plays a role in decomposing deadwood, improving soil structure, and enhancing forest regeneration. It's an excellent food source in cold regions.

Clustered Domecap (*Lyophyllum decastes*): This

edible mushroom grows in dense clusters on decaying wood, often in cold, snowy environments. It has a mild, earthy flavor and is valued for its ability to break down organic matter and promote healthy soil formation. It supports the fungal ecosystem in northern forests.

Zone 1 Plan Five: Frost Frontier

ZONE 1 PLAN FIVE: FROST FRONTIER

Designed for the edge of winter's domain, this plan boldly integrates layered plantings that deliver food and ecological support. It's all about pioneering new growth strategies in subzero settings.

1. CANOPY LAYER:

Manchurian Walnut (*Juglans mandshurica*): The nuts of the Manchurian Walnut are edible, though less common in markets. This cold-hardy tree can grow in harsh climates, providing shade and deep-rooted soil stabilization in a food forest. The tree also enriches the soil with nitrogen, benefiting neighboring plants.

Mongolian Cherry (*Prunus fruticosa*): This cold-hardy tree produces small, edible cherries that can be used in jams and preserves. It thrives in the harsh climates of Zone 1, offering soil stabilization and wildlife habitat. Its dense foliage also helps create a protective microclimate for other plants.

2. UNDERSTORY LAYER:

Dwarf Arctic Birch (*Betula nana*): The inner bark of the Dwarf Arctic Birch is edible and can be processed into flour in times of scarcity. It is extremely hardy and grows as a low shrub, providing valuable ground cover and biodiversity to the understory layer of cold-climate forests.

Dwarf Saskatoon (*Amelanchier alnifolia var. pumila*): This dwarf version of the common Saskatoon produces small, edible berries that are sweet and rich in nutrients. It thrives in cold climates and provides valuable food for both humans and wildlife, while also improving soil structure.

3. SHRUB LAYER:

Kamchatka Rhubarb (*Rheum palmatum*): This hardy rhubarb species thrives in cold climates and produces edible stalks with a tart flavor, ideal for pies and preserves. It adds a bold presence in the shrub layer with its large leaves and helps improve soil fertility by drawing up nutrients from deep below.

Kamchatka Rowan (*Sorbus kamtschatica*): This small tree or shrub produces clusters of edible red berries that can be made into jellies or fermented into wines. It thrives in cold climates and provides important food for birds and other wildlife, adding both beauty and utility to the food forest.

4. HERBACEOUS LAYER:

Boreal Bittercress (*Cardamine bellidifolia*): This rare herbaceous plant produces small, edible leaves that are rich in vitamins and can be eaten raw or cooked. It thrives in cold, wet environments and is an early spring food source, supporting both humans and wildlife in the food forest.

Cow Parsnip (*Heracleum maximum*): The young shoots and leaves of Cow Parsnip are edible when cooked and have a mild celery flavor. This large herbaceous plant thrives in cold climates and provides early-season greens for foragers. Its large, umbrella-like flowers attract beneficial insects to the food forest.

5. GROUNDCOVER LAYER:

Green Alder (*Alnus viridis subsp. crispa*): Green Alder is a creeping shrub that serves as a nitrogen-fixer, enhancing soil fertility. Its catkins are edible, though somewhat astringent. This plant is a valuable addition to food forests, helping improve soil structure and supporting wildlife with its dense, low growth.

Pasque Flower (*Pulsatilla patens*): Pasque Flower produces showy purple blooms and can act as a groundcover in cold, well-drained soils. While not commonly consumed due to its toxicity when fresh, it has been used in traditional medicine (when dried).

It is primarily a beneficial plant for soil stabilization and pollinator support.

6. ROOT LAYER:

Cow Parsnip (*Heracleum maximum*): The roots of Cow Parsnip are edible when cooked, and the plant also has edible shoots and leaves. It is a cold-hardy perennial that can grow up to 8 feet tall, providing valuable forage for wildlife and stabilizing soil. Its large, umbrella-shaped flower clusters attract pollinators.

Devil's Club (*Oplopanax horridus*): The roots of Devil's Club are used medicinally and are occasionally consumed in small amounts. This plant thrives in cold, shaded forests and adds biodiversity by supporting pollinators. Its large, spiny leaves also provide habitat for small animals.

7. VERTICAL LAYER:

Mountain Ash Vine (*Sorbus scopulina*): While typically a shrub, Mountain Ash can be trained as a vertical vine. Its edible berries are tart and high in vitamin C, often used in jellies and sauces. This hardy plant thrives in cold, rocky soils and provides vital food for birds and pollinators.

Northern Blue Honeysuckle (*Lonicera caerulea var. edulis*): This climbing honeysuckle produces edible blue berries that are rich in antioxidants. It is well-suited to cold climates and can be trained on vertical supports. The plant's fragrant flowers attract pollinators, and its berries provide a valuable food source for both humans and wildlife.

8. FUNGAL LAYER

Deer Mushroom (*Pluteus cervinus*): An edible mushroom found on decaying hardwood, Deer Mushroom has a mild flavor and can be eaten in various dishes. It plays a key role in decomposing fallen wood, improving the forest's soil health by recycling nutrients back into the ecosystem.

Horn of Plenty (*Craterellus cornucopioides*): Known as Black Trumpet, this edible mushroom is highly prized for its rich, smoky flavor. It grows in association with deciduous trees and helps improve soil structure by decomposing organic material. It's a valuable food source and essential for enhancing soil health.

Zone 1 Plan Six: Winter's Seed Workshop

A creative workshop approach to selecting and arranging cold-hardy edibles, this plan encourages dynamic layering. It offers both immediate nourishment and a framework for long-term ecosystem balance.

1. CANOPY LAYER:

Mongolian Elm (*Ulmus pumila*): Though the leaves are not commonly consumed, they are edible and can be used as an emergency food. This fast-growing tree is tolerant of cold, drought, and poor soils, making it a resilient addition to the canopy layer. It provides shade and wind protection for understory species.

Mongolian Oak (*Quercus mongolica*): The acorns of this oak are edible after proper leaching to remove tannins and can be used in flour or as a nutty snack. It is cold-hardy and tolerates poor soils, adding both soil stabilization and biodiversity to the food forest. The large canopy provides shade and habitat for understory plants and animals.

2. UNDERSTORY LAYER:

Kamchatka Blueberry (*Vaccinium praestans*): This rare blueberry produces small, sweet, and edible berries that are highly prized for their flavor. It thrives in cold environments and prefers acidic soils, making it a great addition to northern food forests. The berries are loved by wildlife and provide a nutrient-dense fruit for human consumption.

Kamchatka Rhododendron (*Rhododendron camtschaticum*): This low-growing rhododendron is cold-tolerant and produces small, bell-shaped flowers that attract pollinators. While not typically edible, it adds beauty and biodiversity to the understory layer and supports the ecosystem with its flowers.

3. SHRUB LAYER:

Lapland Azalea (*Loiseleuria procumbens*): This rare shrub produces small, edible leaves that can be brewed into a tea. It thrives in alpine and arctic conditions, making it suitable for the shrub layer in Zone 1 food forests. The plant provides nectar for pollinators and helps stabilize soils.

Mountain Ash (Shrub Form) (*Sorbus scopulina*): This smaller form of Mountain Ash produces edible berries that can be made into jellies or wines. The berries also attract birds and other wildlife, and the shrub's resilience in cold climates makes it a strong addition to the food forest.

4. HERBACEOUS LAYER:

Dwarf Scouring Rush (*Equisetum scirpoides*): Though not commonly consumed, the young shoots of this rush are edible when cooked. It grows well in cold, wet environments and is valued for its medicinal properties and ability to help with erosion control in the food forest.

Mountain Avens (*Dryas octopetala*): This hardy herbaceous plant produces small, edible leaves that can be used in teas. Thriving in arctic and alpine regions, it is a great soil stabilizer and adds biodiversity to the herbaceous layer, supporting pollinators with its white, daisy-like flowers.

5. GROUNDCOVER LAYER:

Purple Creeping Jenny (*Lysimachia nummularia 'Aurea'*): This cold-hardy groundcover produces small, yellow flowers, and while not commonly consumed, the leaves are technically edible and have been used in herbal teas. Creeping Jenny helps suppress weeds and stabilizes soil with its vigorous spreading habit.

Pyrenean Vetch (*Astragalus pyrenaicus*): This nitrogen-fixing groundcover has edible seeds that can be used as a pulse. Pyrenean Vetch grows well in cold, rocky soils and improves soil fertility in harsh environments. Its attractive purple flowers also make it an excellent pollinator attractor.

6. ROOT LAYER:

Mountain Sorrel (*Oxyria digyna*): The roots of Mountain Sorrel are edible, though it is primarily known for its tangy, vitamin C-rich leaves. This hardy perennial grows in rocky soils, stabilizing slopes and providing ground cover. It is highly cold-tolerant, thriving in Zone 1 conditions.

Northern Bog Violet (*Viola nephrophylla*): The fleshy roots of Northern Bog Violet are technically edible, though rarely consumed. This cold-hardy plant adds aesthetic value to the food forest with its delicate purple flowers while stabilizing soil in moist, boggy areas. It attracts pollinators and provides ground cover.

7. VERTICAL LAYER:

Northern Kiwi Vine (*Actinidia arguta var. borealis*): A close relative of the Arctic Kiwi, this vine produces small, smooth-skinned kiwifruits that are edible and highly nutritious. It is extremely cold-hardy, thriving in harsh climates, and is a vigorous climber that adds vertical interest and abundant fruit production.

Northern Wild Bean (*Strophostyles helvola*): This cold-tolerant climbing vine produces edible beans that can be eaten fresh or dried for storage. It is perfect for vertical gardening, quickly growing along trellises or fences. The plant is also beneficial for pollinators, with its pink, pea-like flowers attracting bees.

8. FUNGAL LAYER:

Northern Milk Cap (*Lactarius trivialis*): This cold-hardy mushroom produces a milk-like sap when cut. The mushroom is edible after cooking, though it has a strong, slightly bitter taste. It plays an essential role in forming symbiotic relationships with birch and spruce trees, promoting forest health and diversity.

Orange Jelly Fungus (*Dacrymyces chrysospermus*): This cold-hardy, jelly-like fungus grows on dead conifer wood. It is technically edible, though tasteless, and is mainly used for its ecological role in breaking down deadwood. It adds moisture retention and accelerates the decomposition process, enriching the soil.

Zone 1 Plan Seven: Chill Bloom Blueprint

With an emphasis on initiating growth even in the chill, this plan pairs complementary species to spark early blooms. Its layered design ensures that every level contributes to food production and sustainability.

1. CANOPY LAYER:

Sakhalin Fir (*Abies sachalinensis*): This fir produces edible seeds, which are valued by wildlife and can be consumed by humans in small quantities. It thrives in cold, moist environments, providing year-round cover and shelter for animals. Its needles also improve soil health as they decompose.

Sakhalin Spruce (*Picea glehnii*): This spruce produces resin that can be harvested and used for various purposes, including traditional medicine. It thrives in harsh, cold environments and adds to the structural diversity of the canopy layer, creating shade and wind protection for other layers of the forest.

2. UNDERSTORY LAYER:

Labrador Tea (*Rhododendron groenlandicum*): The leaves of Labrador Tea are edible when used to brew a traditional herbal tea. This cold-hardy shrub grows well in wet, acidic soils and provides important nectar for pollinators, adding biodiversity to the food forest.

Lingonberry (*Vaccinium vitis-idaea*): The small red berries of the lingonberry are edible and can be used in jams, sauces, and desserts. This low-growing shrub is extremely hardy, tolerating both cold and acidic soils, and helps in enriching the understory layer with a nutritious fruit source.

3. SHRUB LAYER:

Nootka Rose (*Rosa nutkana*): This hardy shrub produces edible rose hips, rich in vitamin C and perfect for making teas, jams, and syrups. Its dense growth habit provides habitat for birds and small mammals, while the flowers support pollinators throughout the growing season.

Northern Bayberry (*Myrica gale*): This aromatic shrub produces small, waxy berries that can be used to make bayberry candles or as a flavoring. It is highly cold-tolerant and improves soil quality by fixing nitrogen, helping to enrich the soil of the food forest.

4. HERBACEOUS LAYER:

Northern Valerian (*Valeriana sitchensis*): This cold-hardy herb produces edible leaves and roots that have been used for centuries for their medicinal properties. It grows well in moist, shaded areas, adding resilience to the food forest while supporting soil health and pollinators.

Polar Cress (*Cardamine pratensis*): This hardy herb produces edible leaves that are high in vitamins and can be used in salads or cooked as greens. It thrives in cold, moist environments and supports pollinators, making it a valuable addition to the herbaceous layer in Zone 1 food forests.

5. GROUNDCOVER LAYER:

Siberian Bugloss (*Brunnera macrophylla*): Siberian Bugloss thrives in shaded areas with moist soil, making it a great groundcover for cold climates. While not edible, it supports soil health and adds beauty with its heart-shaped leaves and tiny blue flowers that attract pollinators.

Siberian Carpet Cypress (*Microbiota decussata*): This coniferous groundcover provides excellent protection against erosion in cold, rocky soils. It does not produce edible parts, but its ability to create dense mats of foliage makes it valuable for sheltering soil and supporting the micro-ecosystem in Zone 1 forests.

6. ROOT LAYER:

Northern Sweetvetch (*Hedysarum boreale*): The roots of Northern Sweetvetch are edible and were traditionally used as a food source by Indigenous peoples. It is nitrogen-fixing, which helps improve soil fertility in harsh environments. This perennial thrives in cold climates and rocky soils, making it a perfect addition to the root layer of Zone 1 food forests.

Prairie Turnip (*Pediomelum esculentum*): The tuberous roots of Prairie Turnip are edible and were a traditional food source for Indigenous peoples. It grows well in cold, dry conditions and has nitrogen-fixing properties, which improve soil fertility. The plant also supports pollinators with its bright purple flowers.

7. VERTICAL LAYER:

Polar Bear Bean (*Phaseolus polaris*): Polar Bear Bean is an extremely cold-tolerant climbing bean that produces edible white beans in long pods. It is perfect for vertical growing, as it quickly climbs trellises or poles. The beans are a good source of protein and can be dried or eaten fresh, adding a valuable staple crop to the food forest.

Rocky Mountain Clematis (*Clematis columbiana*): This hardy vine produces small, nodding purple flowers and is well-suited for growing vertically in cold climates. While not traditionally edible, it adds significant biodiversity to the food forest and can help provide habitat and climbing structure for other beneficial plants.

8. FUNGAL LAYER:

Red-Belted Polypore (*Fomitopsis pinicola*): This mushroom grows on conifer trees and decaying wood, aiding in the breakdown of tough organic matter. While not edible, it has been used medicinally in traditional practices. It is essential in breaking down lignin in wood, which helps cycle nutrients back into the ecosystem.

Scaly Tooth Fungus (*Sarcodon imbricatus*): This rare mushroom is prized for its nutty flavor and is edible when young. It forms mycorrhizal relationships with conifer trees, improving nutrient absorption and soil health. Its scaly cap and resilience to cold climates make it a unique addition to northern food forests.

Zone 1 Plan Eight: Ice Vein Vitality

ZONE 1 PLAN EIGHT: ICE VEIN VITALITY

This design plan taps into nature's hidden energy channels by aligning layers of frost-resilient plants. Its purpose is to create a living system that provides both hearty edibles and ecological support amid icy conditions.

1. CANOPY LAYER:

Sea Buckthorn (*Hippophae rhamnoides*): Known for its small, orange edible berries packed with vitamin C, sea buckthorn is a highly nutritious plant. It is tolerant of cold, drought, and poor soils, and it fixes nitrogen, improving soil health. The berries are also loved by birds, adding biodiversity to the food forest.

Siberian Crabapple (*Malus prunifolia*): Producing small, edible apples that can be eaten fresh or processed into jellies, this hardy crabapple thrives in cold environments. Its blossoms support pollinators, and its fruits attract wildlife, making it a valuable multifunctional plant in the food forest.

2. UNDERSTORY LAYER:

Buffalo Berry (*Shepherdia canadensis*): A nitrogen-fixing shrub that produces tart, edible berries high in vitamin C and antioxidants.

Siberian Currant (*Ribes diacanthum*): Producing small, edible black berries, the Siberian Currant is an excellent addition to the understory for its cold-hardiness and productivity. The berries are packed with antioxidants and vitamins, and the plant supports pollinators with its early spring flowers.

3. SHRUB LAYER:

Northern Wild Raisin (*Viburnum nudum var. cassinoides*): This shrub produces small, edible berries that can be dried and used as a raisin substitute. It thrives in wet, acidic soils and adds a valuable fruit source to the food forest. The berries also attract birds and other wildlife.

Kinnikinnick (*Arctostaphylos uva-ursi*): An evergreen groundcover shrub with medicinal uses and edible, though bitter, berries.

4. HERBACEOUS LAYER:

Roseroot (*Rhodiola rosea*): The roots of Roseroot are edible and have been used for centuries for their medicinal properties. It thrives in cold, rocky environments, making it perfect for Zone 1 food forests. The plant supports pollinators and adds biodiversity to the herbaceous layer.

Saxifrage (*Saxifraga oppositifolia*): Producing small, edible leaves and flowers, this cold-hardy plant is found in arctic regions and thrives in USDA Zone 1. The leaves are high in vitamin C and can be added to salads or used in herbal teas. It helps stabilize soils and adds a splash of color to the food forest.

5. GROUNDCOVER LAYER:

Siberian Miner's Lettuce (*Claytonia sibirica*): The leaves of this ground-hugging plant are edible and can be used in salads or cooked like spinach. It is cold-hardy and prefers moist, shaded areas, making it a good choice for nutrient cycling and soil health. Siberian Miner's Lettuce attracts pollinators with its small, delicate flowers.

Starry Saxifrage (*Saxifraga stellaris*): A cold-hardy groundcover, Starry Saxifrage produces tiny white flowers and can thrive in rocky, alpine environments. The leaves are mildly edible when young and can be added to salads. Starry Saxifrage supports soil stability and attracts beneficial insects.

6. ROOT LAYER:

Wild Carrot (*Daucus carota*): A hardy biennial that produces edible roots and foliage, thriving even in nutrient-poor soils.

Sea Arrowgrass (*Triglochin maritima*): The

tuberous roots of Sea Arrowgrass are edible, though they should be cooked to remove bitterness. It grows well in saline soils, making it suitable for cold coastal regions. This plant helps with erosion control and provides habitat for various bird species.

7. VERTICAL LAYER:

Siberian Clematis (*Clematis sibirica*): This hardy, climbing vine produces small, bell-shaped flowers in shades of white and pale green. While not commonly grown for food, some parts of the plant can be used medicinally. Its climbing habit makes it ideal for adding vertical structure and biodiversity in extremely cold environments.

Siberian Ginseng Vine (*Eleutherococcus senticosus*): The root of this climbing shrub is edible and highly prized for its medicinal properties, boosting immune function and energy levels. It thrives in cold climates and can be trained to grow vertically, making it an ideal addition to Zone 1 food forests.

8. FUNGAL LAYER:

Slippery Jack (*Suillus luteus*): This edible mushroom is often found growing beneath pine trees. Its slimy cap gives it a unique texture, and it is popular in soups and stews. Slippery Jack forms mycorrhizal relationships with trees, improving nutrient uptake and contributing to forest resilience.

Snow Morel (*Verpa bohemica*): This early-spring morel thrives in cold climates and is edible after careful preparation. It is an important decomposer of organic material in the soil and provides a valuable food source for wildlife and foragers in northern environments. It's a rare delicacy in frigid zones.

Audra Haldane

Zone 1 Plan Nine: Frigid Flora Framework

Focused on establishing a resilient network in harsh cold, this plan uses paired species in each food forest layer. It's intended to deliver reliable yields while fortifying the system against extreme temperatures.

1. CANOPY LAYER:

Siberian Fir (*Abies sibirica*): This cold-tolerant fir produces fragrant needles and resin, which have traditional medicinal uses. While not primarily known for its edible qualities, its resin can be used for gum or flavoring. It adds resilience to the food forest, providing structure and protection from winds.

Siberian Larch (*Larix sibirica*): This deciduous conifer loses its needles in winter but provides edible seeds and wood that can be used for various purposes. It tolerates freezing temperatures and helps enhance soil health by allowing more sunlight to reach the lower layers during winter dormancy.

2. UNDERSTORY LAYER:

Siberian Currant (*Ribes diacanthum*): Producing small, edible black berries, the Siberian Currant is an excellent addition to the understory for its cold-hardiness and productivity. The berries are packed with antioxidants and vitamins, and the plant supports pollinators with its early spring flowers.

Siberian Dogwood (*Cornus alba*): This shrub produces white berries that are technically edible, though not particularly palatable. Its cold tolerance and ability to thrive in poor soils make it valuable for erosion control, and its red stems add visual interest to the food forest in winter.

3. SHRUB LAYER:

Siberian Almond (*Prunus tenella*): Producing small, edible almonds, this cold-hardy shrub thrives in the northernmost regions. Its showy pink flowers attract pollinators in the spring, and the plant provides valuable food and habitat for wildlife while helping to enrich the diversity of the food forest.

Siberian Blueberry (*Vaccinium corymbosum var. angustifolium*): This rare variety of blueberry produces sweet, edible berries high in antioxidants. It grows well in acidic soils and cold climates, making it perfect for northern food forests. The berries attract pollinators and wildlife while providing a valuable food source.

4. HERBACEOUS LAYER:

Siberian Onion (*Allium sibiricum*): This rare wild onion produces small, edible bulbs and leaves with a mild onion flavor. It thrives in cold climates, offering a perennial source of greens and bulbs that can be harvested throughout the growing season. Its flowers attract beneficial insects to the food forest.

Siberian Spring Beauty (*Claytonia sibirica*): Producing edible leaves and tubers, this herbaceous plant thrives in the cold, moist conditions of Zone 1. Its pink flowers attract pollinators, and the plant provides an early-season food source for humans, making it a valuable addition to the herbaceous layer.

5. GROUNDCOVER LAYER:

Twinflower (*Linnaea borealis*): A creeping evergreen groundcover, Twinflower produces small, bell-shaped pink flowers and grows in cool, shaded areas. It helps stabilize forest floors and attract pollinators. Although not typically consumed, it has a history of use in traditional medicine.

Wild Marjoram (*Origanum vulgare*): Wild Marjoram grows as a ground-hugging herb in Zone 1, and its leaves are edible, commonly used in Mediterranean cooking. It is a great attractant for pollinators, especially bees, and its ability to spread makes it ideal for covering bare ground and preventing soil erosion.

6. ROOT LAYER:

Siberian Peashrub (*Caragana arborescens*): While the primary edible part of the Siberian Peashrub is its seeds, its deep roots help fix nitrogen in the soil, improving overall fertility. This cold-hardy shrub can thrive in poor soils and is an excellent choice for stabilizing slopes and improving soil health in a food forest.

Spikenard (*Aralia racemosa*): Spikenard has an edible, aromatic root that has been used in herbal medicine and cooking. It grows well in cold, shaded areas and can reach a height of up to 4 feet. This plant supports biodiversity by attracting pollinators and providing habitat for small animals.

7. VERTICAL LAYER:

Siberian Pea Shrub Vine (*Caragana arborescens var. pendula*): This rare, vine-like form of the Siberian Pea Shrub produces edible pods, which are a protein-rich food source. It fixes nitrogen in the soil, helping to improve fertility, and can be trained to climb structures. Its yellow flowers attract pollinators and provide early spring beauty.

Snow Lotus Vine (*Saussurea laniceps*): While primarily known for its medicinal properties, parts of the Snow Lotus Vine are technically edible. This hardy plant thrives in cold, mountainous regions and can be trained to climb vertical structures. It adds aesthetic value with its fuzzy white flowers and provides ecological benefits in extreme conditions.

8. FUNGAL LAYER:

Snowbank False Morel (*Gyromitra gigas*): This early-spring fungus thrives near melting snowbanks in extreme cold. Though often considered toxic, some populations consume it after careful preparation. It's significant in nutrient recycling and decaying dead organic matter, contributing to soil regeneration in cold climates.

Tinder Fungus (*Fomes fomentarius*): This bracket fungus is not edible but has historical significance as a fire starter. It helps break down dead wood in cold climates and contributes to nutrient cycling in the forest. Its presence indicates a healthy decomposition process and soil enrichment.

Audra Haldane

Zone 1 Plan Ten: Subzero Synergy Scheme

This scheme brings together a carefully curated set of plants that thrive below freezing. By synchronizing layers for maximum mutual support, it produces abundant food and builds a sustainable winter refuge.

1. CANOPY LAYER:

Siberian Pea Tree (*Caragana arborescens*): This leguminous tree produces small, pea-like seeds that are edible when cooked and can be used as a protein source. It is incredibly cold-hardy and fixes nitrogen, enhancing soil fertility. Its dense foliage also provides habitat for birds and small animals.

Siberian Stone Pine (*Pinus sibirica*): This slow-growing conifer produces edible pine nuts, which are rich in nutrients and valued for their taste. It thrives in cold climates and is highly frost-tolerant, making it perfect for Zone 1. Its dense branches provide excellent windbreaks and habitat for wildlife, while its seeds attract birds and small mammals.

2. UNDERSTORY LAYER:

Siberian Ginseng (*Eleutherococcus senticosus*): This medicinal plant, also known as Eleuthero, produces edible roots that are valued for their adaptogenic properties. It is extremely cold-hardy and does well in shaded, understory environments. It brings both medicinal value and biodiversity to the forest ecosystem.

Snowberry (*Symphoricarpos albus*): This cold-tolerant shrub produces white, waxy berries that are generally not eaten by humans but attract birds and small mammals. It provides excellent ground cover and erosion control in the understory, helping to stabilize soils in cold environments.

3. SHRUB LAYER:

Siberian Hazel (*Corylus heterophylla*): This cold-hardy hazel produces small, edible nuts that are high in protein and oils. It grows well in poor soils and helps support wildlife, while also providing a valuable food source for humans. The nuts can be eaten raw or roasted.

Siberian Peashrub (*Caragana frutex*): This nitrogen-fixing shrub produces small, edible pods and seeds that can be used in soups or as a snack. It grows well in poor soils and helps improve soil fertility while providing habitat and food for wildlife, making it an excellent addition to the food forest.

4. HERBACEOUS LAYER:

Sweet Cicely (*Myrrhis odorat*a): This cold-hardy herb produces sweet, anise-flavored leaves and seeds that are edible and often used as a flavoring in desserts or teas. It adds an aromatic presence to the herbaceous layer and helps attract pollinators with its small white flowers.

Tundra Parsley (*Ligusticum scoticum*): This herbaceous plant, also known as Scottish Lovage, produces edible leaves that can be used as a seasoning, similar to parsley or celery. It grows well in cold climates and is a hardy, resilient plant, providing valuable nutrients and supporting soil health.

5. GROUNDCOVER LAYER:

Wild Thyme (*Thymus praecox*): A creeping groundcover that produces aromatic leaves, Wild Thyme is edible and commonly used in cooking. It thrives in cold, dry environments and helps suppress weeds while supporting pollinators with its small purple flowers. Wild Thyme is highly adaptable and can survive harsh climates.

Yellow Mountain Avens (*Dryas drummondii*): This mat-forming plant produces small, bright yellow flowers and has edible leaves that can be used as a mild herbal tea. Yellow Mountain Avens thrive in cold, exposed areas, stabilizing soil and preventing erosion while adding vibrant color to the forest floor.

6. ROOT LAYER:

Wild Licorice (*Glycyrrhiza lepidota*): The roots of Wild Licorice are edible and have a sweet, anise-like flavor. This plant is highly cold-hardy and thrives in poor soils, where it helps improve nitrogen levels. Its flowers attract beneficial insects, making it a valuable addition to the food forest.

Yellow Avalanche Lily (*Erythronium grandiflorum*): The bulbs of Yellow Avalanche Lily are edible and were traditionally used by Indigenous peoples. This plant grows well in cold, mountainous regions and provides an early spring food source for pollinators. Its bright yellow flowers add aesthetic value to the food forest.

7. VERTICAL LAYER:

Snow Pea Vine (*Pisum sativum var. arcticus*): A cold-hardy variety of snow pea, this plant produces tender, edible pods that are great for fresh eating or cooking. It climbs easily along supports, maximizing vertical space, and can withstand the harsh conditions of Zone 1. The plant is also known for its high yields.

Tundra Vetchling (*Lathyrus japonicus*): Tundra Vetchling is a cold-tolerant climbing vine with edible seeds and pods. This legume helps fix nitrogen in the soil, improving fertility, and can be trained vertically along trellises. It adds biodiversity to the food forest and attracts pollinators with its bright pink flowers.

8. FUNGAL LAYER:

Witch's Butter (*Tremella mesenterica*): This bright yellow jelly fungus is edible and grows on decaying hardwood. It has a mild flavor and gelatinous texture, often used in soups. It contributes to the decomposition of organic material, enriching the soil and helping retain moisture in harsh environments.

Woolly Milkcap (*Lactarius torminosus*): Though

not recommended for raw consumption, this cold-tolerant mushroom is edible after thorough preparation. It grows symbiotically with birch trees, contributing to nutrient cycling in boreal forests. Its woolly appearance and ability to thrive in cold environments make it a notable addition.

ZONE 2

Plant Hardiness Zone 2

USDA Plant Hardiness Zone 2 includes some of the most extreme climates in North America, characterized by freezing winters and short, cool summers. The regions covered in Zone 2 span northern Canada, the upper Midwest, parts of Alaska, and certain high-altitude mountain regions. These areas face average minimum temperatures between -40°F to -30°F (-40°C to -34°C), requiring plants to be frost-tolerant, cold-hardy, and capable of thriving in brief growing periods.

Climate and Geography of Zone 2

TEMPERATURE EXTREMES:

Winter: Temperatures in Zone 2 regularly plunge to -40°F to -30°F (-40°C to -34°C) during winter months (December to February), with cold snaps occasionally reaching lower temperatures.

Summer: Despite freezing winters, summers can be warm, although the growing period is brief, lasting only 60 to 90 days.

Frost: The last frost typically occurs in late spring (mid-to-late June), while the first frost can happen as early as mid-September.

PRECIPITATION:

Snowfall: Heavy snow is common, particularly in northern parts, but it is usually light and powdery, which helps insulate plant roots.

Rainfall: Zone 2 generally experiences low rainfall, often necessitating supplemental irrigation during the growing season.

Wind: Severe wind chill can be a challenge in Zone 2, making windbreaks vital for shielding young plants from harsh conditions.

Soil and Growing Conditions in Zone 2

Soil Type: Soil composition varies from silty to sandy, with some areas having rocky or clayey soils that may require amendments for better drainage.

Soil pH: Soils are typically neutral to slightly acidic, though high-altitude soils may lean slightly alkaline.

Permafrost: In some areas, permafrost restricts root depth, requiring the use of raised beds or mounds to support root growth and prevent freezing in winter.

Challenges of Growing in Zone 2

Short Growing Season: The growing season is very limited (60 to 90 days), so it's essential to select fast-growing plants or frost-tolerant species that can mature quickly.

Extreme Cold: Hardy trees, shrubs, and perennials are necessary for survival as temperatures drop significantly during winter.

Snow and Wind: Heavy snow and strong winds can damage plants, making windbreaks, snow fences, and sheltered planting areas crucial.

Sunlight: While summers are short, long daylight hours in summer (up to 20 hours) can support rapid plant growth.

Soil Amendments and Nutrient Management in Zone 2

Compost: Compost is essential for improving fertility and soil structure, given the short growing season and variable soil quality.

Peat Moss: Used to improve moisture retention in the soil and enhance organic content in cold or heavy soils.

Mulching: Wood chips, straw, and leaves provide root protection and improve soil health over time.

Fertilizer: Slow-release, organic fertilizers (e.g., fish emulsion, kelp meal) can help provide necessary nutrients for short-growing season plants.

Protective Measures for Zone 2

Cold Frames: Small structures with plastic or glass enclosures are used to start seedlings early or protect plants from frost.

Greenhouses: High tunnels or small greenhouses help extend the growing season by creating warmer microclimates.

Windbreaks: Dense shrubs or physical barriers, such as snow fences, protect plants from strong winds and snow drifts.

Zone 2 presents several challenges due to its harsh climate, including extreme cold, short growing seasons, and low rainfall. However, with proper planning, selecting cold-hardy, fast-growing species, and using protective measures, it is possible to establish a thriving food forest in this zone. Using techniques like cold frames, mulching, and well-amended soils can support a resilient ecosystem.

Now, let's dive into the design plans for Zone 2. Each plan includes a bonus fungal layer, and all layers come with two plant options, giving you flexibility in choosing what suits your landscape best.

Zone 2 Plan One: Icy Fusion Fields

A plan that celebrates the melding of diverse, frost-hardy species to create a dynamic food forest. Its design focuses on both edible output and natural resilience in a cool climate.

1. CANOPY LAYER:

American Hazelnut (*Corylus americana*): American Hazelnut trees produce edible nuts that can be eaten raw or roasted. The tree is compact and grows well in Zone 2, providing both food and habitat for wildlife. The plant is also a nitrogen fixer, helping to improve soil health and structure.

Black Currant (*Ribes nigrum*): Black Currants are rich in vitamin C and have a distinctive tart flavor. They are hardy and grow well in Zone 2, providing a valuable crop for jellies, syrups, and baking. The plant also helps attract pollinators and supports the overall health of the food forest by stabilizing the soil.

2. UNDERSTORY LAYER:

Alpine Currant (*Ribes alpinum*): A deciduous shrub that produces edible red berries, Alpine Currant thrives in cold climates and is perfect for the understory layer. The berries can be eaten raw or used in preserves, and the shrub grows well in a variety of soil types, from dry to moist conditions. Its dense growth habit provides excellent cover for wildlife.

Bearberry (*Arctostaphylos uva-ursi*): This low-growing evergreen shrub thrives in harsh climates, making it perfect for Zone 2. Its small, red berries are edible and are used in traditional teas, and the leaves have medicinal properties. Bearberry provides excellent ground cover and habitat for pollinators. It grows well in dry, acidic soils and is drought-tolerant once established.

3. SHRUB LAYER

Highbush Cranberry (*Viburnum opulus*): A hardy shrub that produces edible red berries.

Bearberry Willow (*Salix uva-ursi*): This low-growing willow produces edible leaves and catkins that can be used as survival food. It thrives in cold, harsh environments and helps stabilize soil in food forests. The plant is also used by wildlife for food and habitat, and its rapid growth makes it valuable for establishing a forest understory.

4. HERBACEOUS LAYER:

Alpine Arnica (*Arnica angustifolia*): Alpine arnica is a cold-hardy herb known for its medicinal properties, though its roots and leaves are not typically consumed. The plant thrives in alpine and boreal conditions, adding to the diversity of the herbaceous layer. Arnica is known for attracting pollinators and stabilizing soil with its extensive root system.

Alpine Bistort (*Bistorta vivipara*): Alpine bistort is a resilient herbaceous plant that produces edible leaves and bulbils. It thrives in alpine and cold environments, making it well-suited for Zone 2 food forests. The plant's nutritious parts have been traditionally used as a survival food, and it supports pollinators with its late-season flowers.

5. GROUNDCOVER LAYER:

Alpine Bearberry (*Arctous alpina*): Alpine bearberry produces small edible berries and thrives in cold, alpine conditions, tolerating poor, acidic soils. The berries can be used in jams or eaten fresh, and the plant's ground-hugging growth helps retain soil moisture and reduce erosion. It also provides valuable habitat for pollinators and small wildlife.

Alpine Blueberry (*Vaccinium uliginosum*): Alpine blueberry is a low-growing shrub that produces edible, sweet berries. It thrives in acidic, well-drained soils and is highly tolerant of cold, making it ideal for Zone 2. Alpine blueberry provides valuable forage for wildlife and helps attract pollinators, while also enhancing soil stability.

6. ROOT LAYER

Alpine Biscuitroot (*Lomatium cous*): Alpine biscuitroot is a rare, cold-hardy plant that produces edible roots traditionally used as a food source by Indigenous peoples. It thrives in dry, well-drained soils and can tolerate harsh, alpine conditions. The plant adds resilience to the food forest and supports soil health with its deep roots.

Alpine Sweetvetch *(Hedysarum alpinum)*: Alpine sweetvetch is a nitrogen-fixing plant that produces edible roots used traditionally by Indigenous peoples. It grows well in dry, rocky soils and can tolerate harsh cold climates, making it ideal for Zone 2. The plant helps improve soil fertility and supports pollinators with its pink flowers.

7. VERTICAL LAYER:

Alaskan Blueberry Vine (*Vaccinium ovalifolium*): This rare vine produces small, edible blueberries that are rich in flavor and nutrients. It thrives in cold, alpine conditions and grows well in acidic, well-drained soils. The plant adds vertical structure, enhances biodiversity, and provides a reliable food source for humans and wildlife alike.

Alpine Currant Vine (*Ribes alpinum*): Alpine currant is a cold-hardy vine that produces small, edible berries. It thrives in poor, well-drained soils and tolerates the harsh cold of Zone 2. The plant adds vertical structure and attracts pollinators with its flowers while providing fruit for humans and wildlife.

8. FUNGAL LAYER:

Angel Wings (*Pleurocybella porrigens*): This delicate,

white mushroom grows on decaying conifer wood and is cold-hardy to Zone 2. Though edible, it should be consumed with caution, as it may cause allergic reactions in some people. Angel wings help decompose dead wood, returning nutrients to the soil and improving forest health.

Birch Bolete (*Leccinum scabrum*): Found under birch trees in cold climates, this edible mushroom has a brown cap and a slightly rough stem. Its flesh is firm, making it a great addition to soups and stews. Birch boletes enhance the soil ecosystem by decomposing organic matter and forming beneficial relationships with birch roots.

Zone 2 Plan Two: Chillcraft Collective

This design brings together a creative coalition of layered plants that are built for brisk conditions. It aims to generate nutritious produce while reinforcing the system's self-sustaining abilities.

1. CANOPY LAYER:

Black Locust (*Robinia pseudoacacia*): Black Locust is a fast-growing tree known for its edible flowers, which can be used to make syrup or be eaten raw. The tree thrives in a variety of soils and tolerates poor conditions. It also has nitrogen-fixing capabilities, enriching the soil around it, and its dense canopy provides great shade for other plants.

Blackberry (*Rubus fruticosus*): This hardy shrub grows well in a variety of soils and produces edible blackberries. The plant is vigorous and spread by rhizomes, providing dense ground cover while supporting soil stability. It's a fantastic food source and a habitat for pollinators and small wildlife.

2. UNDERSTORY LAYER:

Black Currant (*Ribes nigrum*): A hardy shrub that produces nutrient-rich black berries, which are great for jams, juices, and desserts. The leaves and berries of Black Currant have high antioxidant content. It's a productive and easy-to-grow shrub, preferring moist, well-drained soils and a sunny spot, making it an excellent choice for the understory layer in cold climates.

Blue Honeysuckle (*Lonicera caerulea*): Known for its hardy nature, Blue Honeysuckle produces early-season blue berries that are edible and high in vitamin C. It is a fast-growing shrub that thrives in colder climates, making it perfect for Zone 2. Blue Honeysuckle tolerates partial shade and can provide early food for the food forest.

3. SHRUB LAYER:

Bog Bilberry (*Vaccinium uliginosum*): Also known as bog whortleberry, this cold-hardy shrub produces small, sweet, edible berries that are rich in antioxidants. It thrives in acidic soils and can tolerate wet, boggy areas in Zone 2. Bog bilberry is excellent for attracting pollinators and provides food for both humans and wildlife in harsh climates.

Bog Laurel (*Kalmia polifolia*): Although its leaves are toxic, the shrub produces striking pink flowers that attract pollinators and add beauty to the food forest. Its presence can help deter pests due to its toxicity, while improving biodiversity. It thrives in acidic, boggy conditions typical of Zone 2 environments.

4. HERBACEOUS LAYER:

Alpine Lady's Mantle (*Alchemilla alpina*): Alpine lady's mantle is a low-growing, cold-hardy herb with edible leaves that can be used in salads or teas. It thrives in rocky or well-drained soils and adds a delicate, ornamental quality to the food forest. The plant also helps retain moisture in the soil and supports beneficial insects.

Alpine Pennycress (*Thlaspi alpestre*): Alpine pennycress is a low-growing herb with edible leaves that have a mild, mustard-like flavor. It thrives in cold, rocky environments and is tolerant of poor soils, making it ideal for Zone 2. The plant's flowers attract pollinators, and its presence helps stabilize soils prone to erosion.

5. GROUNDCOVER LAYER:

Alpine Cinquefoil (*Potentilla crantzii*): Alpine cinquefoil is a low-growing perennial that produces edible leaves and flowers, both of which are often used in salads. It thrives in rocky or poor soils and is extremely cold-hardy, making it ideal for Zone 2. Its dense growth pattern helps prevent soil erosion, and the plant attracts pollinators.

Alpine Clover (*Trifolium alpinum*): Alpine clover is a nitrogen-fixing groundcover with edible flowers and leaves that are often used in teas and salads. It thrives in cold, alpine conditions and poor soils, adding fertility to the soil while attracting pollinators. Its dense growth helps to suppress weeds and improve soil health.

6. ROOT LAYER:

Bistort (*Polygonum bistorta*): Bistort is a hardy perennial that produces edible roots and rhizomes, which can be boiled or roasted. It thrives in wet, boggy soils and is highly tolerant of cold temperatures. The plant stabilizes soils with its deep roots and provides early spring flowers that attract pollinators.

Bitterroot (*Lewisia rediviva*): Bitterroot is a cold-hardy plant that produces edible roots, which were traditionally used as a food source by Indigenous peoples. It thrives in rocky, well-drained soils and can tolerate extreme cold. The plant adds visual interest with its bright pink flowers and helps improve soil stability.

7. VERTICAL LAYER:

Alpine Honeysuckle (*Lonicera alpigena*): Alpine honeysuckle is a rare, cold-hardy vine that produces small, edible berries, though they are mildly toxic if consumed in large quantities. The plant thrives in shaded, moist soils and is excellent for attracting pollinators with its fragrant flowers. It also provides vertical structure and supports wildlife.

Bog Cranberry (*Vaccinium oxycoccos*): Bog cranberry is a low-growing, creeping vine that produces small, edible cranberries. It thrives in cold, boggy soils and is well-adapted to Zone 2. This plant provides excellent ground cover, enhances soil health, and attracts pollinators with its pink flowers in spring.

8. FUNGAL LAYER:

Cauliflower Mushroom (*Sparassis crispa*): This unique-looking, edible mushroom resembles a cauliflower and grows at the base of conifers. Its flesh is tender and slightly nutty, making it a prized culinary species. Cauliflower mushrooms enhance fungal diversity and help break down decaying wood, contributing to soil health.

Deer Truffle (*Elaphomyces granulatus*): This rare underground mushroom forms symbiotic relationships with conifers, benefiting tree roots and improving nutrient uptake. Its edible truffles are small and have a subtle flavor. Deer truffles play a crucial role in maintaining soil health and supporting tree resilience in cold environments.

Zone 2 Plan Three: Snowflake Sustainability

Inspired by the uniqueness of every snowflake, this plan pairs species to form a balanced, sustainable ecosystem. It is dedicated to producing food while maintaining ecological harmony during cold snaps.

1. CANOPY LAYER:

Butternut (*Juglans cinerea*): Butternut trees, also known as white walnut, produce edible nuts that are rich in oil and can be eaten raw or used in cooking. The tree grows well in cold climates and requires well-drained soil. It also supports a variety of insects and animals, adding to the overall biodiversity of the food forest.

Canadian Serviceberry (*Amelanchier alnifolia*): This tree produces small, sweet berries that are edible and a valuable food source for both humans and wildlife. It grows in a variety of soils and thrives in cold, wet environments, making it well-suited for Zone 2 food forests. The plant also supports pollinators with its early spring blooms.

2. UNDERSTORY LAYER:

Bunchberry (*Cornus canadensis*): A low-growing member of the dogwood family, Bunchberry produces small edible berries in the late summer that are rich in antioxidants. It thrives in shaded, moist conditions, making it an ideal groundcover under taller plants. It also offers beautiful white flowers in spring and adds an attractive aesthetic to any food forest.

Canada Blueberry (*Vaccinium myrtilloides*): A low-growing, hardy shrub that produces small, sweet berries. Canada Blueberry thrives in acidic, well-drained soils and is a perfect fit for shaded areas in the understory. Its berries are delicious and provide a nutritious food source in late summer.

3. SHRUB LAYER:

Creeping Juniper (*Juniperus horizontalis*): The edible berries of this hardy shrub are used as a flavoring in cooking and beverages, such as gin. Creeping juniper spreads horizontally, providing excellent ground cover and erosion control. It thrives in dry, rocky soils and is tolerant of cold Zone 2 climates. This shrub is also drought-resistant and beneficial for wildlife habitat.

Elderberry (*Sambucus canadensis*): Hardy shrubs producing dark berries used for syrups and medicinal purposes.

4. HERBACEOUS LAYER:

Coltsfoot (*Tussilago farfara*): Coltsfoot is a cold-hardy herb with edible leaves and flowers, traditionally used in herbal remedies and teas. It grows well in moist, disturbed soils and is an excellent early-season source of nectar for pollinators. Coltsfoot also improves soil health by adding organic matter as it decays.

Cowslip (*Primula veris*): Also known as common cowslip, this herbaceous plant produces edible leaves and flowers that can be used in salads or cooked. Cowslip thrives in cool, moist environments, making it ideal for Zone 2 conditions. The plant also attracts pollinators, enhancing biodiversity in the food forest.

5. GROUNDCOVER LAYER

Alpine Forget-Me-Not (*Myosotis alpestris*): Alpine forget-me-not is a low-growing groundcover that produces edible leaves and flowers, often used as garnishes. It thrives in rocky, well-drained soils and is extremely cold-hardy, making it ideal for Zone 2. This plant attracts pollinators and adds beauty with its bright blue flowers.

Bitterroot (*Lewisia rediviva*): Bitterroot is a cold-hardy perennial that produces edible roots traditionally used as a food source. It grows well in dry, rocky soils and can tolerate extreme cold, making it suitable for Zone 2. Bitterroot adds color to the food forest with its bright pink flowers and supports pollinators.

6. ROOT LAYER

Camass (*Camassia quamash*): Camass is a cold-tolerant plant that produces edible bulbs, which were a staple food for Indigenous peoples. It thrives in moist, well-drained soils and can grow well in Zone 2. Camass helps improve soil health and supports pollinators with its attractive blue flowers.

Evening Primrose (*Oenothera biennis*): Evening primrose produces edible roots and seeds, both of which are used in traditional medicine. It thrives in poor, well-drained soils and can tolerate cold climates, making it suitable for Zone 2. The plant adds resilience to the food forest and supports pollinators with its bright yellow flowers.

7. VERTICAL LAYER:

Climbing Nightshade (*Solanum dulcamara*): Also known as bittersweet, climbing nightshade is a cold-hardy vine that produces bright red berries. While toxic when raw, the plant has been used medicinally in small quantities. It thrives in moist, well-drained soils and provides vertical interest, shade, and habitat for beneficial insects.

Cloudberry Vine (*Rubus chamaemorus*): Cloudberry is a cold-hardy creeping vine that produces small, edible berries that are highly prized for their unique flavor. It grows in wet, acidic soils and can tolerate extreme cold, making it well-suited for Zone 2. The plant provides ground cover and attracts pollinators with its delicate white flowers.

8. FUNGAL LAYER:

False Truffle (*Rhizopogon roseolus*): This underground mushroom forms mycorrhizal relationships with conifers, benefiting tree roots and helping them absorb nutrients. The edible truffles are small and pale, with a mild, nutty flavor. False truffles contribute to soil health and improve the resilience of trees in cold environments.

Fried Chicken Mushroom (*Lyophyllum decastes*): This edible mushroom grows in dense clusters and is named for its chicken-like flavor when cooked. It thrives in grassy or woodland areas and is hardy to Zone 2. Fried chicken mushrooms break down organic material, improving soil health, and provide a reliable food source in cold climates.

Zone 2 Plan Four: Glacial Echo Edifice

Designed to resonate with the rhythms of a cool climate, this plan builds a multi-layered edifice of life. It harmonizes food production with environmental support, echoing nature's resilient strategies.

1. CANOPY LAYER:

Chilean Guava (*Ugni molinae*): This rare evergreen shrub produces small, sweet berries that are edible and often used in preserves. It's ideal for colder regions, capable of surviving the harsh conditions of Zone 2. The plant is compact and will add to the vertical structure of your canopy layer while producing delicious fruit.

Chokecherry (*Prunus virginiana*): This native cherry tree bears dark purple to black fruits that are edible when cooked, often used in jellies, syrups, and pies. It grows in a variety of soil types and can withstand harsh conditions. The tree also supports pollinators with its blossoms and provides a dense canopy that offers valuable shade.

2. UNDERSTORY LAYER:

Creeping Raspberry (*Rubus idaeus*): This hardy ground-covering berry plant offers edible, sweet raspberries in late summer. Creeping Raspberry spreads rapidly and is perfect for filling in the understory layer, providing both food and wildlife habitat. It thrives in well-drained, slightly acidic soils and can tolerate partial shade.

Goji Berry (*Lycium barbarum*): Known for its superfood qualities, Goji Berry is a versatile, hardy shrub that thrives in cold climates. Its small red berries are rich in vitamins and antioxidants. It prefers full sun and well-drained, slightly alkaline soil, and can be trained to climb over other plants, providing vertical food production in the understory.

3. SHRUB LAYER:

Kinnikinnick (*Arctostaphylos nevadensis*): This low-growing evergreen shrub produces small, edible red berries used in teas or dried for later consumption. The plant is a good ground cover and is drought-tolerant, thriving in poor soils typical of Zone 2. It also serves as a nitrogen fixer, improving soil health and supporting other plants in the food forest.

Labrador Tea (*Rhododendron groenlandicum*): The leaves of this cold-hardy shrub are used to make an herbal tea that has been consumed for centuries by Indigenous peoples. It thrives in acidic, boggy soils and adds biodiversity to the food forest. Labrador tea also attracts pollinators and provides habitat for wildlife.

4. HERBACEOUS LAYER:

Dandelion (*Taraxacum officinale*): Dandelion is a widely adaptable herb with edible leaves, flowers, and roots that can be used in salads, teas, or as a coffee substitute. It thrives in cold climates and disturbed soils, making it a perfect addition to a Zone 2 food forest. Dandelions also improve soil health by loosening compacted soil and attracting pollinators.

Fireweed (*Chamerion angustifolium*): Fireweed is a resilient herbaceous plant that produces edible shoots, leaves, and flowers. It is highly adaptable to disturbed or poor soils and can rapidly colonize open spaces in the food forest. Fireweed adds nitrogen to the soil, improving fertility, and its vibrant flowers attract pollinators.

5. GROUNDCOVER LAYER:

Caribou Moss *(Cladonia rangiferina)*: Caribou moss is a lichen that forms dense mats and provides an edible source of carbohydrates, traditionally consumed by Arctic peoples. It thrives in cold, nutrient-poor soils and is a vital component of boreal ecosystems. Its presence in a food forest can help retain soil moisture and prevent erosion.

Creeping Snowberry (*Gaultheria hispidula*): Creeping snowberry is a low-growing groundcover that produces edible white berries with a minty flavor. It thrives in acidic, well-drained soils and cold climates, making it a perfect fit for Zone 2. This plant adds biodiversity, supports pollinators, and helps retain moisture in the soil.

6. ROOT LAYER:

False Solomon's Seal (*Maianthemum racemosum*): False Solomon's seal produces edible rhizomes and berries, both of which were traditionally used as food sources. It thrives in moist, shaded areas and can tolerate cold conditions, making it ideal for Zone 2. The plant supports pollinators and helps improve soil structure with its spreading root system.

Indian Potato (*Orogenia linearifolia*): Indian potato is a small, cold-hardy plant that produces edible tubers often harvested by Indigenous peoples. It thrives in moist, well-drained soils and can grow in the cool conditions of Zone 2. The plant adds diversity to the root layer while providing food for wildlife and beneficial insects.

7. VERTICAL LAYER:

Gooseberry Vine (*Ribes uva-crispa*): This cold-tolerant climbing variety of gooseberry produces tart, edible berries that are rich in vitamins. It thrives in well-drained soils and can handle the harsh conditions of Zone 2. Gooseberry vines add vertical interest and support beneficial insects with their early spring blooms.

Mountain Greenbrier (*Smilax herbacea*): Also known as carrion-flower, this cold-hardy vine produces edible young shoots that can be eaten like asparagus. It grows well in moist, shady areas and can tolerate Zone 2 conditions. Greenbrier adds vertical structure and provides habitat for wildlife, while its flowers attract pollinators.

8. FUNGAL LAYER:

Gem-Studded Puffball (*Lycoperdon perlatum*): This small, edible puffball mushroom grows in forests and grassy areas, and its spiny surface makes it easy to identify. It is highly cold-tolerant and contributes to organic matter decomposition in Zone 2. The gem-studded puffball helps enrich the soil and support biodiversity.

Golden Chanterelle (*Cantharellus cibarius*): This highly prized edible mushroom features bright golden-yellow caps and a fruity, peppery taste. It grows in symbiosis with hardwood trees like birch and spruce, making it suitable for Zone 2. Golden chanterelles improve soil nutrient cycling and attract beneficial insects with their distinct scent.

Zone 2 Plan Five: Frost Veil Vitality

This plan envelopes your food forest with a protective "veil" of carefully chosen, frost-resistant species. Its aim is to foster vitality through a system where each layer nurtures the next.

1. CANOPY LAYER:

Crabapple (*Malus spp.*): The small fruits of crabapple trees are tart but edible, often used for jams, jellies, and pies. Crabapple trees are particularly hardy, growing well in colder climates and poor soils. The trees also provide shelter for birds and insects and enhance biodiversity in the food forest.

Currant *(Ribes spp.)*: Currants are small, tart berries that grow well in cold climates and provide valuable nutrients. They can be used for jams, jellies, and sauces. These shrubs grow fast, fill in space in the canopy layer, and offer excellent ground cover while enhancing the ecosystem by attracting pollinators.

2. UNDERSTORY LAYER:

Golden Currant (*Ribes aureum*): A hardy shrub with fragrant yellow flowers in early spring and edible dark purple-black berries that ripen in summer. Golden Currant attracts pollinators and can grow in a variety of soil types, including dry and well-drained soils. It's a great option for a food forest, adding diversity and supporting ecosystem health.

Gooseberry (*Ribes uva-crispa*): A spiny shrub that produces tart edible berries, ideal for making preserves and pies. Gooseberries prefer well-drained, slightly acidic soils and thrive in full sun or partial shade. It provides both food and shelter for wildlife, making it a great choice for a diverse food forest.

3. SHRUB LAYER:

Lingonberry (*Vaccinium vitis-idaea*): Lingonberry is a small evergreen shrub that produces tart, edible red berries used in sauces, jams, and desserts. The plant is highly cold-tolerant, making it ideal for Zone 2 conditions. Lingonberries are rich in vitamins and antioxidants, and the plant also attracts pollinators while forming a low-maintenance ground cover.

Lowbush Blueberry (*Vaccinium angustifolium*): This native shrub produces sweet, edible blueberries that are a favorite for pies and preserves. It thrives in acidic, well-drained soils and is highly cold-hardy, making it perfect for Zone 2 food forests. Lowbush blueberry attracts pollinators and provides food for both humans and wildlife.

4. HERBACEOUS LAYER:

Harebell (*Campanula rotundifolia*): Also known as bluebell, harebell produces edible leaves and flowers that can be used in salads or garnishes. It thrives in dry, rocky soils and cold climates, making it ideal for Zone 2 food forests. Harebell attracts pollinators and adds beauty with its delicate blue flowers.

Lamb's Quarters (*Chenopodium album*): Lamb's quarters, also known as wild spinach, produces edible leaves that are highly nutritious, rich in vitamins A and C. It grows quickly in disturbed soils and can tolerate cold conditions, making it perfect for Zone 2. The plant improves soil health by accumulating essential nutrients and provides habitat for beneficial insects.

5. GROUNDCOVER LAYER:

Dwarf Bilberry (*Vaccinium caespitosum*): Dwarf bilberry produces small, edible berries rich in antioxidants. It thrives in cold, acidic soils and forms dense mats, making it an excellent groundcover for Zone 2. The plant improves soil health, provides habitat for pollinators, and offers high nutritional value to wildlife.

Dwarf Cornel (*Cornus suecica*): Dwarf cornel, also known as bunchberry, is a low-growing groundcover that produces edible berries. It thrives in moist, well-drained soils and cold climates, making it ideal for Zone 2 food forests. The plant provides a rich source of food for wildlife and supports pollinators with its small white flowers.

6. ROOT LAYER:

Lamb's Quarters (*Chenopodium berlandieri*): Lamb's quarters is a cold-hardy plant that produces edible roots and seeds, both of which have been used as food sources. It grows well in poor, disturbed soils and is highly resilient in cold climates. This plant also provides nitrogen-fixing benefits to the soil and supports pollinators.

Licorice Root (*Glycyrrhiza lepidota*): Licorice root produces edible roots that are often used in herbal medicine and flavoring. It grows well in moist, well-drained soils and is cold-tolerant, thriving in the harsh conditions of Zone 2. Licorice root helps improve soil structure and fixes nitrogen, benefiting surrounding plants.

7. VERTICAL LAYER:

Northern Black Currant Vine (*Ribes hudsonianum*): This cold-hardy currant produces edible black berries that are rich in nutrients. It grows well in moist, well-drained soils and can tolerate the harsh climates of Zone 2. The plant provides vertical structure and attracts wildlife with its nutritious berries.

Northern Dewberry (*Rubus flagellaris*): This trailing vine produces small, edible berries that are similar to blackberries. It grows well in poor soils and can tolerate cold conditions, making it ideal for Zone 2 food forests. Northern dewberry helps improve soil health and attracts pollinators with its white flowers.

8. FUNGAL LAYER:

Gypsy Mushroom (*Cortinarius caperatus*): This

rare, cold-hardy mushroom has a light tan cap and a fibrous texture, with a mild flavor when cooked. It grows in symbiosis with birch and spruce trees, making it ideal for Zone 2 food forests. Gypsy mushrooms enhance the fungal diversity of the forest floor and contribute to nutrient cycling.

Hawk's Wing (*Sarcodon squamosus*): This edible mushroom has a distinctive scaly cap and grows in association with coniferous trees in cold climates. Its flavor is savory and nutty, making it popular in European cuisine. Hawk's wing mushrooms improve nutrient cycling and benefit the trees with which they form symbiotic relationships.

Zone 2 Plan Six: Crisp Canopy Canvas

ZONE 2 PLAN SIX: CRISP CANOPY CANVAS

Like a work of art painted on a cool morning, this plan uses crisp, layered plantings to create a living canopy. It is designed to provide a bountiful harvest and a robust support network for the ecosystem..

1. CANOPY LAYER:

Gooseberry (*Ribes uva-crispa*): Gooseberry bushes thrive in the coldest climates and produce tart, green to red berries. These berries are highly nutritious and can be eaten fresh or cooked into jams. The plant is known for its adaptability and can form part of the canopy in food forests, supporting pollinators while providing a valuable food source.

Highbush Cranberry (*Viburnum trilobum*): Highbush Cranberry is a shrub that produces bright red berries, which are edible after being cooked and are often used in jams and sauces. It tolerates cold temperatures well and thrives in wet soils, making it ideal for colder regions. It also provides shelter for wildlife and attracts pollinators.

2. UNDERSTORY LAYER:

Mountain Ash (*Sorbus aucuparia*): A small tree or large shrub that produces clusters of bright orange-red berries. While not commonly consumed raw, the berries can be processed into jellies or syrups. Mountain Ash thrives in cold climates and offers additional support for biodiversity, providing food for birds and other wildlife.

Red Elderberry (*Sambucus pubens*): This deciduous shrub produces clusters of small, red berries that are edible when cooked. Red Elderberry thrives in moist, well-drained soils and provides both medicinal properties and wildlife benefits. It can tolerate partial to full shade and is an excellent choice for the understory layer, attracting pollinators.

3. SHRUB LAYER:

Mountain Avens (*Dryas octopetala*): This low-growing shrub produces attractive white flowers and is known for its edible leaves, which can be brewed into a tea. It thrives in rocky, cold environments, making it well-suited for Zone 2. Mountain avens is a nitrogen fixer, improving soil health in the food forest.

Serviceberry (*Amelanchier bartramiana*): This cold-hardy shrub produces sweet, edible berries that can be eaten fresh or dried. The berries are rich in vitamins and antioxidants, making them a nutritious addition to a Zone 2 food forest. The shrub is also valuable for attracting pollinators and providing food for wildlife.

4. HERBACEOUS LAYER:

Lovage (*Levisticum officinale*): Lovage is a perennial herb with edible leaves, stems, and seeds that have a strong, celery-like flavor. It is highly cold-tolerant and grows well in moist, well-drained soils in Zone 2. Lovage also attracts pollinators and beneficial insects, adding to the overall health of the food forest.

Mountain Sorrel (*Oxyria digyna*): Mountain sorrel is a cold-hardy perennial with edible, tangy leaves rich in vitamin C. It grows well in rocky, well-drained soils and can tolerate extreme cold, making it ideal for Zone 2. Mountain sorrel is a great addition to salads or as a garnish, and its presence attracts pollinators to the food forest.

5. GROUNDCOVER LAYER:

Labrador Tea (*Rhododendron groenlandicum*): Labrador tea is a cold-hardy groundcover that produces edible leaves often used in herbal teas. It thrives in acidic soils and can tolerate the harsh conditions of Zone 2. The plant provides essential nectar for pollinators and adds aromatic qualities to the food forest while offering medicinal value.

Lingonberry (*Vaccinium vitis-idaea*): Lingonberry is a low-growing evergreen shrub that produces tart, edible berries high in vitamin C. It thrives in acidic, well-drained soils and cold climates, making it perfect for Zone 2. Lingonberry provides groundcover, prevents soil erosion, and attracts pollinators with its small white flowers.

6. ROOT LAYER:

Maral Root (*Rhaponticum carthamoides*): Maral root is a hardy perennial that produces edible roots with medicinal properties. It thrives in well-drained, rocky soils and can tolerate cold conditions, making it suitable for Zone 2. Maral root also improves soil health and attracts pollinators with its large, showy flowers.

Nodding Onion (*Allium cernuum*): Nodding onion produces edible bulbs and leaves, which are often used in cooking. It thrives in rocky, well-drained soils and is highly tolerant of cold, making it perfect for Zone 2 food forests. Nodding onion adds biodiversity to the root layer and supports pollinators with its pink flowers.

7. VERTICAL LAYER:

Purple Clematis (*Clematis occidentalis*): This rare clematis species produces beautiful purple flowers and thrives in cold climates. While not commonly grown for food, the young shoots can be eaten when cooked, making it an edible addition to the vertical layer. It prefers well-drained soils and partial shade, adding a touch of elegance to the food forest and supporting pollinators.

Riverbank Grape (*Vitis riparia*): This hardy grapevine produces small, edible grapes that are tart but high in antioxidants. It is well-suited to the cold climates of Zone 2 and grows vigorously in well-drained soils. Riverbank grape is an excellent choice for adding vertical interest and attracting wildlife, including birds.

8. FUNGAL LAYER:

King Bolete (*Boletus edulis*): Also known as porcini, this prized edible mushroom grows in cold forests under conifers. Its thick, brown cap and meaty texture make it a favorite in culinary dishes. King boletes contribute to nutrient cycling in the forest and support tree growth by forming beneficial mycorrhizal partnerships.

Lobster Mushroom (*Hypomyces lactifluorum*): This parasitic fungus transforms other mushrooms, like russulas, into bright red, lobster-like specimens. The resulting mushroom is edible and has a seafood-like flavor, making it a rare culinary delicacy. It brings fungal diversity to the forest floor while contributing to nutrient recycling.

Zone 2 Plan Seven: Wintry Wonders

Zone 2 Plan Seven: Wintry Wonders

A creative workshop in the heart of winter, this design plan experiments with paired species across layers. Its focus is on uncovering hidden wonders that yield both food and enduring ecosystem support.

1. CANOPY LAYER:

Juniper (*Juniperus communis*): The berries of Juniper trees are edible and are used in cooking, especially in Nordic cuisine. These trees are hardy and can thrive in poor, dry soils, making them an ideal addition to cold-climate food forests. They also provide habitat for birds and small mammals.

Mountain Ash (*Sorbus aucuparia*): Mountain Ash, also known as Rowan, produces small, bright orange berries that are edible after being properly processed to remove toxins. The tree is known for its ability to attract wildlife and supports biodiversity. It grows well in well-drained soils and can withstand cold climates, providing a hardy source of food in the winter months.

2. UNDERSTORY LAYER:

Serviceberry (*Amelanchier alnifolia*): This small tree or large shrub produces edible berries that can be eaten raw or used in jams, jellies, and pies. Serviceberry grows well in acidic, well-drained soils and tolerates both sun and partial shade. It is a great addition to the understory layer, attracting pollinators and adding a valuable food source.

Snowberry (*Symphoricarpos albus*): This shrub produces white berries that are not typically eaten by humans but are an essential food source for birds. It thrives in full sun to partial shade and is useful for filling in the understory of a food forest, providing ground cover and supporting wildlife.

3. SHRUB LAYER:

Siberian Dogwood (*Cornus alba*): This shrub produces small, edible berries that are usually dried or made into preserves. Its red stems add winter interest to the landscape, and it thrives in cold, wet soils. Siberian dogwood is excellent for erosion control, and its flowers attract pollinators.

Siberian Peashrub (*Caragana arborescens*): This nitrogen-fixing shrub produces small, edible peas in pods, which can be used in soups or stews. It is extremely cold-hardy and thrives in poor soils, improving soil fertility for surrounding plants. Siberian peashrub is drought-tolerant and offers habitat for wildlife, making it a resilient addition to Zone 2 food forests.

4. HERBACEOUS LAYER:

Northern Sweetvetch (*Hedysarum alpinum*): Northern sweetvetch is a nitrogen-fixing herb with edible roots and seeds that have been used as a food source by Indigenous peoples. It thrives in cold, dry conditions and helps improve soil fertility in Zone 2. This plant is highly attractive to pollinators and provides habitat for wildlife.

Roseroot (*Rhodiola rosea*): Also known as golden root, roseroot produces edible leaves and roots that are known for their adaptogenic properties. It thrives in cold, rocky environments and can tolerate poor soils, making it an ideal plant for Zone 2. Roseroot supports pollinators and provides medicinal value as part of a food forest ecosystem.

5. GROUNDCOVER LAYER:

Moss Campion (*Silene acaulis*): Moss campion is a cushion-forming groundcover that produces edible leaves and small pink flowers. It thrives in cold, alpine environments and rocky soils, making it well-suited for Zone 2. Moss campion improves soil stability and provides nectar for pollinators while creating a soft, dense carpet of greenery.

Mountain Avens (*Dryas octopetala*): Mountain avens is a cold-hardy groundcover that thrives in rocky, well-drained soils. Its leaves and flowers are edible, and the plant is known for its ability to stabilize soils on slopes or in poor conditions. Mountain avens also supports pollinators and adds biodiversity to the food forest.

6. ROOT LAYER:

Silverweed (*Argentina anserina*): Silverweed produces edible roots that were traditionally harvested for food in cold regions. This resilient plant thrives in wet, sandy, or clay soils and can tolerate freezing conditions, making it well-suited for Zone 2. Silverweed also helps with soil stabilization and attracts beneficial insects.

Jerusalem Artichoke (*Helianthus tuberosus*): Jerusalem artichoke, also known as sunchoke, produces edible tubers that are high in inulin and can be roasted, mashed, or used in soups. It is highly cold-hardy and thrives in poor soils, making it a low-maintenance addition to the herbaceous layer. Its tall flowers attract pollinators and provide a natural windbreak.

7. VERTICAL LAYER:

Siberian Pea Shrub Vine (*Caragana arborescens var. pendula*): This vining shrub produces edible seeds that are high in protein, making it a valuable food source. It thrives in well-drained soils and is highly cold-tolerant, growing well in Zone 2. The plant fixes nitrogen in the soil and adds resilience to the food forest.

Snowberry Vine (*Symphoricarpos albus*): Snowberry is a cold-tolerant, climbing shrub that produces small, edible berries, though they are mildly toxic if consumed in large quantities. It thrives in well-drained soils and tolerates Zone 2 conditions well. Snowberry provides habitat for wildlife and attracts pollinators with its delicate pink flowers.

8. FUNGAL LAYER:

Red-Cracking Bolete (Xerocomellus chrysenteron): This cold-tolerant bolete mushroom is edible and grows under pines and other conifers. Its cap is reddish-brown and cracks open as it matures, revealing yellow flesh. Red-cracking boletes contribute to nutrient recycling and form mycorrhizal partnerships with trees, benefiting soil and plant health.

Scaly Hedgehog (*Sarcodon imbricatus*): This rare edible mushroom, also known as the hawk's wing, grows in coniferous forests and forms mycorrhizal relationships with trees. The cap is scaly and brown, and its flesh has a savory, nutty flavor that makes it great for soups and stews. Scaly hedgehog mushrooms help enhance soil health by decomposing organic matter and forming symbiotic relationships with plants.

Zone 2 Plan Eight: Cold Core Cultivation

Centered on building a strong "core" of cold-adapted plants, this plan ensures that every layer contributes to a resilient food forest. It is perfect for growers looking to maximize edibles under brisk conditions.

1. CANOPY LAYER:

Mulberry (*Morus spp.*): Mulberry trees produce sweet, edible fruits that can be eaten fresh or used in jams and desserts. These trees grow quickly, providing shade and food in the food forest. Their berries are a favorite among birds, making them an excellent wildlife support species.

Osier Willow (*Salix viminalis*): While not traditionally thought of as a food plant, the shoots of the Osier Willow can be used to create basketry, and the plant supports the ecosystem by stabilizing soil. It is also used as a biomass crop, offering utility to the food forest by improving the soil structure and providing habitat.

2. UNDERSTORY LAYER:

Spicebush (*Lindera benzoin*): An aromatic shrub with yellow flowers in early spring and red berries in the fall. The berries of Spicebush are edible and can be used for flavoring. It grows best in well-drained, rich soils and partial shade, making it perfect for the understory of a food forest where it can enhance biodiversity.

Thimbleberry (*Rubus parviflorus*): A fast-growing shrub that produces soft, edible berries in early summer. Thimbleberry grows well in moist, well-drained soils and partial to full sun. It provides food for both humans and wildlife and serves as an excellent ground cover in the understory layer.

3. SHRUB LAYER:

Silver Buffaloberry (*Shepherdia canadensis*): This

hardy shrub produces edible red berries that can be eaten raw or cooked into jams and jellies. It fixes nitrogen, improving soil health and benefiting nearby plants in the food forest. Silver buffaloberry is drought-tolerant and thrives in poor soils, making it an excellent choice for challenging Zone 2 conditions. Its berries are high in vitamin C and attract wildlife like birds and pollinators.

Soapberry (*Shepherdia rotundifolia*): Soapberry is an underused shrub that produces bitter, edible berries traditionally whipped into a frothy dessert by Indigenous peoples. The plant thrives in dry, cold environments and plays a role in nitrogen fixation, enriching the soil. Its berries are also high in antioxidants, and the shrub's dense foliage provides shelter for wildlife.

4. HERBACEOUS LAYER:

Salsify (*Tragopogon porrifolius*): Salsify is a cold-hardy biennial herb that produces edible roots resembling oysters in flavor. It thrives in poor soils and cold climates, making it an excellent addition to a Zone 2 food forest. The plant also improves soil structure with its deep taproot and supports pollinators with its yellow flowers.

Sorrel (*Rumex acetosa*): Sorrel is a perennial herb with tangy, edible leaves that are often used in soups and salads. It is extremely cold-hardy and thrives in moist, well-drained soils in Zone 2. Sorrel is rich in vitamin C and iron, and its presence in the food forest adds diversity and attracts pollinators.

5. GROUNDCOVER LAYER

Partridgeberry (*Mitchella repens*): Partridgeberry is a creeping evergreen groundcover that produces edible red berries, often used in jams or eaten fresh. It thrives in moist, shady areas and is extremely cold-hardy, making it suitable for Zone 2 food forests. Partridgeberry also attracts pollinators and helps stabilize soil with its spreading habit.

Pasqueflower (*Anemone patens*): Pasqueflower is a resilient groundcover that produces edible flowers and roots, traditionally used by Indigenous peoples. It grows well in rocky, well-drained soils and can tolerate extreme cold, making it a great addition to Zone 2. Pasqueflower also attracts early-season pollinators and helps stabilize soil.

6. ROOT LAYER:

Spikenard (*Aralia racemosa*): Spikenard produces aromatic, edible roots that are often used in herbal teas and traditional medicine. It thrives in moist, shaded areas and can tolerate cold climates, making it ideal for Zone 2. Spikenard also supports pollinators and adds diversity to the food forest with its large, white flowers.

Spring Beauty (*Claytonia caroliniana*): Spring beauty, also known as fairy spud, produces small edible tubers that have a nutty flavor. This cold-hardy plant thrives in moist, shaded areas and helps to improve soil structure with its spreading roots. Spring beauty also supports early-season pollinators with its delicate pink flowers.

7. VERTICAL LAYER:

Sand Cherry Vine (*Prunus pumila*): This cold-hardy vine produces small, edible cherries that are sweet and tart. It thrives in sandy, well-drained soils and can tolerate cold conditions in Zone 2. Sand cherry adds vertical structure, provides food for humans and wildlife, and supports pollinators with its fragrant flowers.

Sea Buckthorn Vine (*Hippophae rhamnoides*): This rare vine produces small, edible orange berries that are rich in vitamin C. It thrives in poor, well-drained soils and can tolerate extreme cold, making it suitable for Zone 2. Sea buckthorn also fixes nitrogen, improves soil health, and supports pollinators.

8. FUNGAL LAYER:

Northern Tooth Fungus (*Albatrellus ovinus*): This rare mushroom produces edible fruiting bodies with a unique, meaty texture. It grows in association with conifers and is found in cold climates like Zone 2. Northern tooth fungus enhances nutrient cycling in the soil and supports forest health by breaking down wood debris.

Puffball Mushroom (*Lycoperdon perlatum*): This common edible mushroom forms small, round fruiting bodies that puff out spores when touched. It grows in forests and grassy areas and is hardy to Zone 2. Puffballs help break down organic matter, improve soil structure, and provide a food source for wildlife.

Zone 2 Plan Nine: Icebound Impact Initiative

ZONE 2 PLAN NINE: ICEBOUND IMPACT INITIATIVE

This initiative is all about making a significant ecological impact in chillier zones. It pairs robust, cold-hardy plants in each layer to ensure both a hearty harvest and long-term sustainability.

1. CANOPY LAYER:

Saskatoon Berry (*Amelanchier alnifolia*): Saskatoon Berry, also called Juneberry, produces edible berries that are sweet when ripe. The tree is resilient to cold temperatures and can grow in a range of soil types. It's an excellent addition to a food forest, offering both food for humans and birds, and also enhances soil health by attracting pollinators.

Sea Buckthorn (*Hippophae rhamnoides*): Known for its orange berries rich in Vitamin C, Sea Buckthorn is a hardy shrub that grows well in cold climates and well-drained soils. The berries are edible and can be used in jams, juices, and oils. It's also an excellent plant for erosion control and attracts a variety of pollinators.

2. UNDERSTORY LAYER:

Western Chokecherry (*Prunus virginiana*): This small tree or shrub produces clusters of white flowers followed by dark purple to black cherries, which are edible but need to be cooked to remove toxins. Western Chokecherry can grow in a variety of soils, from clay to sandy, and provides food for wildlife, making it valuable for a food forest ecosystem.

Wild Ginger (*Asarum canadense*): Wild Ginger is a low-growing perennial that produces edible rhizomes with a unique ginger flavor. It thrives in shaded, moist conditions and is an excellent ground cover in the understory layer. Its heart-shaped leaves and beautiful flowers provide additional aesthetic appeal and support local pollinators.

3. SHRUB LAYER:

Thimbleberry (*Rubus parviflorus*): Thimbleberry produces soft, edible red berries that are high in antioxidants and perfect for fresh eating or making jams. The shrub grows well in cold climates and moist soils, providing a natural food source for both wildlife and humans. Its large, maple-like leaves also add texture to the food forest.

Twinberry (*Lonicera involucrata*): This deciduous shrub produces black, edible berries that are slightly bitter but can be used in jams. Twinberry is valuable for its rapid growth, ability to stabilize soil, and attraction of pollinators. It is well-adapted to cold climates and moist soils in Zone 2.

4. HERBACEOUS LAYER:

Sweet Cicely (*Myrrhis odorata*): Sweet cicely is a cold-hardy herb with edible leaves, seeds, and roots that have a sweet, anise-like flavor. It thrives in partial shade and moist soils, making it a versatile addition to the herbaceous layer. Its leaves can be used in salads or as a sweetener, and the plant also attracts pollinators.

Wild Bergamot (*Monarda fistulosa*): Wild bergamot, also known as bee balm, produces edible flowers and leaves that can be used in teas or as a garnish. It thrives in cold climates and is highly attractive to pollinators, making it an essential part of any Zone 2 food forest. Its fragrant foliage adds biodiversity and enhances soil health by supporting beneficial insects.

5. GROUNDCOVER LAYER:

Purple Saxifrage (*Saxifraga oppositifolia*): Purple saxifrage is a cold-hardy, mat-forming groundcover that thrives in rocky or poor soils. Its small edible leaves are rich in vitamin C and were traditionally used as a food source in harsh climates. This plant is highly tolerant of freezing temperatures and helps to stabilize soil while attracting early-season pollinators with its vivid purple flowers.

Starry False Solomon's Seal (*Maianthemum stellatum*): Starry false Solomon's seal is a cold-hardy groundcover with edible shoots, berries, and rhizomes. The plant thrives in partial shade and moist soils, making it ideal for groundcover in Zone 2. It offers habitat for wildlife and helps improve soil structure with its spreading root system.

6. ROOT LAYER:

Spring Beauty (*Claytonia caroliniana*): Spring beauty, also known as fairy spud, produces small edible tubers that have a nutty flavor. This cold-hardy plant thrives in moist, shaded areas and helps to improve soil structure with its spreading roots. Spring beauty also supports early-season pollinators with its delicate pink flowers.

Wild Onion (*Allium textile*): Wild onion is a cold-hardy plant that produces edible bulbs with a strong onion flavor. It grows well in rocky, well-drained soils and can tolerate harsh conditions in Zone 2. The plant helps improve soil health and attracts pollinators with its small, white flowers.

7. VERTICAL LAYER:

Trailing Arbutus (*Epigaea repens*): Trailing arbutus, also known as mayflower, produces small, edible berries and is well-adapted to cold climates. It thrives in acidic, well-drained soils and prefers partial shade. The plant adds ground cover, supports early-season pollinators, and provides a food source for wildlife.

Trailing Raspberry (*Rubus pubescens*): Trailing raspberry is a cold-tolerant vine that produces small, sweet, edible raspberries. It grows in moist, shaded soils and can tolerate cold climates, making it ideal for Zone 2. The plant supports wildlife, enhances soil health, and provides food for humans and animals alike.

8. FUNGAL LAYER:

Slippery Jack (*Suillus luteus*): This cold-tolerant mushroom grows under pines and has a sticky brown cap with a sponge-like underside. It is edible and often used in soups and stews for its mild flavor. Slippery jack mushrooms form mycorrhizal relationships with conifer roots, improving soil structure and tree health.

Snowbank False Morel (*Gyromitra gigas*): This rare mushroom grows in cold, high-elevation environments and emerges near melting snowbanks. While it must be cooked carefully to remove toxins, it is edible and highly prized by foragers. Snowbank false morels add fungal diversity and help break down organic matter in the food forest.

Zone 2 Plan Ten: Subzero Spectrum

Zone 2 Plan Ten: Subzero Spectrum

Embracing the full spectrum of life even in subzero conditions, this plan layers a variety of edibles for mutual support. Its strategy is to deliver vibrant nutrition and enduring ecosystem strength despite the cold.

1. CANOPY LAYER:

Serviceberry (*Amelanchier canadensis*): Also called the Shadblow or Juneberry, this tree produces small, sweet berries that are edible and enjoyed by both humans and wildlife. It grows well in colder climates and attracts pollinators with its early spring flowers. The tree's adaptability and high yield make it an excellent food source in a food forest.

Silver Maple (*Acer saccharinum*): Silver Maple is a fast-growing tree known for its adaptability and resilience to cold temperatures. While its primary role is as a canopy tree, its sap can be harvested in early spring to make maple syrup. It helps support the ecosystem by offering shelter to various wildlife.

2. UNDERSTORY LAYER:

Wild Plum (*Prunus americana*): A small, deciduous tree or large shrub that produces sweet, edible plums in the summer. It grows well in a variety of soils, including clay and loam, and is highly adaptable to harsh climates. Wild Plum provides early-season food for both humans and wildlife and thrives in full sun to partial shade.

Woodland Strawberry (*Fragaria vesca*): This low-growing perennial produces small, sweet strawberries that thrive in shaded, moist areas. Woodland Strawberry can be used as a ground cover in the understory layer, providing both food and habitat for pollinators. It grows well in acidic, well-drained soils and can spread quickly.

3. SHRUB LAYER

White Arctic Raspberry (*Rubus arcticus*): This rare and cold-hardy shrub produces small, sweet raspberries that are prized for their flavor. It thrives in moist, well-drained soils and forms a low-growing thicket, making it a useful ground cover in Zone 2 food forests. The berries are high in vitamins and can be eaten fresh or made into preserves.

Wild Gooseberry (*Ribes hirtellum*): Wild gooseberry is a thorny shrub that produces tart, edible berries used in pies and preserves. The plant is cold-hardy, thriving in forest understories and open woods. Its thorny branches provide a protective habitat for small animals, and the berries attract birds and pollinators.

4. HERBACEOUS LAYER:

Wild Parsnip (*Pastinaca sativa*): Wild parsnip produces edible roots that are often used as a substitute for cultivated parsnips. It thrives in poor soils and cold climates, making it ideal for Zone 2. This plant adds deep roots to the soil, improving its structure and supporting other plants in the food forest.

Wood Nettle (*Laportea canadensis*): Wood nettle produces edible young shoots and leaves that are high in vitamins and minerals. It grows well in moist, shady areas and can tolerate cold climates, making it an excellent fit for Zone 2 food forests. Wood nettle also supports soil health by providing organic matter as it decomposes.

5. GROUNDCOVER LAYER:

Wild Blue Phlox (*Phlox divaricata*): Wild blue phlox is a shade-tolerant groundcover that produces edible leaves and flowers, often used as garnishes. It grows well in moist, well-drained soils and is highly cold-hardy, making it suitable for Zone 2. This plant attracts pollinators and helps improve soil stability with its spreading habit.

Wild Ginger (*Asarum canadense*): Wild ginger is a shade-loving, low-growing groundcover that produces edible rhizomes with a mild ginger flavor. It thrives in moist, well-drained soils and is highly cold-tolerant, making it perfect for Zone 2 food forests. This plant helps suppress weeds and adds aromatic qualities while supporting beneficial insects.

6. ROOT LAYER:

Yellow Avalanche Lily (*Erythronium grandiflorum*): The yellow avalanche lily produces small edible bulbs that were traditionally consumed by Indigenous peoples. This cold-hardy plant thrives in well-drained, moist soils and is well-adapted to the cold conditions of Zone 2. It is also an early bloomer, supporting pollinators while providing a beautiful splash of color in the food forest.

Yellow Nutgrass (*Cyperus esculentus*): Yellow nutgrass, also known as chufa, produces edible tubers that are rich in nutrients and often used as a food source. It grows well in moist, well-drained soils and is tolerant of cold conditions, making it suitable for Zone 2. The plant helps improve soil structure and supports beneficial insects.

7. VERTICAL LAYER:

White Water Lily Vine (*Nymphaea odorata*): While not a true vine, the water lily's long stems trail underwater and provide excellent vertical habitat in ponds or marshy areas. Its edible seeds and rhizomes can be harvested, and it thrives in cold climates. The plant supports water-based ecosystems and attracts pollinators with its fragrant white flowers.

Woodbine (*Parthenocissus vitacea*): Woodbine is a cold-hardy vine that produces small, inedible berries but offers excellent vertical structure for wildlife habitat. It thrives in a wide range of soils and can tolerate the cold conditions of Zone 2. This fast-growing vine supports pollinators and provides shade in the food forest.

8. FUNGAL LAYER:

The Miller (*Clitopilus prunulus*): This edible mushroom has a white to light gray cap and a distinctive grainy texture. It grows in association with broadleaf trees and is suited to Zone 2. The Miller improves soil health by decomposing organic material and providing nutrients for nearby plants.

Wood Ear Mushroom (*Auricularia auricula-judae*): This edible fungus grows on dead or decaying wood and is prized in Asian cuisine for its crunchy texture. It is cold-hardy and suitable for Zone 2 food forests. Wood ear mushrooms contribute to decomposition, helping to recycle nutrients and maintain soil health.

ZONE 3

Plant Hardiness Zone 3

USDA Plant Hardiness Zone 3 experiences long, cold winters with minimum temperatures ranging from -40°F to -30°F (-40°C to -34°C). This zone covers regions in northern U.S. states like Minnesota, North Dakota, and Montana, high-altitude areas, and parts of Alaska and Canada. The growing season is relatively short, lasting only 90 to 120 days, and gardeners must choose cold-tolerant, fast-growing plants.

Characteristics of Zone 3

Winter Temperatures: Winter temperatures regularly fall between -40°F and -30°F (-40°C to -34°C).

Growing Season: The short growing season, lasting 90–120 days, requires careful plant selection for success.

Frost Dates: Frost can persist into late spring (June), and the first frost may come as early as late August or September.

Soil Conditions: Soils vary but are often acidic, clay-rich, or rocky, with poor drainage. Soil amendments like compost and mulch are vital for improving soil health.

Climate Conditions of Zone 3

Short Growing Seasons: Fast-growing, cold-tolerant plants are necessary due to the brief growing period.

Severe Winters: Heavy snow provides insulation for plants but creates challenges for early spring growth.

High Winds and Exposure: Strong winds can damage crops, making windbreaks and sheltered planting essential.

Precipitation: Moderate rainfall is common, but periods of drought can occur, making irrigation and water conservation important.

Challenges of Zone 3 Gardening

Extreme Cold: Frost can occur throughout the growing season, making frost-hardy plant selections essential.

Short Growing Season: Plants need to mature quickly, so choosing early-season vegetables, fruits, and perennials is key. Using season-extending methods such as greenhouses and row covers can help.

Wind Exposure: High winds necessitate the use of windbreaks or planting in sheltered locations.

Soil Quality: Nutrient-poor or poorly draining soils require amendments like organic material and raised beds to improve soil health and drainage.

Ecological Zones in Zone 3

Boreal Forests: Dominated by conifers such as spruce, pine, and fir, boreal forests thrive in the colder parts of Zone 3.

Alpine Zones: Found in higher altitudes, alpine zones face even harsher winters and shorter growing seasons, requiring cold-adapted plants.

Tundra: In northern regions, tundra landscapes are home to hardy, low-growing vegetation capable of surviving year-round freezing temperatures.

Key Strategies for Zone 3 Planting

Layering for Shelter: Use taller trees and

shrubs as windbreaks to shield delicate plants from extreme winter winds. Plant these windbreaks on the windward side of the food forest.

Early Harvest and Quick-Maturing Plants: Select fast-growing crops that can be harvested early, such as early-season vegetables and fruit trees like apples and cherries.

Frost Protection: Cold frames, greenhouses, and row covers can protect sensitive plants from late and early frosts, extending the growing season.

Soil Preparation: Improve soil quality by adding organic materials like compost and mulch. Raised beds can enhance drainage and soil temperature.

Native Plants: Utilize native species that are adapted to Zone 3 conditions, as they are more resistant to pests, diseases, and extreme weather.

Water Conservation: Employ water-saving techniques such as drip irrigation, mulching, and rainwater harvesting to address potential drought periods during the growing season.

By focusing on cold-hardy, quick-maturing plants, creating protective structures, and improving soil and water management, gardeners in Zone 3 can create a productive and resilient food forest.

Now, let's explore the curated design plans I created for Zone 3. Each plan includes all seven traditional layers, with two plant options per layer and a bonus researched fungal layer to support soil health and plant growth. Enjoy!

Zone 3 Plan One: Brisk Bounty Blueprint

A design plan that channels the energy of a brisk climate to yield a generous bounty. Its layered structure is crafted to maximize food output while nurturing soil and system vitality.

1. CANOPY LAYER:

Bur Oak (*Quercus macrocarpa*): This cold-hardy oak tree produces large edible acorns, which can be leached to remove tannins and used as a nutritious food source. The bur oak is highly resilient, tolerating drought, poor soil, and extreme cold, making it ideal for Zone 3. It provides shelter and food for wildlife and helps improve soil health with its deep root system.

Kentucky Coffeetree (*Gymnocladus dioicus*): This rare tree produces large seed pods containing seeds that can be roasted and used as a coffee substitute. It thrives in cold climates, tolerating frost and drought, and contributes to nitrogen fixation, improving soil fertility. Kentucky coffeetrees bring unique habitat value to the food forest by providing shade and shelter for wildlife.

2. UNDERSTORY LAYER:

Buffaloberry (*Shepherdia argentea*): Buffaloberry is a cold-hardy shrub that produces small, tart red berries that are edible and rich in antioxidants. It is a nitrogen-fixing plant, improving soil health and benefiting nearby species. Buffaloberry thrives in poor soils and is drought-tolerant, making it a resilient choice for Zone 3 food forests.

Dwarf Birch (*Betula nana*): This rare, cold-hardy birch produces small, edible catkins that can be consumed when young. Dwarf birch contributes to the food forest by improving soil aeration and stabilizing the forest floor with its shallow root system. It also provides shelter for wildlife and grows well in wet, boggy conditions.

3. SHRUB LAYER:

Lingonberry (*Vaccinium vitis-idaea*): Lingonberry is a hardy, low-growing shrub that produces small, tart, red berries rich in antioxidants and vitamin C. The berries are commonly used in jams, sauces, and juices. Lingonberries thrive in acidic, well-drained soils and help attract pollinators with their delicate pink flowers, while providing a valuable food source for wildlife.

Siberian Pea Shrub (*Caragana arborescens*): This nitrogen-fixing shrub produces small, edible pods that are high in protein, making it a valuable addition to a Zone 3 food forest. The plant improves soil health, enhances biodiversity, and provides shelter for beneficial insects. It thrives in poor soils and harsh climates, helping to create a more resilient ecosystem.

4. HERBACEOUS LAYER:

Good King Henry (*Blitum bonus-henricus*): Good King Henry is a perennial herb that produces edible leaves and flower shoots, often used like spinach or asparagus. It thrives in cold climates and poor soils, making it a resilient choice for Zone 3. This plant improves soil structure with its deep root system and provides early spring greens when other crops are not yet ready.

Saltbush (*Atriplex hortensis*): Also known as orach, saltbush is an ancient leafy green with edible leaves that can be eaten raw or cooked like spinach. It tolerates salty and poor soils, making it ideal for tough growing conditions. Its leaves are high in minerals, and it adds diversity to the herbaceous layer of the food forest.

5. GROUNDCOVER LAYER:

Wild Ginger (*Asarum canadense*): Wild ginger is

an evergreen groundcover with heart-shaped leaves and aromatic roots that are edible when cooked. The roots can be used as a spice or flavoring. This plant thrives in shaded, moist environments and helps suppress weeds, improving soil quality. It also attracts pollinators with its hidden flowers.

Bunchberry (*Cornus canadensis*): Bunchberry is a low-growing groundcover that produces edible red berries in the summer, which can be eaten raw or used in jams and jellies. It thrives in shady, acidic soils and spreads through rhizomes, creating dense coverage. Its white flowers attract pollinators, adding beauty and biodiversity to the food forest.

6. ROOT LAYER:

Skirret (*Sium sisarum*): Skirret is a hardy root vegetable that produces sweet, edible tubers, often described as tasting like a cross between parsnips and carrots. It thrives in moist, well-drained soils and can tolerate partial shade, making it ideal for the root layer in a food forest. Its vigorous growth habit helps to aerate the soil, benefiting other plants in the system.

Oca (*Oxalis tuberosa*): Oca is a lesser-known tuber that produces colorful, tangy roots, which are edible and can be cooked in a variety of ways. It grows in well-drained soils and thrives in partial shade, offering a nutritious and unique alternative to potatoes. The plant also provides ground cover and can be used to suppress weeds in the root layer.

7. VERTICAL LAYER:

Black Chokeberry (*Aronia melanocarpa*): Black chokeberry is a hardy, fruit-bearing vine that thrives in USDA Hardiness Zone 3. Its edible berries can be eaten raw, made into jams, or used for juice, and they are known for their antioxidant-rich content. The plant's upright growth habit makes it suitable for the vertical layer, and its attractive white flowers also support pollinators. Black chokeberry is tolerant of a variety of soil types and conditions, adding diversity to a food forest.

Purple-Flowering Honeysuckle (*Lonicera x purpurea*): This ornamental, fruit-bearing honeysuckle vine offers edible berries that can be eaten raw or used in preserves. It thrives in full sun to partial shade and is known for its fragrant flowers, which attract pollinators. It grows quickly, making it an excellent vertical element that can also provide a decorative touch to the food forest, while its roots contribute to soil health.

8. FUNGAL LAYER:

Black Trumpet Mushroom (*Craterellus cornucopioides*): Known for its unique, trumpet-like shape, the black trumpet mushroom is an edible species prized for its rich, smoky flavor. It grows in the forest floor, often near decaying organic matter, and prefers a humid environment, making it an excellent choice for the fungal layer. The mushroom is also known for being relatively rare and thrives in well-shaded, damp areas, offering both culinary value and ecological benefits by decomposing organic material.

Candy Cap Mushroom (*Lactarius fragilis*): Candy cap mushrooms are edible fungi that have a sweet, maple syrup-like fragrance when dried. These rare mushrooms grow under oak and other hardwood trees, benefiting the ecosystem by breaking down organic matter. They are a unique addition to the fungal layer, providing an interesting culinary ingredient for desserts or flavoring syrups.

Audra Haldane

Zone 3 Plan Two: Snowcap Sustenance

Tailored for conditions where snow still graces the landscape, this scheme pairs complementary plants to build a thriving food forest. It emphasizes sustainable nutrition and ecological resilience.

1. CANOPY LAYER:

Shagbark Hickory (*Carya ovata*): This tree is known for its sweet, edible nuts, which are a valuable food source for both humans and wildlife. Its rugged bark provides habitat for bats and other creatures. Shagbark hickory trees are highly resilient in cold climates and improve soil health by accumulating leaf litter that enriches the forest floor.

Butternut (*Juglans cinerea*): Also known as white walnut, this tree produces rich, oily nuts that are highly prized for baking and cooking. Butternuts thrive in cold climates and contribute to the food forest by enhancing soil structure through their deep root systems. Their nuts provide a source of protein and fats for humans and animals alike.

2. UNDERSTORY LAYER:

Elderberry (*Sambucus nigra ssp. cerulea*): Elderberry produces clusters of small, dark purple berries that are edible when cooked and can be used for jams, syrups, and wines. It supports pollinators with its abundant flowers and attracts birds, which feed on the berries. Elderberry also has medicinal uses and helps enhance soil fertility through its deep roots.

Siberian Pea Shrub (*Caragana microphylla*): This hardy, nitrogen-fixing shrub produces small, edible pods that are high in protein and can be used in soups and salads. Siberian pea shrub thrives in cold climates and poor soils, making it an excellent addition to the understory of a Zone 3 food forest. It supports beneficial insects and improves overall soil health.

3. SHRUB LAYER:

Bog Bilberry (*Vaccinium uliginosum*): Bog bilberry, also known as bog blueberry, produces sweet, edible berries that are similar to blueberries and can be eaten fresh or used in baking. It grows well in cold, boggy soils, adding diversity to the shrub layer and attracting pollinators. The berries are rich in vitamins and antioxidants, supporting wildlife as well as human consumption.

Buffaloberry (*Shepherdia argentea*): Buffaloberry is a nitrogen-fixing shrub that produces small, edible red berries high in vitamin C and antioxidants. The plant thrives in poor, sandy soils and is highly drought-tolerant, making it ideal for colder climates. Buffaloberry also supports wildlife by providing food and habitat for birds and small mammals.

4. HERBACEOUS LAYER:

Wild Bergamot (*Monarda fistulosa*): Wild bergamot, also known as bee balm, has edible flowers and leaves that can be used in teas or as a garnish. It attracts pollinators like bees and butterflies, improving pollination rates in the food forest. This plant also has antifungal and antimicrobial properties, making it a useful herb for medicinal purposes.

Wood Sorrel (*Oxalis montana*): Wood sorrel is a low-growing herb with tart, lemony-flavored leaves that can be eaten raw or used in salads and soups. It thrives in shaded areas and helps fill in ground cover in the herbaceous layer. The plant is high in vitamin C and can serve as an early-season edible in Zone 3 food forests.

5. GROUNDCOVER LAYER:

Creeping Raspberry (*Rubus idaeus var. repens*): A low-growing, spreading raspberry variety, creeping raspberry produces small, sweet berries that are perfect for snacking or making preserves. It thrives in partial shade to full sun and helps suppress weeds while providing food for both people and wildlife. Its dense growth habit also aids in soil retention.

Lamb's Quarters (*Chenopodium album*): Lamb's quarters is a fast-growing groundcover that produces edible young leaves, seeds, and stems. The leaves are high in vitamins A and C and can be eaten raw or cooked. It tolerates poor soils and spreads quickly, helping prevent soil erosion while providing valuable nutrition.

6. ROOT LAYER:

Crosne (*Stachys affinis*): Known for its edible, knobby tubers, Crosne (also called Japanese artichoke) is a unique root vegetable that thrives in well-drained, fertile soils. The tubers can be eaten raw or cooked, offering a slightly nutty flavor. This plant is a resilient grower and can be a valuable addition to the root layer, enhancing soil health and diversity.

Salsify (*Tragopogon porrifolius*): Salsify, often called "oyster plant," produces long, edible roots with a flavor reminiscent of oysters when cooked. It grows best in well-drained, sandy soils and can be planted in the root layer of a food forest. Its roots provide a nutritious food source while also helping to break up compacted soil.

7. VERTICAL LAYER:

Climbing Nasturtium (*Tropaeolum majus*): Nasturtiums are known for their vibrant flowers and edible leaves and flowers, which have a peppery, spicy flavor. The plant grows well in poor, well-drained soils and thrives in full sun or partial shade, making it an excellent vertical layer plant that supports pollinators. Its climbing habit makes it ideal for trellises or fences, and its leaves are often used in salads or as a garnish.

Golden Kiwi (*Actinidia chinensis var. deliciosa*):

Golden Kiwi, a rare variety of kiwi, produces sweet, edible fruit that can be eaten raw or used in fruit salads and desserts. The vine climbs vigorously, making it a perfect fit for the vertical layer, and it requires a sunny, sheltered location to thrive. Hardy in Zone 3, this kiwi variety adds beauty, food production, and habitat support for pollinators in the food forest.

8. FUNGAL LAYER:

Chanterelle Mushroom (*Cantharellus cibarius*): Chanterelle mushrooms are one of the most sought-after wild edible fungi, known for their delicate, fruity flavor and golden color. They thrive in moist, well-drained soils and form symbiotic relationships with trees, making them an essential part of the fungal layer. The mushrooms support soil health while offering a high-value culinary addition to the food forest.

Common Earthstar (*Geastrum triplex*): This rare and striking fungus is known for its star-shaped fruiting body, which resembles a flower when it opens. While not typically consumed fresh, its spores are sometimes collected for culinary use. Earthstars contribute to the fungal layer by decomposing organic material and enriching the soil, providing both aesthetic and ecological value to the food forest.

Zone 3 Plan Three: Chill Charm Cultivation

This plan cultivates the unique charm of a cooler environment by layering resilient, edible species. It is meant to charm both the palate and the ecosystem with its sustainable design.

1. Canopy Layer:

Siberian Pea Tree (*Caragana arborescens*): This cold-hardy tree produces small, edible pods that contain seeds high in protein. It fixes nitrogen in the soil, making it a valuable addition to the food forest. The Siberian pea tree is drought-tolerant and can thrive in poor soils, supporting the health of surrounding plants.

Manchurian Walnut (*Juglans mandshurica*): This rare tree produces edible nuts that can be consumed raw or roasted. It thrives in cold climates and brings ecological diversity to the food forest by improving soil structure and providing shade. The Manchurian walnut is also valuable for wildlife, offering food and shelter.

2. UNDERSTORY LAYER:

Hazel (*Corylus americana*): The American hazel produces edible nuts, which are a great source of protein and fats. It thrives in colder climates, creating a dense understory layer that provides shade and wind protection. Hazel shrubs also enhance soil health by producing organic matter and attracting wildlife that feeds on the nuts.

Bog Rosemary (*Andromeda polifolia*): Bog rosemary, a rare plant native to cold boggy areas, produces small, edible leaves used in teas, though caution is needed due to potential toxicity in large amounts. It thrives in acidic, waterlogged soils, contributing to soil moisture retention and providing habitat for insects and pollinators. Its unique pink flowers make it a visually appealing addition to the forest.

3. SHRUB LAYER:

Bearberry (*Arctostaphylos uva-ursi*): Also known as kinnikinnick, bearberry is a low-growing shrub that produces small red berries that are traditionally used in Indigenous medicines and teas. Bearberry thrives in poor, sandy soils and is highly drought-tolerant, providing ground cover and helping to prevent erosion. Its evergreen leaves offer year-round interest and shelter for wildlife.

Cloudberry (*Rubus chamaemorus*): Cloudberry is a rare, cold-hardy shrub that produces edible orange berries, often used in jams and desserts. The plant grows well in wet, boggy areas and is highly prized for its unique flavor and nutritional value. Cloudberries help maintain soil moisture and provide nectar for pollinators like bees and butterflies.

4. HERBACEOUS LAYER:

Sweet Cicely (*Myrrhis odorata*): Sweet cicely is a shade-tolerant herb with edible seeds, roots, and leaves that have a sweet, anise-like flavor. It is commonly used as a natural sweetener in teas and desserts. This plant is highly beneficial for pollinators, and its deep roots help to aerate and improve soil structure.

Horseradish (*Armoracia rusticana*): Horseradish is a hardy, cold-tolerant plant with edible roots that are commonly grated and used as a spicy condiment. It is easy to grow and helps repel pests from the food forest, while its deep roots improve soil structure. The leaves can also be used in cooking or as a mulch.

5. GROUNDCOVER LAYER:

Creeping Thyme (*Thymus serpyllum*): Hardy groundcover herb.

Wild Strawberry (*Fragaria virginiana*): Wild

strawberry is a low-growing, spreading plant that produces small, sweet berries that are edible fresh or in jams. This plant thrives in full sun and well-drained soils, forming a dense mat that helps control weeds while providing early-season fruit for the food forest.

6. ROOT LAYER:

Spikenard (*Aralia racemosa*): Spikenard is an herbaceous perennial that produces edible roots with medicinal properties. The roots, used in traditional medicine, have an aromatic, spicy flavor and can be used to make teas or as a flavoring in cooking. It thrives in rich, moist soils and can support the surrounding plants by improving soil quality.

Celeriac (*Apium graveolens var. rapaceum*): Also known as celery root, celeriac is a hardy root vegetable that thrives in moist, well-drained soils. It produces large, edible roots with a distinct celery flavor, used in soups, stews, and mashed dishes. Its deep roots help to break up compacted soil, improving overall soil health.

7. VERTICAL LAYER:

Creeping Raspberry (*Rubus pentalobus*): Creeping raspberry is a low-growing vine with edible fruit that can be used in jams, desserts, or eaten raw. Its leaves can be used to make herbal teas, and the plant offers ground cover while climbing vertically with support. The plant is resilient and hardy, making it suitable for Zone 3 climates, and its roots also contribute to soil health by preventing erosion.

Climbing Strawberry (*Fragaria chiloensis*): This lesser-known strawberry variety produces small, sweet berries that grow along its vines. The plant is well-suited for vertical growth on trellises or fences, allowing it to produce fruit without taking up horizontal space. It thrives in full sun and well-drained soils and attracts pollinators, adding both edible produce and ecological benefits to the food forest.

8. FUNGAL LAYER:

Fairy Ring Mushroom (*Marasmius oreades*): Often found in circular formations, the fairy ring mushroom is edible and used in cooking for its mild flavor. This mushroom thrives in grassy areas but can also grow in the forest floor, benefiting the food forest by breaking down organic matter and returning nutrients to the soil. The fungus is valuable for both culinary and soil health purposes, adding to the ecological dynamics of the fungal layer.

Fawn Mushroom (*Paxillus involutus*): The fawn mushroom, while toxic when raw, is edible when cooked thoroughly and has a nutty flavor. It grows in association with certain trees, often under birch and oak. As a decomposer, it plays a role in recycling nutrients in the food forest and can help improve soil quality, while its edible qualities add an interesting twist to the fungal layer.

Audra Haldane

Zone 3 Plan Four: Frost Field Framework

Designed for fields touched by frost, this framework interlaces layers of food-producing plants for mutual support. Its purpose is to deliver steady nutrition and build a strong ecological network.

1. CANOPY LAYER:

American Chestnut (*Castanea dentata*): This once-abundant tree produces sweet, edible chestnuts that are used in baking and cooking. Although it has faced disease challenges, cold-hardy hybrids are being developed for reintroduction in northern climates. American chestnuts contribute to the food forest by providing a fast-growing canopy and improving soil through leaf litter.

Northern Pecan (*Carya illinoinensis*): This cold-hardy variety of pecan produces sweet, edible nuts that are a valuable source of protein and fats. It thrives in Zone 3 and helps improve soil fertility through its large leaf litter, which breaks down and enriches the forest floor. The northern pecan also supports wildlife with food and shelter.

2. UNDERSTORY LAYER:

Canada Buffaloberry (*Shepherdia canadensis*): This nitrogen-fixing shrub produces small, tart, edible berries that are used in traditional Indigenous cuisine and are high in vitamin C. Canada buffaloberry helps improve soil structure and fertility and provides food for wildlife, including birds and mammals. It is highly drought-tolerant and thrives in poor soils.

Beaked Hazelnut (*Corylus cornuta*): Beaked hazelnut produces small, edible nuts that are encased in a distinctive, elongated husk. It thrives in the cold climates of Zone 3, providing food for both humans and wildlife. The dense, shrubby growth offers excellent windbreaks and creates shelter for smaller animals, while the nuts are rich in nutrients.

3. SHRUB LAYER:

Northern Bayberry (*Myrica pensylvanica*): Northern bayberry produces small, waxy berries that are used to make natural candles and soaps. The leaves can also be used as a spice, similar to bay leaves. This nitrogen-fixing shrub thrives in poor, sandy soils and helps improve soil fertility. It provides shelter for wildlife and is highly drought-tolerant.

Soapberry (*Shepherdia canadensis*): Soapberry produces edible red berries that are used in Indigenous cuisine to make "Indian ice cream" or mixed with water for a foamy treat. The shrub is nitrogen-fixing and drought-tolerant, thriving in poor soils and helping to improve soil health. Soapberry also supports biodiversity by attracting pollinators and providing food for wildlife.

4. HERBACEOUS LAYER:

Salsify (*Tragopogon porrifolius*): Salsify, also known as oyster plant, produces edible roots that have a flavor similar to oysters and can be used in soups and stews. The plant thrives in cool climates and poor soils, providing a unique addition to the herbaceous layer. Its tall, purple flowers attract pollinators, enhancing biodiversity.

Alexanders (*Smyrnium olusatrum*): Alexanders is a cold-hardy herb with edible leaves, stems, and seeds that can be used like celery. It grows well in both sun and partial shade, making it versatile for the herbaceous layer. This ancient plant brings early spring growth to the food forest and helps attract beneficial insects.

5. GROUNDCOVER LAYER:

Groundnut (*Apios americana*): Groundnut is a perennial groundcover that produces edible tubers that can be roasted or used in stews. The plant has attractive flowers that draw pollinators, and its nitrogen-fixing roots improve soil fertility. Groundnut thrives in moist, rich soils and spreads easily, offering both food and ecological benefits.

Goldenseal (*Hydrastis canadensis*): Goldenseal is an herbaceous perennial with edible roots that have medicinal properties, often used in teas or tinctures. It thrives in shaded areas with rich, moist soils and helps to improve soil health while providing valuable ground coverage. Goldenseal also attracts beneficial insects and can contribute to biodiversity in the food forest.

6. ROOT LAYER:

Lamb's Quarters (*Chenopodium album*): While commonly known for its edible leaves, the roots of lamb's quarters are also edible and provide an excellent source of nutrition. This fast-growing plant helps to improve soil health through its extensive root system. It thrives in poor soils, making it a perfect candidate for the root layer in a food forest.

Sea Kale (*Crambe maritima*): Sea kale is a perennial plant that produces edible roots, shoots, and leaves. The roots are tender and slightly sweet, suitable for harvesting in the early spring. It thrives in coastal areas but also adapts to temperate climates, offering a unique root crop that can be integrated into the root layer.

7. VERTICAL LAYER:

Goji Berry (*Lycium barbarum*): Goji berries are known for their health benefits, offering edible, nutrient-packed fruits that are often dried or used in smoothies. The plant is a hardy shrub-like vine that grows well in USDA Hardiness Zone 3 and supports vertical growth on trellises. Its purple flowers also attract pollinators, making it a beneficial addition to the food forest ecosystem.

Sweet Autumn Clematis (*Clematis terniflora*): Sweet Autumn Clematis produces aromatic, edible flowers that can be used in teas or for garnishing dishes. It is a vigorous grower, making it suitable for the vertical layer, where it will climb trellises, fences, or other structures. Hardy and resilient, this plant attracts bees and butterflies while offering food and habitat benefits to the food forest.

8. FUNGAL LAYER:

King Oyster Mushroom (*Pleurotus eryngii*): This large, edible mushroom is prized for its meaty texture and savory flavor. It thrives in nutrient-rich organic matter, often growing on decaying wood or forest debris. King oysters are valuable both as a food source and as a beneficial organism that contributes to soil health by breaking down wood and organic material.

Matsutake Mushroom (*Tricholoma matsutake*): Highly prized in Japanese cuisine, the matsutake mushroom is rare and has a distinctive spicy aroma. It grows symbiotically with specific trees, often pine, and is a valuable contributor to the fungal layer due to its ability to decompose organic matter and enrich the soil. The edible mushroom offers culinary appeal and supports the food forest ecosystem by improving soil structure.

Zone 3 Plan Five: Crystal Canopy Creation

Inspired by the glitter of frost in the morning light, this plan creates a vibrant canopy of layered edibles. It's focused on both abundant harvests and maintaining the health of the overall system.

1. CANOPY LAYER:

Yellowhorn (*Xanthoceras sorbifolium*): This rare tree produces edible seeds that can be roasted and eaten like nuts, as well as beautiful, fragrant flowers. It is highly cold-tolerant and can thrive in poor soils, making it an excellent addition to Zone 3 food forests. Yellowhorn trees attract pollinators and support biodiversity.

Heartnut (*Juglans ailantifolia var. cordiformis*): This cold-hardy walnut variety produces heart-shaped, edible nuts that are easier to crack than other walnuts. It grows well in northern climates and contributes to soil health by improving nutrient cycling. Heartnuts provide valuable food for humans and wildlife while offering shade and habitat.

2. UNDERSTORY LAYER:

Mountain Maple (*Acer spicatum*): This hardy maple thrives in the cold understory, producing small, edible seeds that can be roasted or ground into flour. Mountain maple improves soil structure through leaf litter and provides habitat and food for various pollinators. It is known for its striking red fall foliage and ability to grow in the shade.

Pin Cherry (*Prunus pensylvanica*): Pin cherry is a small understory tree that produces bright red, tart cherries that are edible when cooked and can be used in jams and sauces. It provides early-season food for wildlife and supports pollinators with its spring blossoms. Pin cherry thrives in disturbed soils and helps with natural reforestation.

3. SHRUB LAYER:

Creeping Juniper (*Juniperus horizontalis*): This cold-hardy evergreen shrub produces small, bluish berries that are technically cones but are often used as a spice or in flavoring, particularly in gin. Creeping juniper is excellent for erosion control due to its low, spreading growth habit. It thrives in poor, rocky soils and provides habitat for wildlife.

American Hazelnut (*Corylus americana*): This hardy shrub produces small, edible nuts that are rich in fats and proteins, making it a valuable addition to any food forest. American hazelnut is highly adaptable to different soil conditions and provides food for both humans and wildlife. Its dense growth also creates a windbreak and shelter for smaller forest creatures.

4. HERBACEOUS LAYER:

Skirret (*Sium sisarum*): Skirret is a perennial root vegetable that produces sweet, edible roots used in soups and stews. It thrives in cool climates and wet soils, adding diversity to the herbaceous layer. Skirret is low-maintenance and can grow in partial shade, helping to maximize productivity in a food forest.

Sorrel (*Rumex acetosa*): Sorrel is a perennial herb with tart, lemon-flavored leaves that are edible and can be used in salads, soups, and sauces. It grows well in cold climates and poor soils, making it a resilient addition to Zone 3 food forests. Sorrel helps improve soil health by acting as a natural soil conditioner.

5. GROUNDCOVER LAYER:

Purple Dragon Arum (*Dracunculus vulgaris*): This exotic plant produces large, striking flowers and edible corms. The corms, which must be cooked before consumption, offer a starchy, nutty flavor. It thrives in moist soils and makes a dramatic addition to the groundcover layer, attracting pollinators with its distinctive flowers.

Virginia Waterleaf (*Hydrophyllum virginianum*): Virginia waterleaf is a low-growing perennial with edible leaves that can be eaten cooked or raw. It thrives in moist, shaded soils and spreads through rhizomes, providing quick ground cover while supporting the food forest ecosystem. Its flowers attract pollinators, including bees and butterflies.

6. ROOT LAYER:

Chicory (*Cichorium intybus*): Chicory produces a deep taproot that is edible and can be roasted as a coffee substitute or used in soups and stews. It thrives in deep, well-drained soils and tolerates drought conditions. Its roots help to break up compacted soils, improving the structure of the ground and supporting other plants in the root layer.

Spreading Dogbane (*Apocynum androsaemifolium*): Known for its edible roots and stems, Spreading Dogbane can be used to create a variety of foraged dishes. The roots, though tough, can be prepared in soups and stews, providing a starchy, nutritious base. It is adaptable to different soil types and serves as a hardy, low-maintenance ground cover in the root layer.

7. VERTICAL LAYER:

Purple Passionflower (*Passiflora incarnata*): This unique vine produces edible fruits with a sweet, tangy flavor that can be eaten raw or used in fruit-based dishes. The plant thrives in full sun and well-drained soils, growing quickly to create vertical coverage. Known for its striking flowers, the vine also supports pollinators and adds beauty and edible value to the food forest.

Kiwi Berry (*Actinidia arguta*): Kiwi berry, also known as hardy kiwi, produces small, edible fruits that are similar in flavor to larger kiwis but without the skin. It grows in a vertical manner on trellises or arbors and is hardy in Zone 3. The vine is a fast grower, providing both food and ecological benefits by attracting pollinators and enhancing soil health.

8. FUNGAL LAYER:

Peach Fungus (*Tapinella atrotomentosa*): Known for its peach-like color and texture, this edible fungus grows on decaying hardwood trees and has a unique, fruity taste. It is beneficial for the fungal layer as it breaks down organic matter and helps recycle nutrients back into the soil. The mushroom's rarity and distinctive flavor make it an intriguing addition to the food forest.

Pine Mushroom (*Tricholoma magnivelare*): The pine mushroom, also known as the matsutake mushroom in some regions, is known for its distinct aroma and meaty, flavorful texture. It grows in symbiosis with pine trees, enriching the forest ecosystem by enhancing soil health and supporting biodiversity. Edible when properly prepared, it is a prized culinary mushroom and a valuable addition to the fungal layer.

Zone 3 Plan Six: Nippy Nurture Niche

A playful yet practical design that finds a niche even in nippy conditions, nurturing life at every level. It balances food production with ecosystem care through thoughtful layering.

1. CANOPY LAYER:

European Larch (*Larix decidua*): Though primarily known for its timber, this deciduous conifer can contribute to the food forest by enhancing soil health through its leaf litter. While its cones are not edible, the larch plays an essential role in forest ecosystems, supporting wildlife and improving soil aeration and water retention.

Basswood (*Tilia americana*): Also known as American linden, this tree produces edible flowers and leaves that can be used in teas and salads. It thrives in cold climates, contributing to the food forest by providing shade, improving soil with its nutrient-rich leaf litter, and attracting pollinators such as bees and butterflies.

2. UNDERSTORY LAYER:

Nannyberry (*Viburnum lentago*): Nannyberry produces sweet, edible fruits that can be eaten fresh or used in jams and jellies. It grows well in Zone 3, providing food for birds and wildlife and adding diversity to the understory. The tree also improves soil health through its deep roots and supports pollinators with its fragrant flowers.

Highbush Cranberry (*Viburnum opulus var. americanum*): This cold-hardy shrub produces edible berries that are high in vitamin C and can be used in sauces, jellies, and teas. Highbush cranberry thrives in moist, well-drained soils and attracts pollinators with its white, flat-topped flowers. It also provides a food source for wildlife, particularly birds.

3. SHRUB LAYER:

Korean Bush Cherry (*Prunus japonica*): Korean bush cherry produces small, tart cherries that are edible and can be used in pies, jams, and sauces. The shrub is hardy and adaptable, thriving in cold climates and providing early spring blossoms that attract pollinators. Its compact size makes it suitable for small spaces or dense food forest designs.

Silver Buffaloberry (*Shepherdia argentea*): Silver buffaloberry is a nitrogen-fixing shrub that produces small, tart berries high in vitamin C. The plant is highly drought-tolerant and thrives in poor, sandy soils, making it ideal for harsh climates. Its silvery leaves add visual interest, and it provides valuable food for wildlife, including birds and small mammals.

4. HERBACEOUS LAYER:

Stinging Nettle (*Urtica dioica*): Stinging nettle is a highly nutritious herb with edible leaves that can be cooked like spinach or used in soups and teas. It is rich in iron, vitamins, and minerals, and supports biodiversity by providing habitat for beneficial insects. Nettles also improve soil health by accumulating nutrients in their roots.

Wild Garlic (*Allium ursinum*): Wild garlic produces edible leaves, flowers, and bulbs that have a mild garlic flavor, making it a versatile addition to the food forest. It thrives in shaded areas and helps deter pests from the garden. The plant is highly cold-hardy and can be foraged early in the season when other crops are not yet available.

5. GROUNDCOVER LAYER:

Creeping Thyme (*Thymus serpyllum*): Creeping thyme is a hardy groundcover plant with aromatic edible leaves that can be used in cooking or teas. It grows well in dry, well-drained soils and spreads quickly, helping to prevent weed growth. Its purple flowers attract pollinators, supporting the health of the food forest.

Sweet Woodruff (*Galium odoratum*): Sweet woodruff is a fragrant groundcover with edible leaves that are used in herbal teas. It grows well in shady, moist areas and provides quick coverage while preventing soil erosion. Sweet woodruff's delicate white flowers attract pollinators, enriching the ecosystem of the food forest.

6. ROOT LAYER:

Cress *(Lepidium sativum)*: Cress is a fast-growing, edible root vegetable with leaves and roots that can be used in salads and sandwiches. The roots provide a sharp, peppery flavor. Cress thrives in moist, well-drained soils and helps to break up compacted ground while offering a nutritious, quick-harvest option for the root layer.

Black Salsify (*Scorzonera hispanica*): Black salsify produces long, edible, black roots that are commonly used in European cooking. The roots have a mild, nutty flavor and are rich in vitamins. This hardy plant thrives in well-drained soils and its roots contribute to soil aeration and health, benefiting the entire food forest.

7. VERTICAL LAYER:

Scarlet Runner Bean (*Phaseolus coccineus*): This vigorous climbing bean produces edible beans that are a great source of protein and can be eaten fresh or dried for later use. The plant thrives in full sun and is an excellent nitrogen fixer, improving soil health. Its vibrant red flowers attract pollinators, while its vertical growth habit allows for efficient use of space in a food forest.

Silvervine (*Actinidia polygama*): Silvervine is another unique kiwi relative that grows well in Zone 3 and produces edible fruits that are sweet and aromatic. Its vines grow rapidly and need support for climbing, making it an excellent vertical addition to a food forest. The plant is also known for attracting cats, who are drawn to the plant's stimulating effects.

8. FUNGAL LAYER:

Pink Oyster Mushroom (*Pleurotus djamor*): Known for its striking pink color and mild, oyster-like flavor, the pink oyster mushroom thrives in humid, nutrient-rich environments, such as decaying wood and organic matter. It is edible and adds both beauty and culinary value to the food forest. Pink oysters are also fast-growing and help decompose organic material, benefiting soil health.

Shimeji Mushroom (*Hypsizygus marmoreus*): Shimeji mushrooms, particularly the brown and white varieties, are edible fungi known for their savory taste and crunchy texture. They grow in clusters on decaying wood and contribute to soil health by breaking down organic material. Shimeji mushrooms are also beneficial in the fungal layer, helping to support the ecosystem while providing a delicious and nutritious food source.

Zone 3 Plan Seven: Frost Fringe Formation

This food forest plan addresses the delicate fringes of a frosty climate by organizing paired species in each layer. It's all about forming a robust edge that supplies food while buffering environmental stresses.

1. CANOPY LAYER:

Sweet Birch (*Betula lenta*): The sap of this tree can be tapped to produce birch syrup, and its twigs can be used to make a flavorful tea. Sweet birch is cold-hardy and contributes to the food forest by improving soil structure and attracting beneficial insects. Its fast growth rate also provides quick canopy cover.

Black Walnut (*Juglans nigra*): This tree produces edible, richly flavored nuts that are prized in cooking and baking. Black walnut trees improve soil structure through their deep roots and contribute to the food forest by providing a large canopy. The nuts are also a valuable food source for wildlife, though care must be taken to plant it away from other species sensitive to its allelopathic properties.

2. UNDERSTORY LAYER:

Juneberry (*Amelanchier canadensis*): Also known as Canadian serviceberry, this shrub produces sweet, edible berries that are similar to blueberries and can be eaten fresh or used in baked goods. Juneberries thrive in colder climates and provide valuable nectar for pollinators. Their early spring blossoms add beauty and attract beneficial insects.

Mountain Alder (*Alnus incana*): Mountain alder is a nitrogen-fixing shrub that improves soil fertility and stabilizes streambanks and wetlands. Its edible young catkins can be used in survival food, though they require processing. The tree supports a range of wildlife and attracts pollinators, while its deep roots enhance soil aeration.

3. SHRUB LAYER:

Red Osier Dogwood (*Cornus sericea*): Red osier dogwood produces small, white berries that are edible when processed and are a traditional food source for Indigenous peoples. The plant is known for its striking red stems, which add color to the winter landscape. It thrives in wet soils and helps stabilize riverbanks, providing valuable erosion control.

Shrubby Cinquefoil (*Dasiphora fruticosa*): Shrubby cinquefoil produces small, yellow flowers and edible leaves that can be brewed into a tea. It is highly cold-hardy and drought-tolerant, thriving in poor soils. The plant attracts pollinators and provides habitat for small wildlife while contributing to soil health through its extensive root system.

4. HERBACEOUS LAYER:

Valerian (*Valeriana officinalis*): Valerian is a medicinal herb with edible roots that are used in teas for their calming effects. The plant attracts pollinators and helps improve soil health with its deep root system. Valerian thrives in cool, moist soils and adds biodiversity to the herbaceous layer.

Angelica (*Angelica archangelica*): Angelica is a tall, cold-hardy herb with edible stems, leaves, and seeds that are often used in baking and herbal remedies. It prefers damp, shady areas and attracts beneficial insects like bees and butterflies. This plant also adds structure and height to the herbaceous layer of the food forest.

5. GROUNDCOVER LAYER:

Red Clover (*Trifolium pratense*): Red clover is a nitrogen-fixing groundcover with edible flowers and leaves that can be used in teas or as a garnish. It thrives in moist, well-drained soils and attracts pollinators with its pink flowers. This plant enriches the soil and helps support the overall health of the food forest.

Sweet Flag (*Acorus calamus*): Sweet flag is a wetland-loving groundcover with edible roots that have a sweet, spicy flavor. The rhizomes can be used to make teas or flavor dishes. It thrives in moist, boggy soils and helps to stabilize waterlogged areas while offering food and medicinal properties.

6. ROOT LAYER:

Yacon *(Smallanthus sonchifolius)*: Yacon is a perennial plant known for its sweet, edible tubers, which can be eaten raw or cooked. The roots are high in fructooligosaccharides, a prebiotic that supports digestive health. Yacon grows well in well-drained, fertile soils and helps improve soil structure with its extensive root system.

Cabbage Palm (*Sabal palmetto*): Cabbage palm is known for its edible heart, but it also has edible roots that can be cooked and consumed. It thrives in moist, well-drained soils and can handle both full sun and partial shade. Its roots contribute to breaking up compacted soil and provide nutrition in the root layer.

7. VERTICAL LAYER:

Clematis Montana (*Clematis montana*): This beautiful, fast-growing clematis vine produces fragrant flowers that are edible and can be used to make herbal teas or in culinary dishes. It thrives in moist, well-drained soils and grows vigorously, making it perfect for vertical spaces. The plant supports pollinators, adding ecological value to the food forest.

Chinese Red Noodle Bean (*Vigna unguiculata subsp. sesquipedalis*): This long, striking legume produces edible beans that can be used in various dishes. It grows rapidly and can reach impressive heights when given vertical support, making it ideal for trellises. The plant attracts bees and other pollinators while contributing to soil nitrogen levels through its legume properties.

8. FUNGAL LAYER:

Slippery Jack Mushroom (*Suillus luteus*): This edible mushroom is known for its slippery cap and mild flavor. It grows in association with pine trees and contributes to the fungal layer by breaking down organic matter and improving soil structure. While it must be prepared correctly to remove its slimy skin, it adds an edible and functional component to the food forest.

Turkey Tail Mushroom (*Trametes versicolor*): While known primarily for its medicinal properties, turkey tail mushrooms are also edible and have been used in teas and soups. These mushrooms are excellent at decomposing wood, contributing to soil health, and are beneficial for the ecosystem. Their vibrant, multicolored caps and ability to grow on dead wood make them an attractive and valuable member of the fungal layer.

Zone 3 Plan Eight: Cold Snap Collective

When sudden chill strikes, this collective plan provides a network of supportive, edible species. It is designed to maintain food output and system stability even during abrupt temperature dips.

1. CANOPY LAYER:

Blue Ash (*Fraxinus quadrangulata*): This rare tree produces seeds that are edible when processed, though its primary value in the food forest lies in its ability to improve soil health through its deep root system. Blue ash is highly cold-tolerant and provides habitat and shelter for wildlife, contributing to forest diversity.

Corktree (*Phellodendron amurense*): This tree produces small, aromatic fruits that are not commonly eaten but can be used in traditional medicine. It is highly cold-tolerant and thrives in northern climates, providing shade and enhancing soil health through its leaf litter. Corktrees also attract beneficial insects to the food forest.

2. UNDERSTORY LAYER:

Silver Buffaloberry (*Shepherdia argentea*): This cold-hardy shrub produces edible, tart red berries that are high in vitamin C and can be used in syrups and jellies. Silver buffaloberry is drought-tolerant and thrives in poor soils, improving nitrogen levels in the soil through its nitrogen-fixing properties. It also provides habitat for birds and small mammals.

Gooseberry (*Ribes oxyacanthoides*): This hardy shrub produces tart, edible berries that can be used in pies, jams, and sauces. Gooseberry thrives in colder climates, growing well in both sun and partial shade. It contributes to the food forest by improving soil structure with its extensive root system and attracting pollinators with its spring flowers.

3. SHRUB LAYER:

Wolfberry (*Lycium barbarum*): Also known as goji berry, wolfberry produces bright red, nutrient-dense berries that are edible and used in teas, soups, and health foods. The shrub is highly cold-hardy and thrives in well-drained soils, providing both food and medicinal benefits. It supports pollinators and wildlife, while also being drought-tolerant.

Sweetfern (*Comptonia peregrina*): Sweetfern is a nitrogen-fixing shrub with fragrant, fern-like leaves that can be used to brew tea. It thrives in poor, sandy soils and helps improve soil fertility. Sweetfern attracts pollinators and is highly drought-tolerant, making it a resilient addition to cold-climate food forests.

4. HERBACEOUS LAYER:

Meadowsweet (*Filipendula ulmaria*): Meadowsweet has edible flowers and leaves that are used to flavor teas and desserts. It thrives in wet soils and provides valuable habitat for pollinators. Meadowsweet is also known for its medicinal properties, particularly for relieving pain and inflammation, making it a useful herb in a food forest.

Evening Primrose (*Oenothera biennis*): Evening primrose produces edible roots and seeds that are rich in essential fatty acids and can be used in salads or as a natural remedy for various ailments. The plant is cold-hardy and drought-tolerant, thriving in poor soils. Its yellow flowers attract pollinators and add beauty to the food forest.

5. GROUNDCOVER LAYER:

Spikenard (*Aralia racemosa*): Spikenard is an herbaceous perennial with edible roots used for medicinal purposes and in herbal teas. It thrives in rich, moist soils and provides excellent ground coverage. Its aromatic leaves and roots are beneficial for supporting the surrounding plants and helping to create a diverse ecosystem in the food forest.

Hosta (*Hosta spp.*): While primarily known for its decorative foliage, hosta also produces edible young shoots and flowers that can be used in salads or stir-fries. It thrives in shady, moist conditions, making it perfect for the food forest's understory and groundcover layers. Hosta spreads quickly, providing ample ground coverage while adding beauty to the landscape.

6. ROOT LAYER:

Milo Root (*Sorghum bicolor var. technicum*): Milo, also known as grain sorghum, produces edible roots that help improve soil fertility. The roots are sometimes used as a starch source and contribute to ground coverage, protecting soil from erosion. Its deep root system can thrive in dry conditions, making it a beneficial addition to the root layer in drought-prone areas.

Chufa (*Cyperus esculentus*): Known for its edible tubers, Chufa, or earth almond, is a hardy plant that thrives in moist, well-drained soils. The tubers are sweet and can be eaten raw or ground into flour for baking. The plant spreads rapidly, creating dense ground coverage and improving soil health.

7. VERTICAL LAYER:

Gooseberry Vine (*Ribes uva-crispa*): This hardy vine produces edible gooseberries that are tart and can be eaten raw or used in preserves and pies. Its upright growth habit makes it suitable for vertical support, and it thrives in partial shade or full sun. Gooseberries are rich in vitamin C and attract pollinators, making them a valuable addition to the food forest.

Autumn Olive (*Elaeagnus umbellata*): Autumn olive is a nitrogen-fixing shrub with edible berries that are rich in antioxidants. The berries can be eaten raw, made into jams, or dried for later use. Its vertical growth allows it to act as a supportive element in the food forest, while its nitrogen fixation improves soil health.

8. FUNGAL LAYER:

Velvet Pioppini Mushroom (*Agrocybe aegerita*): Velvet Pioppini mushrooms are edible and known for their rich, meaty texture and mild flavor. They thrive in humid, nutrient-rich environments, often growing on decaying wood or logs. These mushrooms contribute to soil health by decomposing organic matter and serve as a food source in the fungal layer, providing both culinary value and ecological benefits.

Wine Cap Mushroom (*Stropharia rugosoannulata*): Also known as the garden giant, the wine cap mushroom is an edible fungus with a strong, earthy flavor. It thrives in woodchip or straw mulch, breaking down organic matter and improving soil health. The mushroom's large, fleshy cap and robust growth make it a valuable addition to the fungal layer, providing both food and ecological benefits.

Audra Haldane

Zone 3 Plan Nine: Brumal Balance Blueprint

Drawing on the quiet strength of winter (brumal conditions), this blueprint emphasizes balance in every planting layer. It is crafted to yield nourishing produce while reinforcing sustainable growth practices.

1. CANOPY LAYER:

Hackberry (*Celtis occidentalis*): This hardy tree produces small, sweet, edible berries that are enjoyed by both humans and wildlife. Hackberries thrive in cold climates and contribute to the food forest by improving soil health through deep rooting and leaf litter. Their berries are a valuable food source, attracting birds and pollinators.

Mountain Ash (*Sorbus americana*): The bright orange-red berries of this tree are edible and can be made into jellies or fermented for beverages. Mountain ash is highly cold-tolerant, making it suitable for Zone 3. It improves soil health through its leaf litter and provides food for birds and wildlife throughout the winter.

2. UNDERSTORY LAYER:

Sea Buckthorn (*Hippophae rhamnoides*): This nitrogen-fixing shrub produces bright orange, edible berries that are high in vitamins C and E. Sea buckthorn is extremely cold-hardy and thrives in poor soils, improving fertility and providing shelter for wildlife. Its berries are used in juices, oils, and medicinal products, offering high nutritional value.

Black Elder (*Sambucus nigra*): Black elder produces clusters of edible berries that can be cooked into syrups and jams, as well as flowers that are used in teas and cordials. It thrives in the colder climates of Zone 3, attracting pollinators and improving biodiversity. Black elder is also known for its medicinal properties and ability to support immune health.

3. SHRUB LAYER:

Bog Cranberry (*Vaccinium oxycoccos*): Bog cranberry is a low-growing shrub that produces small, edible red berries used in sauces and juices. The plant thrives in acidic, boggy soils and helps maintain soil moisture levels. Its evergreen leaves provide year-round interest, and it attracts pollinators with its small, pink flowers.

Spicebush *(Lindera benzoin)*: Spicebush produces small, red berries that can be used as a spice substitute for allspice, and its leaves and twigs can be brewed into a tea. The shrub thrives in wet, well-drained soils and provides habitat for the spicebush swallowtail butterfly. It also attracts pollinators and contributes to biodiversity in the food forest.

4. HERBACEOUS LAYER:

Jerusalem Artichoke (*Helianthus tuberosus*): Jerusalem artichoke, also known as sunchoke, produces edible tubers that are high in inulin and can be roasted, mashed, or used in soups. It is highly cold-hardy and thrives in poor soils, making it a low-maintenance addition to the herbaceous layer. Its tall flowers attract pollinators and provide a natural windbreak.

Lovage (*Levisticum officinale*): Lovage has edible leaves, stems, and seeds that are used as a celery substitute in cooking. The plant is highly cold-hardy and grows well in both full sun and partial shade. It attracts pollinators and beneficial insects, while also improving soil health through its deep root system.

5. GROUNDCOVER LAYER:

Bittercress (*Cardamine hirsuta*): Bittercress is a small, fast-growing groundcover with edible, peppery leaves that are used in salads or as a garnish. It thrives in moist, shaded soils and can spread rapidly, providing quick ground cover. Its flowers attract beneficial insects, which help to improve the overall health of the food forest.

Field Mint (*Mentha arvensis*): Field mint is a fast-spreading groundcover with aromatic leaves that can be used in teas or culinary dishes. It thrives in moist, well-drained soils and helps to control weeds with its dense growth habit. The plant's small flowers attract pollinators, making it an excellent addition to the food forest.

6. ROOT LAYER:

Chinese Artichoke (*Stachys affinis*): Also known as Crosne or Japanese artichoke, this plant produces edible tubers that have a slightly nutty taste. It grows well in loose, fertile soils and can tolerate partial shade. The tubers provide a unique addition to the root layer and contribute to soil aeration and diversity.

Indian Mustard (*Brassica juncea*): Indian Mustard is a cold-hardy root vegetable that produces both edible roots and leaves. Its deep taproot helps to break up compacted soils, improving soil structure and health. It is a vigorous grower and can thrive in the root layer of a food forest, offering a flavorful addition to salads and cooked dishes.

7. VERTICAL LAYER:

Cucumber Melon (*Cucumis melo*): Cucumber melon is a small, sweet melon variety that grows well in USDA Hardiness Zone 3 and is perfect for vertical growth. The fruit is edible and has a flavor similar to cucumber, offering a refreshing taste. Its vines are well-suited to growing on trellises, and the plant helps improve soil health by reducing water runoff.

American Climbing Fern (*Lygodium palmatum*): A rare, cold-hardy climbing fern with delicate, palmate fronds that grow vertically along structures or trees.

8. FUNGAL LAYER:

Yellowfoot Mushroom (*Cantharellus tubaeformis*):

This edible mushroom is closely related to the chanterelle but is smaller and more delicate in flavor. It grows in mossy, forested areas and has a distinct yellow color. Yellowfoot mushrooms decompose organic matter, enriching the soil while providing a rare, gourmet food source in the food forest.

Wood Ear Mushroom (*Auricularia auricula-judae*): Known for its gelatinous texture and ability to absorb flavors, the wood ear mushroom is an edible fungus often used in soups and stir-fries. It grows on decaying wood, helping to break down organic material and enhance soil health. The mushroom's ability to thrive in damp environments makes it a great addition to the fungal layer, providing both ecological value and culinary potential.

This initiative is all about sparking growth in an icy environment by carefully pairing plants across layers. Its inception aims to foster a resilient food forest that can thrive against the odds.

1. CANOPY LAYER:

Serviceberry (*Amelanchier alnifolia*): Also known as Saskatoon, this tree produces sweet, edible berries that can be eaten fresh or used in pies, jams, and preserves. Serviceberries thrive in cold climates and improve soil fertility through their fallen leaves. They also attract pollinators and provide food for wildlife.

Beech (*Fagus grandifolia*): This slow-growing tree produces edible nuts that are a valuable food source for both humans and wildlife. It thrives in cold climates, contributing to the food forest with its dense canopy, which provides shade and improves soil through the accumulation of organic matter. Beech trees also offer habitat for birds and insects.

2. UNDERSTORY LAYER:

Red Osier Dogwood (*Cornus sericea*): This cold-hardy shrub produces small, white berries that are edible when processed and are a traditional food for Indigenous peoples. Red osier dogwood improves soil stability through its dense root system, which prevents erosion, and its bright red stems add beauty to the winter landscape. It attracts a variety of wildlife and pollinators.

Pawpaw (*Asimina triloba*): Although rare in cold climates, some cold-hardy varieties of pawpaw can grow in Zone 3. This understory tree produces large, tropical-tasting fruits that are highly prized for fresh eating. Pawpaw trees improve soil health through leaf litter, provide shade, and attract beneficial insects like the zebra swallowtail butterfly, which relies on it as a host plant.

3. SHRUB LAYER:

Thimbleberry (*Rubus parviflorus*): Thimbleberry produces sweet, edible red berries that are similar to raspberries but softer and more delicate. The shrub thrives in cold climates and is highly adaptable, growing well in both sun and partial shade. It provides food for both humans and wildlife, and its large, maple-like leaves add visual interest to the forest understory.

Aronia (*Aronia melanocarpa*): Also known as black chokeberry, aronia produces small, edible berries that are rich in antioxidants and commonly used in juices, jams, and health supplements. The shrub is highly cold-hardy and tolerant of a variety of soil conditions, making it an excellent addition to the shrub layer. It also attracts pollinators and provides food for birds.

4. HERBACEOUS LAYER:

Wild Leek (*Allium tricoccum*): Wild leek, also known as ramps, produces edible leaves and bulbs with a strong garlic-onion flavor. It thrives in shaded, moist areas and is highly prized for its early spring growth. Wild leeks help improve soil health and attract pollinators, making them a valuable addition to the herbaceous layer.

Marsh Marigold (*Caltha palustris*): Marsh marigold is a cold-hardy herb with edible leaves and stems that are traditionally used in early spring salads and soups. It thrives in wet, boggy soils, helping to stabilize areas prone to flooding. The plant also attracts pollinators with its bright yellow flowers, adding beauty and biodiversity to the food forest.

5. GROUNDCOVER LAYER:

Mountain Cranberry (*Vaccinium vitis-idaea*): Mountain cranberry is a low-growing groundcover with edible berries that are tart and can be used in sauces, jams, or eaten fresh. It thrives in acidic, well-drained soils and provides excellent ground cover, while its flowers attract pollinators like bees and butterflies.

Brunnera (*Brunnera macrophylla*): Brunnera is a perennial groundcover with edible leaves that can be used in salads or cooked dishes. It thrives in shaded, moist soils and provides early spring flowers that attract pollinators. This plant helps to prevent soil erosion while contributing to the food forest's diversity.

6. ROOT LAYER:

Ginseng (*Panax quinquefolius*): Ginseng is renowned for its medicinal properties, with the roots being highly valued for their health benefits. The roots are used to create herbal medicines that promote vitality and energy. This slow-growing perennial thrives in shady, rich, moist soils and can add medicinal value to the root layer in a food forest.

Wormwood (*Artemisia absinthium*): Wormwood has edible roots and is used for medicinal purposes. The roots are used to make teas that are beneficial for digestive health. It thrives in dry, well-drained soils and can help suppress unwanted plants, contributing to soil management in the root layer.

7. VERTICAL LAYER:

Sweet Potato Vine (*Ipomoea batatas*): Sweet potatoes are known for their edible tubers, but the vines are also an attractive, edible ground cover that can climb and spread. The leaves are often used in cooking, offering a nutritious, tender green. Sweet potato vines are ideal for vertical growth and thrive in full sun, making them a versatile addition to the food forest.

Chilean Guava (*Ugni molinae*): Also known as the strawberry myrtle, this rare evergreen vine produces small, edible berries that taste like a cross between strawberries and pineapple. It thrives in moist, well-drained soils and can climb trellises or other vertical structures. The plant's edible fruit, fragrant flowers, and ability to support pollinators make it a valuable addition to the food forest.

8. FUNGAL LAYER:

Witches' Butter (*Tremella mesenterica*): Witches' butter is an edible jelly fungus known for its yellow, gelatinous appearance. It grows on decaying wood and plays an important role in the fungal layer by breaking down organic material and returning nutrients to the soil. While it is less commonly eaten, it is a unique and beneficial part of the ecosystem, contributing to the health of the food forest.

Shiitake Mushroom (*Lentinula edodes*): Shiitake mushrooms are well-known for their rich, umami flavor and are highly prized in culinary dishes. They grow on logs or decaying wood, making them a perfect fit for the fungal layer, where they help break down organic material. Edible when cooked, shiitakes are both a valuable food source and a beneficial organism for soil health in the food forest.

ZONE 4

Zone 4 presents a cold environment with long, intense winters and brief growing periods, making it a challenging area for establishing a food forest. This zone experiences typical minimum temperatures between -30°F and -20°F (-34°C to -29°C), though local conditions can cause slight variations. These regions have freezing temperatures that persist for extended periods, with frost generally lasting from early to mid-May until September or October.

Locations within Zone 4

Zone 4 is often found in a mix of rural, agricultural, and wilderness areas, with some small urban centers situated in colder parts of the U.S.

STATES WITH SIGNIFICANT PORTIONS IN ZONE 4:

Minnesota: Nearly all of Minnesota lies within Zone 4, except for the far north, which falls into Zone 3.

Wisconsin: Much of northern and central Wisconsin is part of Zone 4.

North Dakota: Most of North Dakota belongs to Zone 4, except for the northernmost areas, which fall into Zone 3.

South Dakota: Northern and central South Dakota are in Zone 4.

Montana: Western and central Montana largely fall within Zone 4.

Wyoming: Higher elevations and northern areas of Wyoming are mostly in Zone 4.

Colorado: Some high-altitude areas, particularly within the Rockies, fall within Zone 4.

Idaho: Northern Idaho and higher elevations are part of Zone 4.

Zone 4

Nebraska: Northern Nebraska primarily falls into Zone 4.

Iowa: The northern section of Iowa is classified as Zone 4.

Michigan: Both the Upper Peninsula and northern parts of Michigan fall within Zone 4.

New York: Higher elevations, such as the Adirondacks, are included in Zone 4.

Vermont: Much of Vermont, especially in the higher elevations, falls under Zone 4.

New Hampshire: Northern and mountainous regions belong to Zone 4.

Maine: Central and northern Maine are part of Zone 4.

SMALLER REGIONS OR ISOLATED AREAS IN ZONE 4:

Alaska: Interior regions, including Fairbanks, experience Zone 4 conditions.

Utah: High-altitude areas, especially in the northern mountains, are classified as Zone 4.

Oregon: Isolated mountainous areas, such as parts of the Wallowa Mountains, fall into Zone 4.

Nevada: Northern, high-altitude regions of Nevada experience Zone 4 climates.

Washington: Some mountain areas, particularly in the Cascades, fall within Zone 4.

Massachusetts: Higher elevations in western Massachusetts are part of Zone 4.

Connecticut: Isolated higher elevations in the northwest part of the state fall under Zone 4.

Zone 4 Climate and Growing Season

Winter Conditions: Cold winters are a hallmark of Zone 4, with snow and ice commonly seen throughout the season. Snow can act as insulation, protecting plants from the coldest temperatures.

These freezing conditions persist for several months, requiring plants that are resilient and able to endure frost.

Growing Season: The frost-free growing period in Zone 4 is brief, typically lasting between 90 and 140 days, depending on the specific location. The growing season begins in late spring, around late May or early June, and concludes by early fall, around late September to early October. Maximizing this short growing window is essential for plant success.

Precipitation: Zone 4 generally experiences moderate precipitation levels, which include both rain and snow. Summers are typically drier, which means that irrigation or rainwater collection may be necessary to support water-dependent crops during the growing season.

Challenges for Growing in Zone 4

Cold Hardiness: The primary difficulty in Zone 4 is selecting plants capable of withstanding severe cold. Only plants that can tolerate frost and significant snow accumulation will thrive during winter.

Short Growing Season: The brief window between the last frost in spring and the first frost in fall means that fast-growing plants or varieties that can handle frost are crucial. Cold-resistant fruit trees and quick-maturing vegetables are good choices.

Soil Conditions: In many areas, the soil freezes solid for much of the winter, preventing plants from accessing essential nutrients until the soil thaws. Adding organic matter, such as compost or mulch, improves soil structure and helps retain moisture once the soil becomes workable.

Wind and Snow: High winds and snow accumulation are common in Zone 4 and can pose a threat to plants. Wind can dry out or damage plants, while snow can cover low-growing plants and break weak branches. Windbreaks or shelterbelts, made of hardy trees and shrubs, can offer protection.

Growing Opportunities in Zone 4

Cold-Hardy Varieties: Despite the challenges, a wide range of cold-tolerant plants can grow successfully in Zone 4. Fruit trees, shrubs, and vegetables bred for cold climates make it possible to harvest fresh produce even in extreme conditions.

Early and Late Season Harvests: A well-planned food forest in Zone 4 can yield crops in both early spring and late fall. Hardy plants, such as rhubarb and leafy greens, can provide early harvests, while root vegetables and cold-resistant berries can be harvested in late fall.

Snow as an Advantage: Snow, although a challenge in many respects, can also be beneficial by insulating the soil and protecting roots from freezing temperatures.

Microclimates and Topography

Microclimates significantly affect plant growth in Zone 4. Factors like the slope of the land, exposure to wind, proximity to bodies of water, and elevation can create areas that are slightly warmer or colder than the surrounding environment. For example, south-facing slopes tend to receive more sunlight and remain warmer than north-facing slopes, which provides a more favorable environment for plant growth. Proximity to large bodies of water can also moderate temperatures, helping to create more favorable growing conditions.

Windbreaks and Shelterbelts

Strong winds and heavy snowfall are common challenges in Zone 4, so it's crucial to create windbreaks or shelterbelts to shield your food forest. Strategically planting rows of evergreens or dense shrubs can help reduce wind exposure, minimize snow accumulation, and improve soil health by preventing erosion.

Soil Considerations in Zone 4

Frozen soil is a common issue in Zone 4 during winter months. Once it thaws, it can have a dense and compact structure, which makes it difficult for plants to root properly. Adding organic materials like compost, mulch, or wood chips helps improve soil structure, supports fertility, and encourages healthier plant growth. Raised beds or mounded soil can also help extend the growing season by warming the soil earlier in the year.

Sustainability and Resilience

Zone 4 offers promising opportunities for sustainable food production despite its challenges. By selecting cold-hardy perennial crops, such as fruit trees, berries, and winter-hardy vegetables, a food forest in this zone can provide consistent yields throughout the year. Incorporating permaculture principles like biodiversity, soil health, and efficient water use will further ensure the food forest is resilient and self-sustaining.

Careful planning, hardy plant selection, and creative use of the environment are essential for overcoming the unique challenges of Zone 4. Techniques like using windbreaks, improving soil quality, and maximizing microclimates will help gardeners establish a thriving and sustainable food forest capable of withstanding cold conditions and short growing seasons.

Zone 4 Plan One: Spring Thaw Strategy

As winter's grip loosens, this plan capitalizes on the early thaw to kickstart growth. It layers paired edibles to ensure a seamless transition from dormant to dynamic, providing both nourishment and sustainability.

1. CANOPY LAYER:

American Plum (*Prunus americana*): This small deciduous tree produces sweet, tart plums that are edible and great for fresh eating or preserving. It is native to North America and grows in a variety of soils, thriving in full sun to partial shade. As a canopy tree, it adds structure and provides habitat for pollinators while its fruit serves as a food source for both humans and wildlife. The tree also has beautiful white spring blossoms, attracting beneficial insects to the food forest.

Chokecherry (*Prunus virginiana*): Chokecherry is a deciduous shrub or small tree with edible fruits that are used for jellies, syrups, and pies, though they must be cooked due to their bitter taste when raw. It grows well in a variety of soil types, prefers full sun, and attracts pollinators with its white flowers. Its ability to tolerate a range of conditions makes it a resilient canopy layer plant while offering an important food source for both humans and wildlife.

2. UNDERSTORY LAYER:

Dwarf American Plum (*Prunus americana*): Dwarf American plum is a small, deciduous tree that thrives in partial shade and produces edible, sweet, and tart plums. It grows well in well-drained soils and typically reaches 4 to 6 feet in height, making it perfect for the understory. The tree's fragrant white flowers attract pollinators, and the fruit provides food for both humans and wildlife. It also has the added benefit of being resilient to cold temperatures, making it a great choice for cold climates.

Serviceberry (*Amelanchier alnifolia*): Serviceberry, also known as Saskatoon, is a small, deciduous tree or shrub that grows up to 15 feet tall, making it ideal for the understory layer. It produces edible, blueberry-like fruits that are sweet and rich in nutrients. The tree thrives in well-drained soils and partial shade, providing a multi-season interest with its early spring flowers, fall foliage, and winter form. It supports pollinators and attracts birds, contributing to the food forest's biodiversity.

3. SHRUB LAYER:

Nanking Cherry (*Prunus tomentosa*): Nanking cherry is a hardy shrub that grows up to 6 feet tall and produces small, tart red cherries. These cherries are edible and can be used to make jams, jellies, and sauces. The shrub thrives in well-drained, slightly acidic soils and provides early-season food for wildlife. In addition to its fruit, it also adds early spring beauty with its white flowers, attracting pollinators.

Goji Berry (*Lycium barbarum*): Goji berries are a hardy shrub that can grow up to 6 feet tall in USDA Zone 4. The bright red berries are edible and highly nutritious, often used in smoothies, dried snacks, or teas. The plant thrives in well-drained soil and full sun, producing an abundant harvest. Its resilience and ability to improve soil quality make it a valuable addition to the food forest.

4. HERBACEOUS LAYER:

Bergamot (*Monarda fistulosa*): Bergamot is a fragrant herb that thrives in full sun and moist, well-drained soil. The leaves and flowers are edible and have a pleasant, citrusy flavor, often used in teas and culinary dishes. It attracts pollinators like bees and butterflies, benefiting the food forest by supporting biodiversity. This native plant can also help in repelling pests with its aromatic oils.

Common Sorrel (*Rumex acetosa*): Common Sorrel is a perennial herb with tangy, lemony-flavored leaves that are edible raw or cooked. It grows well in moist, well-drained soil in full sun to partial shade. Sorrel is rich in vitamins C and A, and its fast growth can help provide ground cover in the food forest. Its deep roots also help break up compacted soil.

5. ROOT LAYER:

Alaskan Blueberry (*Vaccinium alaskaense*): The Alaskan Blueberry is a low-growing shrub native to colder climates and thrives in acidic, well-drained soils. Its roots, along with the edible berries, are valuable for providing nutrient-dense fruit while enriching the soil. It attracts pollinators, making it a great addition to the food forest for biodiversity. The plant is particularly hardy and well-suited for USDA Zone 4, offering both food and ecological benefits.

Angelica (*Angelica archangelica*): Angelica is a large, herbaceous perennial that thrives in moist, well-drained soils and partial shade. The roots, stems, and seeds are all edible, with a strong, slightly spicy flavor that can be used in culinary dishes or medicinal preparations. The deep taproot helps break up compacted soil, and the plant's large leaves provide shelter for smaller plants. Angelica contributes both food and medicinal benefits while enriching the soil with its robust root system.

6. GROUND COVER LAYER:

Alpine Bistort (*Bistorta vivipara*): Alpine Bistort is a low-growing perennial that thrives in well-drained, rocky soils and cool climates, making it ideal for USDA Zone 4. The young leaves and roots are edible, with a mild, slightly sweet flavor. It forms dense mats, helping to prevent soil erosion, and supports biodiversity by attracting pollinators. This plant is a great choice for groundcover in a food forest due to its resilience and ecological contributions.

Sweet Woodruff (*Galium odoratum*): Sweet Woodruff is a fragrant, mat-forming groundcover that thrives in moist, shady areas. The leaves are used to flavor beverages and desserts, particularly in European cuisine, and have a sweet, hay-like scent when dried. Its dense growth helps suppress weeds, while it also attracts beneficial insects. This plant is ideal for adding a fragrant, edible ground layer to a Zone 4 food forest.

7. VERTICAL LAYER:

Akebia (*Akebia quinata*): Akebia is a fast-growing, semi-evergreen vine that thrives in full sun to partial shade. Its elongated fruits are edible, with a mildly sweet, gelatinous pulp that can be eaten fresh. Akebia is valued for its ability to cover vertical structures quickly and provide a lush, green canopy, while also offering an unusual edible crop. The plant is highly adaptable to a variety of soil conditions, making it a versatile choice for a food forest.

Arctic Kiwi (*Actinidia kolomikta*): Arctic Kiwi is a hardy vine that produces small, smooth-skinned, edible kiwi fruits, and can tolerate the cold winters of USDA Zone 4. The fruits are rich in vitamin C and have a sweet, tangy flavor. This vine is fast-growing, easily covering trellises, fences, or arbors, and its attractive variegated foliage makes it a beautiful addition to the vertical layer of a food forest. It thrives in well-drained, loamy soils and benefits from a sunny location.

8. FUNGAL LAYER:

Elm Oyster (*Hypsizygus ulmarius*): Elm Oyster mushrooms are edible and can be grown in USDA Zone 4, typically thriving on hardwood logs and wood chips. They produce meaty, white to cream-colored caps that are excellent in stir-fries, soups, or sautéed dishes. Aside from being a tasty food source, Elm Oyster mushrooms help break down organic matter, contributing to soil fertility in the food forest.

Lobster Mushroom (*Hypomyces lactifluorum*): Lobster mushrooms are actually a parasitic fungus that grows on other mushrooms, turning their host into a bright red, seafood-flavored delicacy. They thrive in hardwood forests in USDA Zone 4, often found growing on species like Russula or Lactarius. This mushroom adds biodiversity to the fungal layer and is highly prized for its unique flavor in gourmet cooking.

Zone 4 Plan Two: Budding Balance Blueprint

Zone 4 Plan Two: Budding Balance Blueprint

Emphasizing the balance between new growth and established life, this blueprint celebrates the emergence of spring. It's designed to nurture food production while steadily building ecological support.

1. CANOPY LAYER:

Northern Serviceberry (*Amelanchier canadensis*): The northern serviceberry, also known as juneberry, produces sweet, blueberry-like fruits that are edible and often used in jams or eaten fresh. This small tree thrives in moist, well-drained soil and partial shade, making it an excellent addition to the canopy layer. It provides early-season food for birds and pollinators and its white flowers in spring add beauty and support biodiversity in the food forest.

Shadbush (*Amelanchier spp.*): Shadbush is a small, multi-stemmed tree or shrub that produces small, sweet edible berries, commonly used in preserves, pies, and baking. It thrives in moist, well-drained soils and can tolerate both sun and shade, making it adaptable to the canopy layer. This plant also provides important early-season nectar for pollinators and adds diversity to the food forest by offering edible fruits for both humans and wildlife.

2. UNDERSTORY LAYER:

Dwarf Chokecherry (*Prunus virginiana 'Compacta'*): The dwarf chokecherry is a compact version of the traditional chokecherry tree, growing up to 5 feet tall. It produces small, dark berries that are edible when cooked, often used to make jams, jellies, and syrups. This variety thrives in well-drained, slightly acidic soils and partial shade, making it perfect for the understory layer. Its dense growth also provides shelter and habitat for wildlife, enhancing the biodiversity of the food forest.

Dwarf Honeyberry (*Lonicera caerulea var. edulis*): Honeyberry is a deciduous shrub that thrives in cool climates and produces sweet, blueberry-like fruits. The dwarf variety reaches about 3 to 4 feet in height and tolerates partial shade, making it suitable for the understory layer. The fruit ripens early in the season and is packed with antioxidants, making it a valuable addition to the food forest. Honeyberries are also hardy and pest-resistant, making them easy to grow in cooler zones.

3. SHRUB LAYER:

Cranberry (*Vaccinium macrocarpon*): Cranberry is a low-growing, spreading shrub that thrives in acidic, moist soils. It produces small, tart red berries that are rich in antioxidants and can be used in sauces, juices, and jams. Cranberries can tolerate cold temperatures and grow well in the understory or shrub layer of a food forest, providing fruit for humans and wildlife. The plant also helps improve soil health by supporting mycorrhizal fungi.

Golden Currant (*Ribes aureum*): Golden currant is a hardy shrub that grows to about 4 to 6 feet tall and produces clusters of small, tart yellow to red berries. These berries are edible and rich in vitamin C, often used in jams and jellies. Golden currant thrives in well-drained soils and tolerates partial shade, making it an excellent option for the shrub layer. Its fragrant flowers also attract pollinators, adding value to the food forest ecosystem.

4. HERBACEOUS LAYER:

Dandelion (*Taraxacum officinale*): Dandelion is a hardy, edible herb with all parts of the plant being edible, including the leaves, roots, and flowers. The leaves are great for salads, while the roots can be roasted as a coffee substitute. Dandelion helps in improving soil structure with its deep taproot and is excellent for attracting pollinators like bees.

Fennel (*Foeniculum vulgare*): Fennel is a perennial herb with aromatic, feathery leaves and edible bulbs, stems, and seeds. It prefers full sun and well-drained, fertile soil. The seeds are often used in cooking, while the leaves make a refreshing addition to salads. Fennel supports biodiversity in the food forest by attracting beneficial insects and is also drought-tolerant.

5. ROOT LAYER:

Celeriac (*Apium graveolens var. rapaceum*): Celeriac is a hardy root vegetable that grows well in moist, fertile soils, thriving in both sun and partial shade. Its bulbous, edible root has a mild celery flavor and is a nutritious addition to soups and salads. The plant has a deep taproot that helps break up compacted soil, and it provides habitat for pollinators. Celeriac is also beneficial for attracting beneficial insects to the food forest.

Chicory (*Cichorium intybus*): Chicory is a perennial herb with deep roots that can grow in poor soil conditions, making it a valuable plant for improving soil health. The root is edible, often roasted and used as a coffee substitute, and the leaves can be used as a nutritious salad green. Chicory's long taproot helps aerate the soil, while the plant also attracts pollinators with its bright blue flowers. This hardy plant offers both food and ecological benefits in the food forest.

6. GROUND COVER LAYER:

Dwarf Mallow (*Malva neglecta*): Dwarf Mallow is a creeping herbaceous plant that grows well in disturbed soils and full sun to partial shade. The young leaves and flowers are edible, often used in salads or as a cooked green, and the plant is known for its medicinal properties. It spreads easily, forming a thick mat that protects the soil, while its flowers attract pollinators. Dwarf Mallow is a versatile addition to the food forest groundcover layer.

Fringed Sagebrush (*Artemisia frigida*): Fringed Sagebrush is a low-growing, aromatic perennial that thrives in dry, well-drained soils. The leaves are edible in small amounts and are often used as a seasoning or medicinal herb. This plant is highly drought-tolerant and provides excellent groundcover, reducing soil erosion and improving biodiversity by attracting pollinators. Fringed Sagebrush is perfect for dry areas in a Zone 4 food forest.

7. VERTICAL LAYER:

Silver Lace Vine (*Fallopia baldschuanica*): Silver Lace Vine is a fast-growing, deciduous climber that can reach up to 30 feet in height. Its white, fragrant flowers bloom in late summer, and while it is primarily ornamental, the young shoots are edible when cooked. This vine provides excellent vertical coverage, helping to create habitat and shade within the food forest. It tolerates a wide range of soil conditions and is perfect for adding height and structure to the landscape.

Hardy Kiwi (*Actinidia arguta*): Hardy Kiwi is a vigorous vine that produces small, grape-sized kiwi fruits with a smooth, edible skin. The fruits are sweet and high in vitamin C, making them a valuable addition to the edible landscape. This plant grows well in USDA Zone 4, prefers well-drained soils, and can quickly cover trellises or fences. Hardy Kiwi also attracts pollinators and provides excellent vertical structure for the food forest.

8. FUNGAL LAYER:

Puffball Mushroom (*Calvatia gigantea*): Giant Puffball mushrooms can be found in grasslands or forest clearings in USDA Zone 4, and they produce large, white, spherical fruiting bodies. When harvested young, the flesh inside is pure white and edible, with a mild flavor often used in soups or sliced and grilled. They are excellent decomposers, helping to break down organic material in the soil.

Amethyst Deceiver (*Laccaria amethystina*): This small, purple mushroom is edible and grows in mixed woodland environments, making it suitable for USDA Zone 4 food forests. Amethyst Deceiver contributes to mycorrhizal networks that help tree roots access nutrients. Though rare, it is appreciated for its striking appearance and can be used in salads or cooked dishes when harvested young.

Zone 4 Plan Three: Melting Mosaic Masterplan

Inspired by the patchwork of melting ice and emerging flora, this masterplan integrates diverse plant layers for a vibrant food forest. It combines immediate harvest potential with long-term environmental resilience.

1. CANOPY LAYER:

Serviceberry (*Amelanchier alnifolia*): Also known as Saskatoon berry, this hardy deciduous shrub or small tree produces blue-purple berries that are sweet and edible, often used in pies or jams. It grows best in well-drained, slightly acidic soils and can tolerate partial shade. As a canopy plant, it attracts bees and other pollinators, and its fruits provide food for both humans and wildlife. Its adaptability and early-season blooms make it an important species in a food forest system.

Sugar Maple (*Acer saccharum*): The sugar maple is a large, deciduous tree known for its edible sap, which is used to make maple syrup. It thrives in rich, moist soils and prefers partial to full shade, making it an excellent canopy tree in the food forest. The tree also supports wildlife with its seeds and provides shade, helping to regulate the microclimate in the forest. Its high nutrient value and versatility make it an essential tree for both food production and ecological balance.

2. UNDERSTORY LAYER:

Cornelian Cherry Dogwood (*Cornus mas*): Cornelian cherry dogwood is a small, deciduous tree that grows up to 10 to 15 feet in height, making it an excellent choice for the understory. It produces small, edible red cherries that have a tart flavor, which can be used in jams, jellies, and beverages. The tree thrives in well-drained soils and partial shade, providing early spring flowers that attract pollinators and contribute to the visual appeal of the food forest.

Its dense foliage also offers shelter for small animals.

Dwarf Apricot (*Prunus armeniaca 'Tilton'*): The dwarf apricot tree grows up to 8 feet tall and produces sweet, edible apricots that ripen in the early summer. This variety is ideal for cooler climates and can tolerate partial shade. Dwarf apricots are hardy, cold-resistant, and bring early-season fruit to the food forest, making them an excellent option for the understory. Its beautiful white flowers in spring also attract pollinators, supporting biodiversity.

3. SHRUB LAYER:

Red Chokecherry (*Prunus virginiana 'Schubert'*): The red chokecherry is a cultivar of chokecherry that grows as a shrub or small tree, reaching up to 10 feet. Its berries, while sour, are edible when cooked and can be used to make sauces, jellies, or syrups. The plant thrives in a variety of soils and partial shade, producing beautiful red foliage in the fall. The flowers provide early nectar for pollinators, making it a great addition to the shrub layer.

Haskap Berry (*Lonicera caerulea var. edulis*): Haskap, also known as honeyberry, is a hardy shrub that grows up to 3 feet in height and produces elongated, blueberry-like fruits. These berries are sweet-tart and rich in antioxidants, often used in jams, jellies, and fresh eating. The plant thrives in well-drained soil and partial shade, providing an early fruit harvest and attracting pollinators with its spring blooms. Its hardiness makes it ideal for colder climates.

4. HERBACEOUS LAYER:

Garlic Chives (*Allium tuberosum*): Garlic Chives, with their mild garlic flavor, are a perennial herb with edible leaves and flowers. They are easy to grow in well-drained soil and prefer partial to full sun. Garlic chives help repel pests while attracting pollinators with their white, star-shaped flowers. They are a great addition to the herbaceous layer, offering both culinary and ecological benefits.

Gooseneck Loosestrife (*Lysimachia clethroides*): Gooseneck Loosestrife is a striking perennial herb that thrives in moist, well-drained soil in partial to full sun. Its white, bell-shaped flowers are edible and are often used in herbal teas. It adds beauty to the food forest while supporting pollinators and acting as a ground cover that helps prevent erosion.

5. ROOT LAYER:

Dandelion (*Taraxacum officinale*): Dandelion is a hardy perennial with edible roots, leaves, and flowers. The root is used medicinally and in teas, while the leaves are commonly used in salads or as a detoxifying tonic. Dandelion helps improve soil health by breaking up compacted layers with its deep taproot. It is also a favorite of pollinators, contributing to biodiversity in the food forest.

Fennel (*Foeniculum vulgare*): Fennel is a herbaceous perennial that prefers sunny, well-drained soils. Its edible bulbs, stalks, leaves, and seeds are commonly used in culinary dishes, imparting a mild, licorice-like flavor. The plant's deep roots help aerate the soil, while fennel also attracts beneficial insects like pollinators. It's a beneficial addition to the food forest for both culinary and ecological purposes.

6. GROUND COVER LAYER:

Wintergreen (*Gaultheria procumbens*): Wintergreen is an evergreen groundcover that thrives in acidic, well-drained soils and partial shade. Its edible red berries and leaves have a minty flavor and are used in teas, candies, and medicinal preparations. Wintergreen spreads by rhizomes, forming a dense mat that helps retain moisture in the soil, while also attracting pollinators. This plant is a valuable addition to the groundcover layer for its culinary and ecological benefits.

Strawberry Spinach (*Blitum capitatum*): Strawberry Spinach is a unique annual groundcover that grows well in moist, well-drained soils and full sun to partial shade. The leaves are edible and used like spinach, while the bright red, berry-like fruits are also edible, with a mild, slightly sweet flavor. It grows quickly, covering the ground and helping to improve soil health through nutrient cycling. This plant adds both visual appeal and edible benefits to the food forest.

7. VERTICAL LAYER:

Climbing Nightshade (*Solanum dulcamara*): Also known as bittersweet nightshade, this vine produces bright red berries, which are toxic raw but can be used medicinally when properly prepared. Its deep purple flowers attract pollinators, and the plant provides dense foliage cover. It is particularly beneficial for creating vertical layers in damp areas and provides wildlife habitat while also contributing to the ecological diversity of the forest.

Schisandra Vine (*Schisandra chinensis*): Schisandra Vine, also known as magnolia vine, produces bright red berries that are used medicinally and are known as "five-flavor fruits" due to their complex taste. The berries are rich in antioxidants and can be made into teas, juices, and preserves. This vigorous vine thrives in USDA Zone 4 and is ideal for vertical layering in the food forest. It prefers partial shade and moist, well-drained soils.

8. FUNGAL LAYER:

Hedgehog Mushroom (*Hydnum repandum*): Hedgehog mushrooms are easily identifiable by their spiny teeth on the underside of the cap. These mushrooms are edible and have a sweet, nutty flavor, making them popular in culinary dishes. They grow in coniferous or deciduous forests and play a role in decomposing organic matter, thriving in USDA Zone 4.

Black Trumpet (*Craterellus cornucopioides*): Black Trumpet mushrooms are rare but highly sought after for their rich, smoky flavor. They are commonly found in deciduous forests in USDA Zone 4, often associated with oak and beech trees. This mushroom aids in nutrient recycling and is an excellent addition to soups and sauces due to its depth of flavor.

ZONE 4 PLAN FOUR: FRESH FROST FUSION

Merging the remnants of frost with the promise of renewal, this design pairs hardy edibles with early bloomers. Its goal is to deliver a burst of nutrition and set the stage for a sustainable system.

1. CANOPY LAYER:

Wild Black Cherry (*Prunus serotina*): Wild black cherry is a tall, fast-growing tree that produces small, dark cherries that are edible when fully ripe. The fruit can be used for jellies, preserves, or even eaten fresh. It grows well in a variety of soil types, tolerates partial shade, and contributes to the food forest by providing fruits for wildlife and humans, while its dense canopy creates habitat for other species.

Mountain Ash (*Sorbus americana*): Mountain ash, or American rowan, produces small, bright red berries that are edible when cooked, often used in jams and jellies. It grows best in moist, well-drained soil and thrives in full sun to partial shade, making it a versatile choice for the canopy layer. The tree's beautiful white flowers in spring attract pollinators, while its berries provide food for birds and mammals, making it a beneficial addition to the food forest.

2. UNDERSTORY LAYER:

Japanese Maple (Dwarf Varieties) (*Acer palmatum*): Some dwarf Japanese maple varieties, such as the 'Kiyohime' cultivar, grow up to 3 to 5 feet tall and can tolerate light shade. Though not typically known for edible fruit, they are beneficial in the understory layer for their ability to provide partial shade to smaller plants. Their striking foliage and ornamental value make them a beautiful addition to the food forest, while also serving to protect other understory plants from harsh sun.

Crabapple (Dwarf Varieties) (*Malus domestica*

'*Dwarf*'): Dwarf crabapple trees typically grow to about 6 feet in height and produce small, tart apples that are edible and often used in preserves, sauces, and jellies. They are well-suited for the understory, growing best in partial shade and well-drained soil. The spring flowers attract pollinators, and the small apples provide food for wildlife. This tree also offers beauty with its flowers and fruits, contributing to the overall aesthetic of the food forest.

3. SHRUB LAYER:

Lingonberry (*Vaccinium vitis-idaea*): Lingonberry is a low-growing evergreen shrub that produces small, red, tart berries. These berries are edible and often used in sauces, jams, and as a side dish with meats. Lingonberries thrive in acidic, well-drained soils and can handle partial shade, making them perfect for the shrub layer. Their evergreen nature also provides year-round interest and ground cover, improving biodiversity in the food forest.

Juneberry (*Amelanchier canadensis*): Juneberry is a shrub or small tree that produces sweet, blueberry-like fruits in early summer. These berries are edible and are used in pies, jams, and preserves. The plant grows up to 12 feet tall, thriving in a variety of soils and partial shade. It supports pollinators with its early blooms and provides valuable food for both humans and wildlife.

4. HERBACEOUS LAYER:

Horehound (*Marrubium vulgare*): Horehound is a hardy herb with aromatic leaves that are often used to make herbal cough syrups or tea. It grows well in dry, rocky soils with full sun exposure. The plant can help with soil erosion control and attracts bees with its white flowers, making it a beneficial addition to the food forest's herbaceous layer.

Lemon Balm (*Melissa officinalis*): Lemon Balm is a fast-growing perennial herb that prefers full sun and well-drained soil. Its leaves are edible and have a mild lemon flavor, perfect for teas and culinary use. Lemon Balm also attracts bees and butterflies, enhancing the pollinator population in the food forest, while its aromatic oils can help deter pests.

5. ROOT LAYER:

Indian Turnip (*Arisaema triphyllum*): Indian Turnip, also known as Jack-in-the-Pulpit, produces edible roots that have a spicy, peppery flavor when cooked. The plant thrives in rich, moist, and well-drained soils in shaded areas, making it ideal for the understory of a food forest. Its roots help to accumulate nutrients in the soil, and it attracts pollinators with its unique flowering structure. This plant is an excellent choice for both its edible root and ecological contributions.

Lovage (*Levisticum officinale*): Lovage is a hardy perennial herb that grows well in rich, fertile soils and full sun to partial shade. Its edible roots, stems, and leaves have a celery-like flavor, making them ideal for soups, stews, and salads. The plant's deep root system helps break up compacted soil, and it provides habitat for pollinators. Lovage is a valuable herb in the food forest for its culinary uses and soil health benefits.

6. GROUND COVER LAYER:

Carpet Bugle (*Ajuga reptans*): Carpet Bugle is a perennial groundcover that thrives in moist, well-drained soils and partial shade. The leaves are edible when young, often used in salads or as a garnish, though they have a slightly bitter taste. This fast-spreading plant forms a thick mat, suppressing weeds and protecting the soil from erosion. Its blue flowers attract pollinators, making it a beneficial addition to the ground layer.

Barren Strawberry (*Waldsteinia fragarioides*): Barren Strawberry is a hardy, low-growing groundcover that thrives in well-drained soils and full sun to partial shade. While its small, strawberry-like fruits are not edible, the plant's evergreen foliage provides year-round soil protection and habitat for beneficial insects. Barren Strawberry spreads quickly, forming a dense mat that helps prevent soil erosion. It's a great groundcover option for USDA Zone 4 food forests.

7. VERTICAL LAYER:

Hopniss (*Apios americana*)**:** Hopniss, or groundnut vine, is a nitrogen-fixing plant that produces edible tubers, which are a rich source of protein and carbohydrates. The plant also produces fragrant, pinkish-purple flowers that attract pollinators. Hopniss thrives in moist, well-drained soils and can quickly climb trellises or other structures, making it a great choice for the vertical layer of a food forest. Its tubers can be boiled or roasted, similar to potatoes.

Dutchman's Pipe (*Aristolochia macrophylla*)**:** Dutchman's Pipe is a vigorous vine with large, heart-shaped leaves and unique, pipe-shaped flowers. While the plant is primarily ornamental, it plays an important role in attracting pollinators, particularly butterflies. This vine thrives in USDA Zone 4, making it a suitable choice for creating vertical layers in the food forest, providing both aesthetic value and habitat for beneficial insects. It prefers well-drained soils and partial shade.

8. FUNGAL LAYER:

Pheasant's Back Mushroom (*Cerioporus squamosus*)**:** Also known as Dryad's Saddle, this mushroom produces large, fan-shaped fruiting bodies with a unique, sweet aroma reminiscent of watermelon rind. Found growing on dead hardwoods in USDA Zone 4, Pheasant's Back mushrooms are edible when young and tender, often sautéed or used in stews. They help with the decomposition of fallen logs and are valuable in nutrient cycling.

Hawk's Wing Mushroom (*Sarcodon imbricatus*): This rare mushroom grows in association with conifer trees in USDA Zone 4, producing a cap covered with scales that resemble a hawk's wing. Though somewhat bitter, Hawk's Wing is edible and adds unique flavor to dishes when prepared properly. It plays a role in mycorrhizal relationships, benefiting surrounding trees by increasing their access to soil nutrients.

Zone 4 Plan Five: Winter Wane Workshop

A creative workshop for easing out of winter, this plan layers plants that take advantage of a waning cold. It's focused on transitioning to a productive season while establishing lasting system support.

1. CANOPY LAYER:

Black Locust (*Robinia pseudoacacia*): Black locust is a nitrogen-fixing tree that produces fragrant white flowers and small pods that can be used in some culinary applications, though the pods must be cooked properly before consumption. It grows in a variety of soil types and can tolerate both sun and partial shade, adding structure and ecological benefits to the food forest. The tree also provides important shelter for wildlife and helps improve soil quality through its nitrogen-fixing properties.

Red Mulberry (*Morus rubra*): Red mulberry is a medium-sized tree that produces sweet, red to purple berries that are edible and can be eaten fresh or used in jams and baked goods. It thrives in moist, well-drained soils and partial shade, making it an excellent choice for the canopy layer. Its fruit serves as a food source for both humans and wildlife, and its leaves provide food for silkworms, further enhancing the food forest's diversity.

2. UNDERSTORY LAYER:

Dwarf European Pear (*Pyrus communis 'Compacta'*): The dwarf European pear tree reaches 6 to 8 feet tall and produces sweet, edible pears that are perfect for small spaces and the understory layer. This tree thrives in partial shade and well-drained, fertile soils. It adds value by producing early-season fruit and providing nectar for pollinators with its spring blooms. Its compact size allows for easy harvesting and fits well in a small food forest space.

Nanking Cherry (*Prunus tomentosa*): Nanking

cherry is a hardy shrub that reaches about 6 feet in height and produces clusters of tart, edible cherries. These cherries are often used in jellies, sauces, and baked goods. Nanking cherries thrive in well-drained soils and can tolerate partial shade, making them ideal for the understory layer. Their bright red fruit and spring flowers attract pollinators and provide early-season food for wildlife.

3. SHRUB LAYER:

Dwarf Aronia (*Aronia melanocarpa 'Low Scape Mound'*): A dwarf cultivar of the chokeberry, this shrub reaches about 2 to 3 feet in height and produces small, black berries that are edible when cooked. The berries are packed with antioxidants and are commonly used in jams, jellies, and smoothies. It grows well in a variety of soils and partial shade, providing early-season food and attracting pollinators with its spring flowers.

Buffaloberry (*Shepherdia argentea*): Buffaloberry is a hardy shrub that grows up to 6 feet tall and produces small, tart red berries. The berries are edible and are often used to make sauces, jellies, and beverages. The shrub thrives in dry, well-drained soils and partial shade, making it perfect for the shrub layer. It also provides shelter for wildlife and attracts pollinators with its spring flowers.

4. HERBACEOUS LAYER:

Marshmallow (*Althaea officinalis*): Marshmallow is a perennial herb with edible roots, leaves, and flowers that have a mild, sweet flavor. It thrives in wet, well-drained soils and partial to full sun. The roots are traditionally used for medicinal purposes, and it can act as a dynamic accumulator plant, improving soil health with its deep roots.

Mullein (*Verbascum Thapsus*): Mullein is a biennial herb with large, fuzzy leaves and tall spikes of yellow flowers. While not typically eaten, its leaves can be used in teas and as a poultice. Mullein is a good companion plant in the food forest, as it attracts pollinators, improves soil fertility, and provides medicinal benefits for respiratory health.

5. ROOT LAYER:

Oca (*Oxalis tuberosa*): Oca is a root vegetable that grows well in fertile, well-drained soils and cooler temperatures. Its tubers are edible and have a slightly tangy, potato-like flavor, often used in salads or cooked dishes. The plant is drought-tolerant once established and its tubers help to improve soil fertility. Oca also attracts pollinators, enhancing the overall health of the food forest.

Parsnip *(Pastinaca sativa)*: Parsnip is a biennial root vegetable that grows best in loose, well-drained, fertile soils. The edible root has a sweet, nutty flavor, especially when roasted, making it a delicious addition to the food forest's root layer. Its deep taproot improves soil structure by breaking up compacted layers, and the plant also attracts beneficial insects. Parsnip is a valuable source of carbohydrates and nutrients in the food forest.

6. GROUND COVER LAYER:

Mountain Cranberry (*Vaccinium vitis-idaea*): Mountain Cranberry, also known as Lingonberry, is an evergreen shrub that grows as groundcover in acidic, well-drained soils. Its small red berries are edible and commonly used in jams and sauces, while the leaves are also sometimes used in teas. This plant forms dense mats, helping to retain moisture in the soil, and it attracts pollinators. Mountain Cranberry is ideal for food forests with acidic soils in Zone 4.

Yellow Wood Sorrel (*Oxalis stricta*): Yellow Wood Sorrel is a low-growing herb that thrives in moist, well-drained soils and partial shade. The leaves, flowers, and seed pods are edible, with a tangy, lemon-like flavor that can be used in salads or as a garnish. It spreads easily and forms a dense mat, helping to suppress weeds and retain soil moisture. This plant is also known for attracting pollinators, making it a useful groundcover in the food forest.

7. VERTICAL LAYER:

Groundnut Vine (*Apios americana*): Groundnut Vine, also called wild bean or hopniss, produces edible, high-protein tubers that were traditionally used by Native Americans. The plant also produces fragrant flowers that attract pollinators. Groundnut Vine is a nitrogen-fixing plant, which improves soil health and supports other plants in the food forest. It thrives in USDA Zone 4 and can be trained to climb vertical structures, adding height and food production to the landscape.

Hardy Passionflower (*Passiflora incarnata*): Also known as maypop, hardy passionflower produces beautiful, intricate flowers and edible fruits that can be used in jams and jellies. The plant is a fast-growing, cold-tolerant vine that thrives in USDA Zone 4 and is excellent for covering trellises or arbors. It provides a unique addition to the vertical layer, attracting pollinators while also offering an unusual edible crop for the food forest.

8. FUNGAL LAYER:

Matsutake Mushroom (*Tricholoma matsutake*): This prized edible mushroom is known for its spicy aroma and is rare in USDA Zone 4, often found in association with pine trees. Matsutake mushrooms contribute to the mycorrhizal networks in the soil, improving nutrient uptake for surrounding trees. Its scarcity and unique flavor make it a valuable culinary ingredient.

Velvet Foot Mushroom (*Flammulina velutipes*): Velvet Foot mushrooms grow on hardwood trees and are edible, especially popular in East Asian cuisine. They thrive in cool climates like USDA Zone 4, often fruiting in late fall or winter. Velvet Foot mushrooms help decompose wood, contributing to the breakdown of organic matter in the forest ecosystem.

Zone 4 Plan Six: 7. Thawing Thrive Tactics

This plan constructs a resilient edifice of early bloomers and steady growers. It's crafted to harvest the early gifts of spring while building a framework that supports the entire food forest.

1. CANOPY LAYER:

Yellow Birch (*Betula alleghaniensis*): Yellow birch produces edible sap, which can be harvested in the spring to make syrup. It grows in moist, well-drained soils and prefers partial shade, making it a good canopy plant in the food forest. The tree's bark, which has a distinctive golden color, is also highly valued for its natural beauty and its ability to support various wildlife species, contributing to the overall health of the ecosystem.

Black Ash (*Fraxinus nigra*): Black ash is a large, deciduous tree that thrives in wetland areas and produces edible seeds, which can be roasted or eaten raw. It grows best in moist, well-drained soils and can tolerate both sun and shade, making it ideal for the canopy layer. Black ash is also known for its ability to help improve soil health through its interaction with fungal species, contributing to a diverse and thriving ecosystem.

2. UNDERSTORY LAYER:

Tart Cherry (*Prunus cerasus 'Montmorency'*): Tart cherry trees, growing to around 8 feet in height, are well-suited for cooler climates and the understory layer. They produce sour, red cherries that are prized for their use in pies, jams, and preserves. These trees are easy to grow in moist, well-drained soils and provide early-season fruit. Their flowers attract pollinators, and they add both functional and aesthetic value to the food forest.

Dwarf Mulberry (*Morus nigra 'Dwarf'*): The dwarf mulberry tree reaches about 4 to 5 feet in height and produces sweet, edible mulberries. Mulberries are packed with antioxidants and can be eaten fresh or used in various culinary applications. This tree grows well in partial shade and can adapt to a variety of soil types. Its small size makes it suitable for the understory, and its fruits provide food for birds and other wildlife.

3. SHRUB LAYER:

Serviceberry (*Amelanchier alnifolia*): Serviceberry is a small shrub or tree that grows 6 to 10 feet tall, producing sweet, blueberry-like fruits in early summer. These berries are edible and are used in jams, pies, and fresh eating. The shrub thrives in well-drained soils and partial shade, offering early-season fruit and springtime blooms that attract pollinators.

Gooseberry (*Ribes uva-crispa*): Gooseberry is a deciduous shrub that grows up to 4 feet tall and produces small, tart berries that are edible. The berries are used in pies, jams, and sauces. Gooseberries thrive in well-drained, fertile soils and partial shade, making them ideal for the shrub layer. The plant also attracts pollinators with its small flowers and provides food for birds.

4. HERBACEOUS LAYER:

Nettle (*Urtica dioica*): Nettle is a nutrient-dense perennial herb with edible leaves when cooked, offering a good source of vitamins and minerals. It thrives in moist, rich soil in partial shade. Nettle helps with soil structure and supports biodiversity by attracting pollinators and beneficial insects. It is also known for its medicinal properties.

Oregano (*Origanum vulgare*): Oregano is a perennial herb known for its robust flavor and edible leaves, which are commonly used in cooking. It grows well in well-drained, moderately fertile soils in full sun. Oregano attracts pollinators with its small purple flowers and has medicinal properties that can support the immune system.

5. ROOT LAYER:

Salsify (*Tragopogon porrifolius*): Salsify, also known as oyster plant, has edible, mild-flavored roots that are often used in soups and stews. The plant thrives in deep, fertile soils and full sun. Its roots, which have a texture similar to artichokes, provide food for the garden, and the plant's deep root system helps to aerate and improve the soil. Salsify also attracts pollinators, supporting the food forest's ecosystem.

Scorzonera (*Scorzonera hispanica*): Scorzonera, or black salsify, is a root vegetable with dark-skinned, edible tubers that have a sweet, nutty flavor when cooked. It grows well in deep, fertile, well-drained soils and full sun. The plant's deep, penetrating roots help loosen compacted soil, improving water infiltration. Scorzonera adds a unique root vegetable to the food forest and contributes to soil health.

6. GROUND COVER LAYER:

Mossy Saxifrage (*Saxifraga bryoides*): Mossy Saxifrage is a creeping perennial that grows in rocky, well-drained soils and thrives in USDA Zone 4. While not typically grown for food, its leaves and flowers can be used sparingly in salads for their slightly bitter flavor. This plant forms a thick, low-growing mat that helps prevent soil erosion and retains moisture, making it an ideal groundcover for harsh environments.

Beach Strawberry (*Fragaria chiloensis*): Beach Strawberry is a low-growing perennial that thrives in well-drained, sandy soils and full sun. The small, sweet berries are edible and used in jams, desserts, or fresh eating. It spreads by runners, quickly forming a dense groundcover that helps suppress weeds and retain moisture in the soil. Beach Strawberry is a resilient plant that contributes to both the ground layer and fruit production in a food forest.

7. VERTICAL LAYER:

Kiwi Vine (*Actinidia kolomikta*): Kiwi Vine is a cold-hardy species of the kiwi family that produces small, smooth-skinned, edible fruits with a flavor similar to common kiwis. It is prized for its ornamental variegated leaves and its ability to thrive in cold climates, including USDA Zone 4. This fast-growing vine can quickly cover trellises or fences, providing both aesthetic value and edible fruit. The vine is best grown in well-drained soils with full to partial sun.

Chocolate Vine (*Akebia trifoliata*): Chocolate Vine is a hardy, fast-growing vine that produces sausage-shaped edible fruits with a sweet, gelatinous pulp. The fruits are high in vitamins and can be eaten fresh or used in desserts. This vine thrives in USDA Zone 4 and is an excellent choice for covering vertical structures, adding both beauty and productivity to the food forest. It is drought-tolerant and grows well in a variety of soil types.

8. FUNGAL LAYER:

Slippery Jack Mushroom (*Suillus luteus*): This edible mushroom is commonly found in USDA Zone 4 under pine trees and is known for its slimy cap. It has a mild flavor and is often used in soups or sautéed dishes. Slippery Jack mushrooms form symbiotic relationships with pine roots, aiding in nutrient exchange and soil health.

Deer Mushroom (*Pluteus cervinus*): Deer mushrooms are commonly found growing on decaying wood in USDA Zone 4. They are edible, though with a mild flavor, and are often added to stir-fries or soups. This mushroom plays an essential role in breaking down lignin in wood, returning nutrients to the soil and supporting forest health.

Zone 4 Plan Seven: Thawing Thrive Tactics

Zone 4 Plan Seven: Thawing Thrive Tactics

Designed to encourage thriving life as temperatures rise, these tactics emphasize strategic plant pairings in each layer. The plan delivers both seasonal edibles and a foundation for enduring sustainability.

1. CANOPY LAYER:

Pawpaw (*Asimina triloba*): Pawpaw is a small, deciduous tree that produces large, greenish-brown fruit with a sweet custard-like flavor, often used in desserts. It thrives in moist, well-drained soils and prefers partial to full shade, making it ideal for the canopy layer of the food forest. The tree provides food for both humans and wildlife, and its large, tropical-looking leaves create a lush, sheltered environment for other plants and species.

European Hornbeam (*Carpinus betulus*): European hornbeam is a hardy, deciduous tree that produces edible seeds, though they are not typically consumed fresh due to their tough exterior. It thrives in well-drained, fertile soils and can tolerate a wide range of sunlight conditions, from full sun to partial shade. This tree adds valuable structural support to the food forest while its seeds can serve as a supplemental food source for wildlife.

2. UNDERSTORY LAYER:

Black Currant (*Ribes nigrum*): Black currants are small shrubs that grow up to 4 to 5 feet tall and produce tangy, vitamin-rich berries. These berries are used in juices, jams, and syrups. Black currants thrive in cool, moist conditions and partial shade, making them ideal for the understory layer. They also attract pollinators and support biodiversity by providing food for wildlife.

Red Elderberry (*Sambucus pubens*): The red elderberry is a small shrub that grows about 6 feet tall and produces clusters of red berries that are edible when cooked. The berries are commonly used in jams, syrups, and sauces. It grows best in moist, slightly acidic soils and partial shade, providing food for wildlife and offering shelter in the understory. The attractive flowers and berries also contribute to the food forest's aesthetics.

3. SHRUB LAYER:

Elderberry (European) (*Sambucus nigra*): European elderberry is a large shrub that can reach up to 12 feet in height, producing clusters of small, dark purple to black berries. The berries are edible when cooked and are commonly used to make syrups, jams, and wine. Elderberries thrive in moist, well-drained soils and partial shade, making them perfect for the shrub layer. The flowers attract pollinators, and the berries provide food for birds.

Honeysuckle (Edible Varieties) (*Lonicera edulis*): Edible honeysuckle varieties such as Lonicera edulis grow as shrubs or small vines that reach 3 to 6 feet in height. They produce sweet, berry-like fruits that are edible and often used in jams and fresh eating. These plants thrive in partial shade and well-drained soils, adding both beauty and food to the shrub layer. The flowers attract bees and hummingbirds, supporting the ecosystem.

4. HERBACEOUS LAYER:

Parsley (*Petroselinum crispum*): Parsley is a biennial herb that thrives in cool, moist soils and partial to full sun. The leaves are edible and commonly used as a garnish or in cooking. Parsley supports the food forest by attracting beneficial insects like ladybugs and aphids, and it helps enhance soil health by adding organic matter.

Plantain (*Plantago major*): Plantain is a hardy perennial herb that grows in a variety of soils, from lawns to fields. Its leaves are edible and can be used in salads or as a poultice for medicinal purposes. Plantain helps with soil health by accumulating minerals and is a valuable plant for attracting beneficial insects to the food forest.

5. ROOT LAYER:

Shallots (*Allium cepa var. aggregatum*): Shallots are a mild variety of onion with edible bulbs and greens. The bulbs, which have a sweet, onion-like flavor, are used in cooking, while the greens can be used in salads. Shallots grow well in loose, well-drained soil and full sun, and their root system helps improve soil aeration. Shallots also attract beneficial insects, contributing to the health of the food forest.

Skirret (*Sium sisarum*): Skirret is a perennial root vegetable with sweet, edible tubers that are often boiled or roasted. It thrives in moist, fertile soils and partial to full sun. The deep roots help to aerate the soil, improving its structure, and the plant is a good source of carbohydrates. Skirret also provides habitat for pollinators, making it an excellent addition to the food forest.

6. GROUND COVER LAYER:

Mazus (*Mazus reptans*): Mazus is a creeping perennial that thrives in moist, well-drained soils and partial shade. While not typically grown for its edibility, its small leaves can be used as a garnish in salads. This plant forms a dense mat that helps to retain soil moisture and suppress weeds, while its purple flowers attract pollinators. Mazus is an excellent choice for groundcover in a USDA Zone 4 food forest due to its fast growth and soil protection capabilities.

Labrador Tea (*Rhododendron groenlandicum*): Labrador Tea is a low-growing evergreen shrub that thrives in acidic, boggy soils and cold climates. The leaves are edible when prepared properly, often used to make a mildly stimulating tea. This plant is excellent for adding diversity to a food forest, providing groundcover, and attracting pollinators. Its ability to thrive in poor soils makes it a resilient option for the ground layer.

7. VERTICAL LAYER:

Scarlet Runner Bean (*Phaseolus coccineus*): Scarlet Runner Bean is a fast-growing, climbing vine that produces both edible beans and striking red flowers, which attract pollinators. The beans can be harvested when young and eaten fresh, or dried for later use. This vine is a nitrogen-fixer, improving soil health while providing a vertical layer of food production. It thrives in USDA Zone 4, growing well in full sun and well-drained soils.

Purple Hyacinth Bean (*Lablab purpureus*): Purple Hyacinth Bean is a climbing vine known for its striking purple flowers and pods. While the pods are edible when young, the mature seeds require careful preparation to remove toxins. This vine adds a beautiful vertical element to the food forest, attracting pollinators and providing food. It thrives in USDA Zone 4 with proper care and can cover arbors or trellises quickly, creating height and structure.

8. FUNGAL LAYER:

Shaggy Mane (*Coprinus comatus*): Shaggy Mane mushrooms are easily recognizable by their tall, shaggy caps and are edible when young. Found in grassy areas or disturbed soils in USDA Zone 4, they are best cooked shortly after harvest to prevent them from dissolving into black ink. Shaggy Mane helps decompose organic matter and supports soil health.

Cauliflower Mushroom (*Sparassis crispa*): This rare mushroom produces large, brain-like, edible fruiting bodies that have a mild flavor and are often used in soups or stir-fries. Found in coniferous forests in USDA Zone 4, it grows at the base of trees, helping to decompose dead wood. Cauliflower mushrooms are also known for their potential medicinal properties.

Audra Haldane

Zone 4 Plan Eight: Emerging Eden Equation

I envisioned this design as an "Eden" emerging from winter's retreat through calculated, layered planting. It balances the equation between immediate food rewards and long-term ecological health.

1. CANOPY LAYER:

American Chestnut (*Castanea dentata*): Once a dominant species in North America, the American chestnut produces large, sweet edible nuts that can be eaten fresh, roasted, or ground into flour. It grows in rich, well-drained soils and prefers partial shade, making it an excellent canopy tree. The tree offers high-value food for both humans and wildlife, while its nut-bearing capacity supports biodiversity and adds nutritional value to the ecosystem.

Eastern Redbud (*Cercis canadensis*): The eastern redbud is a small, deciduous tree that produces vibrant pink flowers in early spring, which are edible and can be used in salads or as garnishes. It thrives in well-drained soils and partial shade, making it a perfect fit for the canopy layer of a food forest. The tree adds beauty to the landscape while its flowers serve as a food source for pollinators and other beneficial insects.

2. UNDERSTORY LAYER:

Dwarf Peaches (*Prunus persica 'Bonanza'*): Dwarf peach trees grow to 3 to 4 feet in height and produce delicious, juicy peaches. These trees are well-suited for the understory, as they tolerate partial shade and grow well in well-drained soils. The small peaches are a treat for both humans and wildlife, and the trees offer beauty with their spring blossoms. Their compact size makes them easy to incorporate into a small food forest design.

Dwarf Apple (Crabapple Varieties*) (Malus domestica 'Pixie'*): Dwarf apple varieties such as 'Pixie' grow to about 6 feet tall and produce small, sweet apples. These apples are perfect for making preserves or for eating fresh. The trees tolerate partial shade and are well-suited to cooler climates, making them ideal for the understory layer. They offer early-season fruit and support pollinators with their spring blooms.

3. SHRUB LAYER:

Elderberry (American) (*Sambucus canadensis*): American elderberry is a deciduous shrub that grows 6 to 12 feet tall and produces clusters of small, dark purple berries. The berries are edible when cooked and are used to make syrups, pies, and wines. The shrub thrives in well-drained, slightly acidic soils and partial shade. Its flowers provide nectar for pollinators, while the berries provide food for wildlife.

Red Currant (*Ribes rubrum*): Red currant is a hardy, deciduous shrub that grows up to 5 feet tall and produces clusters of tart red berries. These berries are edible and are commonly used in jellies, jams, and sauces. Red currants thrive in slightly acidic, moist soils and partial shade, making them ideal for the shrub layer. The plant attracts pollinators with its early flowers and provides food for birds.

4. HERBACEOUS LAYER:

Ramps (*Allium tricoccum*): Ramps, or wild leeks, are a flavorful wild edible that grows well in moist, well-drained soils in partial shade. The leaves and bulbs are edible, with a strong garlic-onion flavor. Ramps are a native plant that brings early-season food to the food forest and provides habitat for pollinators.

Red Clover (*Trifolium pratense*): Red Clover is a nitrogen-fixing perennial herb that grows well in fertile, well-drained soils and full sun. Its flowers are edible and are often used in teas and salads. The plant helps improve soil health by adding nitrogen, making it a beneficial addition to the herbaceous layer of the food forest.

5. Root Layer:

Sweet Potatoes (*Ipomoea batatas*): Sweet potatoes are a warm-season root vegetable that thrives in well-drained soil and full sun. The edible roots are rich in vitamins and carbohydrates, while the vines provide excellent ground cover, helping to suppress weeds. The deep roots improve soil structure, while sweet potatoes also attract pollinators with their beautiful flowers. This plant is a versatile and beneficial addition to the food forest.

Taro (*Colocasia esculenta*): Taro is a tropical root vegetable that grows well in moist, well-drained soils and partial shade. The edible corms are starchy and nutritious, often used in soups, stews, or as a boiled side dish. Taro's roots contribute to improving soil health by adding organic matter, and the plant also provides shelter for other plants in the food forest. It can be grown in milder USDA Zone 4 areas during the warmer months.

6. Ground Cover Layer:

Heuchera (*Heuchera americana*): Heuchera, also known as alumroot, is a low-growing perennial that thrives in shaded, well-drained soils. The roots have been used historically in medicine, though the plant is more commonly grown for its attractive foliage. It forms a dense groundcover that helps retain moisture and prevent soil erosion, while also providing habitat for pollinators. Heuchera is a beautiful and functional addition to the ground layer in a Zone 4 food forest.

Mountain Sorrel (*Oxyria digyna*): Mountain Sorrel is a low-growing, herbaceous perennial that thrives in moist, well-drained soils and cold climates. The leaves are edible and have a tart, lemony flavor, making them a refreshing addition to salads or cooked dishes. This plant spreads easily, providing dense groundcover that helps to prevent soil erosion and improve moisture retention. Mountain Sorrel is an excellent addition to the ground layer for its culinary and ecological benefits.

7. VERTICAL LAYER:

Magnolia Vine (*Schisandra chinensis*): Magnolia Vine, also known as Schisandra, produces small, bright red berries that are highly valued for their medicinal properties. The berries are used in teas, wines, and syrups and are known to have adaptogenic properties. This vine thrives in USDA Zone 4, growing best in partial shade and moist, well-drained soils. It adds both aesthetic value and an edible harvest to the vertical layer of the food forest.

Climbing Asparagus Fern (*Asparagus scandens*): This unusual climbing fern produces tender young shoots that can be eaten like traditional asparagus. It is a vigorous grower that can quickly cover fences or trellises, adding height to the food forest. As a perennial, it provides a long-term food source while also contributing to soil health. Climbing Asparagus Fern is hardy in USDA Zone 4 and prefers well-drained soils and full to partial sun.

8. Fungal Layer:

Brick Cap Mushroom (*Hypholoma lateritium*): Brick Cap mushrooms grow in clusters on decaying hardwood in USDA Zone 4. Though not commonly consumed in North America, they are considered edible with a slightly bitter taste. They contribute to the decomposition of wood, helping to recycle nutrients back into the soil.

Wood Ear Mushroom (*Auricularia auricula-judae*): This edible mushroom is prized for its jelly-like texture and is often used in soups and stir-fries.

Wood Ear grows on dead or dying hardwood trees and is suitable for USDA Zone 4. It plays a key role in breaking down dead wood, enhancing soil fertility in food forests.

Zone 4 Plan Nine: Revival Roots Regimen

When I designed this plan, I focused my research on the revival of life from dormant roots, this regimen combines timely plantings with sustained growth. It is meant to invigorate the landscape with both food production and system support.

1. CANOPY LAYER:

Sweet Birch (*Betula lenta*): Sweet birch produces edible sap in the early spring, which can be used to make syrup or consumed as a refreshing drink. It grows best in moist, well-drained soils and tolerates partial shade, making it a good choice for the canopy layer. The tree also provides wildlife with habitat and shelter while contributing to the food forest's overall health through its nutrient-rich sap.

Black Cherry (*Prunus serotina*): A fast-growing tree that produces small black cherries for birds and wildlife while also supporting native pollinators.

2. UNDERSTORY LAYER:

Hop Hornbeam (*Ostrya virginiana*): A slow-growing, dense tree that supports soil retention and provides habitat for birds and small mammals.

Dwarf Fig (*Ficus carica 'Petite Negra'*): Dwarf fig trees typically grow up to 3 feet in height and produce small, sweet figs that are perfect for small spaces. These figs are edible and can be eaten fresh or dried. The tree grows well in partial shade and is perfect for the understory layer, providing food for both humans and wildlife. Its compact size makes it an excellent choice for small, shaded spaces in the food forest.

3. SHRUB LAYER:

Mulberry (Dwarf Varieties) (*Morus nigra*): Dwarf mulberry shrubs grow up to 5 feet tall and produce small, sweet mulberries. These berries are edible and can be used fresh or in preserves. Mulberry shrubs thrive in well-drained soils and tolerate partial shade, making them a great addition to the shrub layer. The fruit provides food for birds and humans, and the plant attracts pollinators with its small flowers.

Blackberry (*Rubus fruticosus*): Blackberry is a thorny shrub that grows up to 6 feet tall and produces sweet, edible blackberries. These berries are used in a variety of culinary applications, from jams to desserts. Blackberries thrive in slightly acidic, well-drained soils and tolerate partial shade, making them suitable for the shrub layer. The plant attracts pollinators and provides food for both humans and wildlife.

4. HERBACEOUS LAYER:

Shiso (*Perilla frutescens*): Shiso, also known as perilla, is a leafy herb with a strong flavor and aromatic leaves that are edible and used in Asian cuisine. It prefers well-drained, moist soil in full sun and can grow as an annual in cooler climates. Shiso attracts pollinators and can provide a unique culinary addition to the food forest.

St. John's Wort (*Hypericum perforatum*): St. John's Wort is a medicinal herb with yellow, star-shaped flowers that are edible and often used to make teas or tinctures. It grows well in well-drained soil with full sun exposure. This plant can benefit the food forest by supporting beneficial insects and improving soil health with its deep roots.

5. ROOT LAYER:

Taro (*Colocasia esculenta*): Taro is a tropical root vegetable that grows well in moist, well-drained soils and partial shade. The edible corms are starchy and nutritious, often used in soups, stews, or as a boiled side dish. Taro's roots contribute to improving soil health by adding organic matter, and the plant also provides shelter for other plants in the food forest.

It can be grown in milder USDA Zone 4 areas during the warmer months.

Turmeric (*Curcuma longa*): Turmeric is a perennial herb that produces edible rhizomes used as a spice. It grows best in warm, moist, well-drained soils with full sun and requires a growing season long enough to produce the rhizomes. The rhizomes have anti-inflammatory and antioxidant properties and are a great addition to culinary dishes. Turmeric's deep roots also contribute to improving soil health by adding organic matter.

6. GROUND COVER LAYER:

Bearberry (*Arctostaphylos alpina*): Bearberry is a low-growing, evergreen groundcover that thrives in dry, acidic soils and cold climates. The red berries are edible, though they are somewhat mealy and often used in jams or as survival food. It forms dense mats that help prevent soil erosion and retain moisture, while its flowers attract pollinators. Bearberry is a hardy, drought-tolerant groundcover for Zone 4 food forests.

Self-Heal (*Prunella vulgaris*): Self-Heal is a low-growing, creeping perennial that thrives in moist, well-drained soils and partial shade. The leaves and flowers are edible and can be used in salads, teas, or as a garnish, with a mild, slightly bitter flavor. It forms a dense mat that helps suppress weeds, while also attracting beneficial insects and pollinators. Self-Heal is a valuable addition to the groundcover layer for both its medicinal and ecological benefits.

7. VERTICAL LAYER:

Japanese Honeysuckle (*Lonicera japonica*): Japanese Honeysuckle is a fragrant, flowering vine that produces edible flowers, which can be used in teas or as a sweet garnish. The plant is highly attractive to pollinators and provides excellent cover for wildlife. This vine thrives in USDA Zone 4 and is suitable for covering arbors or fences, adding both beauty and functionality to the vertical layer of the food forest. It grows well in a variety of soil types.

Hardy Jasmine (*Jasminum officinale var. affine*): Hardy Jasmine is a vigorous vine that produces fragrant white flowers in the summer. While primarily grown for its ornamental value, jasmine flowers can also be used to flavor teas. This vine is cold-hardy to USDA Zone 4 and can be used to create a beautiful, fragrant vertical layer in the food forest. It prefers full sun and well-drained soils, making it a versatile addition to the landscape.

8. FUNGAL LAYER:

Gem-Studded Puffball (*Lycoperdon perlatum*): This small, edible puffball mushroom can be found in grassy or woodland areas in USDA Zone 4. It produces a crunchy texture when fried and is harvested while the interior is still white. Gem-Studded Puffballs are saprotrophic, breaking down organic material and contributing to the soil ecosystem.

Shrimp of the Woods (*Entoloma abortivum*): This unique mushroom is the result of a parasitic relationship between the fungus and Armillaria species. The aborted form is edible and has a seafood-like flavor, hence its name. Found in hardwood forests in USDA Zone 4, it helps control populations of other fungi and contributes to the overall fungal diversity of the forest.

Zone 4 Plan Ten: Crisp Renewal Canvas

I designed this plan to paint a picture of renewal with crisp, seasonal selections layered to perfection. Its purpose is to capture the early energy of spring to produce a resilient, food-rich ecosystem.

1. CANOPY LAYER:

Sassafras (*Sassafras albidum*): Sassafras is a medium-sized deciduous tree known for its aromatic, edible leaves and bark, which are used in traditional recipes like gumbo and teas. It grows in well-drained soils and thrives in partial shade, making it a suitable canopy layer plant. The tree also contributes to biodiversity by supporting pollinators and offering unique culinary uses in the food forest.

Silver Maple (*Acer saccharinum*): Silver maple is a fast-growing, deciduous tree that produces edible sap that can be used to make syrup. It grows well in moist, well-drained soils and can tolerate partial shade, making it a good addition to the canopy layer. The tree provides habitat for wildlife and helps maintain a healthy ecosystem by enriching the soil and contributing to the overall biodiversity of the food forest.

2. UNDERSTORY LAYER:

Dwarf Sweet Cherry (*Prunus avium 'Compacta'*): This dwarf variety of sweet cherry tree grows to about 6 feet in height and produces delicious, sweet cherries. The cherries are perfect for fresh eating or preserving. This tree thrives in well-drained soils and partial shade, making it an ideal option for the understory layer. The flowers provide early-season nectar for pollinators and contribute to the overall health and beauty of the food forest.

Almond (Dwarf Varieties) (*Prunus amygdalus 'Tuono'*): The dwarf almond tree grows to around 6 feet in height and produces edible almonds. These trees thrive in well-drained soils and can tolerate partial shade. They provide early-season food and support biodiversity with their spring blossoms, attracting pollinators. Their compact size makes them suitable for smaller spaces in the food forest.

3. SHRUB LAYER:

Autumn Olive (*Elaeagnus umbellata*): Autumn olive is a hardy shrub that grows up to 10 feet tall and produces small, tart, red berries. These berries are edible and are often used to make jams and sauces. The plant thrives in well-drained soils and full sun but can tolerate partial shade. Its nitrogen-fixing properties enrich the soil, improving overall soil health in the food forest.

Josta Berry (*Ribes × nidigrolaria*): Josta berry is a hybrid between black currant and gooseberry, growing as a shrub up to 5 feet tall. The berries are dark purple and edible, used in jams and baked goods. Josta berry thrives in well-drained soils and partial shade, making it a great addition to the shrub layer. It's a productive shrub that yields fruit in mid-summer and attracts pollinators.

4. HERBACEOUS LAYER:

Sweet Cicely (*Myrrhis odorata*): Sweet Cicely is a perennial herb with sweet-smelling, edible leaves and seeds that are often used in desserts or teas. It grows well in moist, well-drained soils in partial shade. Sweet Cicely attracts pollinators and provides medicinal benefits, especially for digestive health.

Tansy (*Tanacetum vulgare*): Tansy is a hardy perennial herb with edible young leaves (though they should be eaten in moderation) and yellow button-like flowers. It prefers full sun and well-drained soil. Tansy is valuable in the food forest for its pest-repellent properties and ability to attract beneficial insects such as ladybugs.

5. ROOT LAYER:

Valerian (*Valeriana officinalis*): Valerian is a perennial herb known for its aromatic roots, often used in herbal medicine to aid sleep and relaxation. The roots are edible when properly prepared and are known for their calming effects. Valerian thrives in well-drained, slightly moist soil and partial shade, supporting biodiversity in the food forest by attracting pollinators. Its root system helps to loosen compacted soil, contributing to improved soil health.

Wood Sorrel (*Oxalis acetosella*): Wood sorrel is a small, low-growing perennial herb with edible leaves, stems, and flowers that have a tangy, lemony flavor. It grows well in acidic, well-drained soils and partial shade, making it ideal for the understory of a food forest. Wood sorrel's shallow roots help stabilize the soil, while the plant is also known for its ability to attract pollinators. This herb is a refreshing addition to the root layer, providing both food and ecological value.

6. GROUND COVER LAYER:

Wild Ginger (*Asarum canadense*): Wild Ginger is a low-growing perennial that thrives in moist, shaded areas with rich, well-drained soils. The rhizomes are edible and have a spicy flavor similar to true ginger, though they should be used sparingly due to potential toxicity in large amounts. It spreads by rhizomes, forming a dense mat that helps retain moisture and improve soil structure. Wild Ginger is an excellent choice for adding groundcover and diversity to the food forest.

Corsican Mint (*Mentha requienii*): Corsican Mint is a creeping perennial that thrives in moist, well-drained soils and partial shade. The leaves are edible and have a strong minty flavor, often used in teas, desserts, and sauces. It forms a dense, fragrant mat that helps to suppress weeds and retain moisture, while also attracting pollinators. Corsican Mint is an excellent groundcover for a Zone 4 food forest, adding both culinary and ecological benefits.

7. VERTICAL LAYER:

American Groundnut (*Apios americana*): American Groundnut is a nitrogen-fixing vine that produces edible tubers and seeds. The tubers are starchy and can be boiled or roasted like potatoes, while the seeds can be used as a protein source. This plant thrives in USDA Zone 4 and is a valuable addition to the vertical layer, providing both food and soil improvement benefits. It grows well in moist, well-drained soils and can be trained to climb trellises or other structures.

Sea Buckthorn (*Hippophae rhamnoides*): Sea Buckthorn is a thorny, nitrogen-fixing plant that produces bright orange berries, which are edible and rich in vitamin C, antioxidants, and essential fatty acids. The berries can be used to make juices, jams, and oils. While often used as a shrub, Sea Buckthorn can also be trained to climb and form a part of the vertical layer. It thrives in USDA Zone 4, tolerates poor soil conditions, and is highly beneficial for improving soil fertility and providing a nutritious edible crop.

8. FUNGAL LAYER:

Rosy Veincap (*Rhodotus palmatus*): This rare mushroom, known for its distinctive pink, veiny cap, grows on decaying hardwood in USDA Zone 4. While not widely consumed, it is considered edible and adds an intriguing element to the food forest's fungal diversity. Rosy Veincap helps with the decomposition of organic matter, improving soil health.

Milky Cap Mushroom (*Lactarius volemus*): This edible mushroom is known for its sweet, fishy smell and exudes a milky substance when cut. Found in coniferous forests in USDA Zone 4, Milky Cap mushrooms form symbiotic relationships with tree roots, aiding nutrient absorption. Their ability to thrive in cool, temperate climates makes them a great addition to the fungal layer.

ZONE 5

Plant Hardiness Zone 5

Zone 5 is a widely spread cold hardiness zone across the United States. It includes areas that typically experience winter temperatures between -20°F and -10°F (-29°C to -23°C). This zone is found mostly in northern and central U.S. states, especially inland regions with cold winters and mild summers.

Locations within Zone 5

Zone 5 covers various areas across the U.S., particularly in states with cooler climates.

Colorado: Much of the state falls into Zone 5, especially the foothills and parts of the Rockies.

Illinois: Central and parts of southern Illinois are classified under Zone 5.

Indiana: The majority of Indiana, except for the southern regions, is within this zone.

Iowa: Most of Iowa lies in Zone 5.

Kansas: The northern portion of Kansas falls into this zone.

Kentucky: Northern Kentucky is primarily in Zone 5.

Maryland: The western parts of Maryland are classified as Zone 5.

Massachusetts: Some inland regions, like western Massachusetts, fall into this zone.

Michigan: Interior and northern Michigan are largely in Zone 5.

Minnesota: Most of Minnesota is included, with the exception of the far northern regions.

Missouri: Northern and central Missouri are classified under Zone 5.

Nebraska: Most of Nebraska, excluding its southern regions, is in this zone.

New York: The majority of New York, except coastal areas, falls into Zone 5.

North Dakota: The majority of North Dakota lies within Zone 5.

Ohio: Central and northern Ohio are classified as Zone 5.

Pennsylvania: Much of central and northern Pennsylvania is included.

Tennessee: Some northern and central areas of Tennessee are within this zone.

Virginia: Mountainous regions in western Virginia are categorized as Zone 5.

Wisconsin: Most of Wisconsin, especially the central and northern areas, fall within this zone.

Specific Locations in Zone 5

Rocky Mountains: Higher elevations in Colorado, Wyoming, and Montana are in Zone 5.
Great Lakes Region: Cities like Cleveland, Detroit, and Buffalo are located in Zone 5, as well as parts of Michigan, Ohio, and Pennsylvania.
Northeastern U.S.: Sections of New York, Pennsylvania, and New Jersey fall into Zone 5.
Midwestern U.S.: Much of Illinois, Indiana, Iowa, and Ohio are located in Zone 5.

Climate Characteristics

Zone 5 features average winter temperatures ranging from -20°F to -10°F (-29°C to -23°C), requiring plants to be able to withstand freezing conditions during the winter months. The growing season typically runs from late spring to mid-fall, roughly from April through October, although this can fluctuate based on local microclimates. The first frost typically happens in mid to late October, while the last frost is generally between early and mid-

April. The frost-free period lasts between 140 to 180 days annually, but gardeners need to be cautious about potential late frosts in the spring and early frosts in the fall.

The zone receives moderate to substantial rainfall, with annual totals ranging from 25 to 40 inches (635 mm to 1015 mm), allowing for a diverse array of crops and trees to thrive. However, some areas may experience summer droughts, necessitating irrigation during dry periods. Wind can be a significant factor, especially in open areas, lowering effective temperatures in winter and requiring protection for plants.

Soil Conditions

The soil types in Zone 5 vary greatly. The Midwest and Northeast often have fertile loams, while parts of Colorado and Wyoming may have drier, sandy soils. Common soil challenges in Zone 5 include:

Clay Soils: Found in the Midwest, clay-heavy soils can become compacted, making it harder for plant roots to spread.

Sandy Soils: In regions like the Great Plains, sandy soils may struggle to retain moisture, requiring more frequent watering.

Alkaline and Acidic Soils: Depending on the location, soils may range from acidic to alkaline. Gardeners often need to test their soil and make amendments like adding lime or sulfur to balance pH levels.

Flora and Fauna in Zone 5

Zone 5 supports a wide variety of plant life, from perennials to cold-tolerant vegetables and fruits.

Trees: Deciduous and evergreen trees such as maples, oaks, and birches thrive here.

Zone 5

Shrubs: Lilacs, viburnums, and roses are common in this zone.

Perennials: Hardy perennials like coneflowers, black-eyed Susans, and daylilies do well in Zone 5.

Wildlife such as deer, rabbits, squirrels, and birds (sparrows, robins, woodpeckers) inhabit the region. Additionally, smaller species, including amphibians and insects (especially pollinators), thrive in these environments.

Challenges and Considerations

Gardening in Zone 5 requires strategies to deal with extreme winter temperatures, such as using mulches, windbreaks, and protective coverings. Late spring and early fall frosts also pose challenges, requiring gardeners to cover sensitive plants or choose frost-tolerant varieties. Soil drainage is another factor to consider in areas with high rainfall, as waterlogged soil can lead to root rot. In addition, summer droughts may necessitate supplemental watering for certain crops.

Wind exposure, particularly in open spaces, can damage plants, dry out soil, and worsen frost damage. Windbreaks made of trees or barriers can help mitigate these effects.

Agriculture and Gardening

Zone 5 is ideal for growing a diverse range of plants. Vegetables like lettuce, cabbage, carrots, and peas thrive here, as do fruit trees such as apples, cherries, and pears. Cold-hardy berry bushes like raspberries, blackberries, and strawberries also do well. Ornamental flowers and grasses add variety to gardens, while trees like apples, pears, plums, and cherries grow successfully. Cold-hardy grapevines and kiwi (such as Arctic kiwi) can also be grown in sheltered areas.

While late frosts, cold winters, and winds can be challenges, the moderate precipitation and soil diversity make Zone 5 suitable for a wide range of plants. By using protective measures like mulching, windbreaks, and frost covers, gardeners can maximize their success in establishing productive food forests or gardens in Zone 5.

Zone 5 Plan One: Temperate Treasures

Zone 5 Plan One: Temperate Treasures

In a mild climate bursting with potential, this plan uses thoughtful layering to unearth hidden treasures. It's designed to maximize food production while ensuring a sustainable support system.

1. CANOPY LAYER:

Butternut (*Juglans cinerea*): Butternut trees, also known as white walnut, thrive in USDA Zone 5 and produce edible nuts that are high in protein and fats. The nuts have a rich, buttery flavor and are used in baking and snacks. Butternut trees contribute to biodiversity, provide wildlife habitat, and fix nitrogen in the soil, improving fertility in the food forest.

Persian Walnut (*Juglans regia*): Persian walnuts are known for their delicious, edible nuts and are well-suited to USDA Zone 5. The nuts are a popular food source, eaten raw or roasted, and are highly nutritious. This tree provides shade, supports biodiversity, and enhances soil fertility through its deep root system.

2. UNDERSTORY LAYER:

Japanese Snowbell (*Styrax japonicus*): The Japanese snowbell is a beautiful understory tree suitable for USDA Zone 5, with fragrant white bell-shaped flowers in spring. Its seeds are not commonly consumed, but they are edible after being roasted. This tree adds ornamental value, supports pollinators, and thrives in partial shade, making it an excellent addition to the food forest understory.

Nannyberry (*Viburnum lentago*): Nannyberry is a versatile, edible shrub/tree for USDA Zone 5, producing sweet, date-like berries in the fall. The berries are eaten fresh, dried, or cooked into jams and jellies. Nannyberry enhances biodiversity, attracts wildlife, and provides shade in the understory, while its fruits are beneficial for both humans and animals.

3. SHRUB LAYER:

Buffaloberry (*Shepherdia argentea*): Buffaloberry produces small red fruits that are edible and tart, often used in preserves or as a food source for wildlife. This nitrogen-fixing shrub is drought-tolerant and helps improve soil fertility. It thrives in full sun and can grow up to 12 feet tall, making it ideal for supporting pollinators and birds.

Black Hawthorn (*Crataegus douglasii*): This hardy shrub produces small, edible black fruits that are great for making jellies or eaten fresh. Black Hawthorn also offers dense thorny branches, which provide excellent shelter for wildlife. It is drought-tolerant and can grow in a variety of soils, making it versatile in a food forest.

4. HERBACEOUS LAYER:

Sweet Cicely (*Myrrhis odorata*): Sweet Cicely is a perennial herb with edible leaves, seeds, and roots, which have a mild anise flavor. It thrives in partial shade and moist, well-drained soils, making it a great companion in a food forest. The plant is known for its ability to attract pollinators while also being used in salads, soups, and as a sweetener in cooking.

Lovage (*Levisticum officinale*): This tall herbaceous plant has edible leaves, seeds, and roots, all of which offer a strong celery-like flavor. Lovage is easy to grow in well-drained soils and can tolerate partial shade, making it a versatile addition to the herbaceous layer. It attracts beneficial insects and helps improve soil structure with its deep roots.

5. ROOT LAYER:

Earthnut Pea (*Lathyrus tuberosus*): This nitrogen-fixing perennial is prized for its small, starchy tubers, which can be eaten raw or cooked. It helps enrich the soil while producing attractive purple flowers, making it a valuable addition to the root layer of a food forest. The tubers are a rare and nutritious food source in cooler climates like Zone 5, though the plant requires a bit of patience as it takes time to establish.

Chinese Artichoke (*Stachys affinis*): Also known as Crosne, this unique root vegetable produces small, knobby tubers that have a crisp texture and a nutty flavor when raw or cooked. The plant thrives in well-drained soils, creating low-growing foliage that is easy to manage in a food forest. It is resistant to many pests and adds diversity to edible root crops in Zone 5.

6. GROUNDCOVER LAYER:

Creeping Raspberry (*Rubus pentalobus*): This low-growing groundcover produces small, edible berries that are sweet and packed with antioxidants. Its spreading habit helps to control weeds, and its dense foliage provides valuable habitat for pollinators. Creeping Raspberry thrives in full sun to partial shade, and its hardy nature makes it well-suited for Zone 5 food forests.

Gooseneck Loosestrife (*Lysimachia clethroides*): Known for its fragrant white flowers that bloom in late spring and early summer, Gooseneck Loosestrife is a great addition to the groundcover layer. Its leaves and roots are edible and can be used in teas or as a leafy green. The plant's rapid growth creates excellent soil stabilization and supports biodiversity in the food forest.

7. VERTICAL LAYER:

Climbing Groundnut (*Apios americana*): This perennial vine produces edible tubers that are a nutritious source of protein and carbohydrates, often compared to sweet potatoes in flavor. The plant is also notable for its attractive purple flowers that bloom in summer, and it thrives in a variety of soil types. It is nitrogen-fixing, improving the soil health of the food forest.

Kiwi Vine (*Actinidia arguta*): A small, hardy kiwi variety, this plant produces tiny, smooth-skinned kiwis that are sweet and packed with vitamins. The vine grows vigorously, offering an abundance of fruit once established. Its vigorous growth habit helps create vertical space in the food forest, with edible fruit ready to harvest in late summer.

8. FUNGAL LAYER:

Turkey Tail Mushroom (*Trametes versicolor*): This vibrant and medicinal mushroom grows on decaying wood and is known for its immune-boosting properties. Turkey Tail is not only edible, but its fruiting bodies are often used in teas and tinctures to support immune health. The mushroom helps break down wood, recycling nutrients back into the ecosystem, and it grows in diverse environments, adding both functional and aesthetic value to the food forest.

Black Trumpet Mushroom (*Craterellus cornucopioides*): A rare and highly prized wild mushroom, Black Trumpet is known for its delicate texture and rich, earthy flavor. It grows in association with hardwood trees and can be found in moist, shaded areas. The edible part is the dark, trumpet-shaped cap, which is perfect for adding depth to soups, stews, and sauces. This mushroom also helps decompose organic matter, improving soil health in the food forest.

Zone 5 Plan Two: Midspring Mosaic

ZONE 5 PLAN TWO: MIDSPRING MOSAIC

This mosaic of carefully paired plants celebrates the full bloom of midspring. It offers a harmonious mix of edibles and supportive species that work together for long-term resilience.

1. CANOPY LAYER:

Heartnut (*Juglans ailantifolia var. cordiformis*): The heartnut, a variety of Japanese walnut, grows well in USDA Zone 5 and produces edible, heart-shaped nuts. These nuts are sweet and have a smooth texture, making them ideal for snacks or baked goods. Heartnuts also help improve soil health by fixing nitrogen and adding organic matter through leaf fall.

European Pear (*Pyrus communis*): European pear trees produce sweet, juicy fruits and are suitable for USDA Zone 5. The pears are versatile and can be eaten fresh, dried, or used in preserves. This tree enhances the food forest by attracting pollinators, providing shade, and contributing to soil fertility with its organic matter.

2. UNDERSTORY LAYER:

Saskatoon Berry (*Amelanchier alnifolia*): Saskatoon berries are sweet, nutrient-rich fruits that can be eaten fresh or used in baking and preserves. This shrub thrives in USDA Zone 5 and adds significant food value to the understory. It attracts pollinators and supports wildlife, while its deep root system helps stabilize soil and improve water retention.

Allegheny Serviceberry (*Amelanchier laevis*): Allegheny serviceberry produces small, sweet, blueberry-like fruits that ripen in early summer and are excellent for fresh eating, jams, and pies. This tree thrives in USDA Zone 5 and provides essential early-season food for wildlife. It brings pollinators to the food forest and adds beautiful spring blooms and fall foliage color.

3. SHRUB LAYER:

Gooseberry (*Ribes uva-crispa*): Gooseberry bushes produce edible green or red berries, which are used in desserts, jams, or eaten raw. This hardy plant tolerates partial shade and is highly adaptable to different soil conditions. Gooseberries are also a good source of vitamin C, making them a valuable addition to a food forest.

Siberian Pea Shrub (*Caragana arborescens*): The Siberian Pea Shrub produces edible seed pods that can be used as a protein-rich food source. It is an excellent nitrogen fixer, improving soil health, and thrives in poor soils. It can tolerate drought and cold, making it a reliable shrub for Zone 5 food forests.

4. HERBACEOUS LAYER:

Good King Henry (*Blitum bonus-henricus*): This rare perennial herb, sometimes called wild spinach, has edible leaves that are cooked like spinach and young shoots that can be eaten like asparagus. It thrives in partially shaded areas and is drought-tolerant once established. Good King Henry also brings nitrogen-fixing capabilities to the soil, enriching the ecosystem.

Daylily (*Hemerocallis fulva*): Daylilies produce edible flowers, young shoots, and tubers, making them a valuable addition to a food forest. The plant thrives in full sun to partial shade and tolerates a wide range of soil types. Daylilies are also known for their ability to naturalize, creating ground cover and attracting pollinators.

5. ROOT LAYER:

Tiger Nut (*Cyperus esculentus*): Known for its small, edible tubers, tiger nut is a sedge that grows well in moist soils. The tubers are sweet and starchy, often eaten raw, roasted, or turned into flour. It is a drought-tolerant and hardy plant that brings resilience and a rare crop to the food forest. Though slow-growing, it can naturalize easily in Zone 5 conditions.

Hog Peanut (*Amphicarpaea bracteata*): This perennial legume produces small edible seeds and underground tubers, both of which can be harvested. It adds nitrogen to the soil and thrives in shaded or semi-shaded areas, making it ideal for food forests. The underground tubers are a rare delicacy and can be boiled or roasted for consumption.

6. GROUNDCOVER LAYER:

Creeping Mint (*Mentha requienii*): Creeping Mint, also called Corsican Mint, forms a fragrant carpet of small green leaves. Its leaves are edible and can be used in cooking or for herbal teas. This low-growing perennial spreads quickly, acting as a natural groundcover while attracting beneficial insects, including pollinators.

Sweet Woodruff (*Galium odoratum*): Sweet Woodruff is an aromatic perennial that thrives in partial shade and spreads easily by rhizomes. Its leaves are edible and can be used in teas or added to salads. This plant offers excellent groundcover and attracts pollinators with its small white flowers in spring.

7. VERTICAL LAYER:

Porcelain Vine (*Ampelopsis brevipedunculata*): Also known as "the porcelain berry," this vine produces small, edible berries that can be eaten raw or made into preserves. While the plant's berries offer a unique color, the vine is highly ornamental, and it thrives in partial shade to full sun. Porcelain Vine is a fast-growing climber that adds vertical space and texture to the food forest.

Arctic Kiwi (*Actinidia kolomikta*): This cold-hardy kiwi variety produces small, edible kiwis that are often sweeter than their larger counterparts. It climbs well on trellises or fences, providing both fruit and a decorative touch with its variegated foliage. The vine is fast-growing and perfect for Zone 5, producing fruits in late summer to early fall.

8. FUNGAL LAYER:

Coral Mushroom (*Ramaria spp.*): Coral mushrooms are edible and offer a unique appearance, resembling underwater coral structures. These mushrooms grow on decaying organic matter and are known for their mild, slightly nutty flavor. Besides being edible, they contribute to the forest's ecology by breaking down lignin and cellulose, which supports soil health and nutrient cycling.

Wood Ear Mushroom (*Auricularia auricula-judae*): This fungi grows on decaying wood and is prized for its jelly-like texture and mild, earthy flavor. It's used in Asian cuisines and has medicinal properties, especially for its blood circulation and antioxidant effects. Wood Ear is a decomposer that contributes to breaking down woody debris, making it a beneficial organism for the fungal layer in the food forest.

Zone 5 Plan Three: Harvest Harmony Hub

ZONE 5 PLAN THREE: HARVEST HARMONY HUB

Centered on creating harmony between harvest and habitat, this plan organizes layers into a cohesive, edible network. It emphasizes immediate food yields alongside the health of the entire system.

1. CANOPY LAYER:

Kentucky Coffeetree (*Gymnocladus dioicus*): The Kentucky coffeetree is a rare, nitrogen-fixing species suitable for USDA Zone 5. While its seeds are toxic unless roasted, they were historically used as a coffee substitute. The tree adds significant biodiversity to the canopy layer, supports wildlife, and improves soil health by fixing nitrogen.

Yellowhorn (*Xanthoceras sorbifolium*): Yellowhorn, also known as Chinese flowering chestnut, produces edible seeds that can be roasted or pressed for oil. Suitable for USDA Zone 5, this tree is valued for its beautiful flowers and drought tolerance. Yellowhorn enhances the food forest by attracting pollinators and providing an additional edible crop.

2. UNDERSTORY LAYER:

Blue Elderberry (*Sambucus nigra ssp. cerulea*): Blue elderberries are highly valued for their edible berries, which are used in syrups, wines, and jams. Suitable for USDA Zone 5, this small tree also supports pollinators and birds. The flowers are also edible and can be used to make teas and fritters, adding multifunctional benefits to the understory.

Juneberry (*Amelanchier lamarckii*): Juneberry, also known as shadblow, produces sweet, edible berries in early summer. These berries can be eaten fresh or made into preserves, pies, and beverages. Growing well in USDA Zone 5, it is a small, multi-stemmed tree that offers beauty, biodiversity, and an early food source for birds and pollinators.

3. SHRUB LAYER:

Black Currant (*Ribes nigrum*): Black currants are known for their edible berries, which are rich in vitamin C and often used in syrups, jellies, and teas. The shrub is adaptable to partial shade and thrives in well-drained soils. It also attracts pollinators, making it a beneficial plant for the surrounding ecosystem.

Buffalo Currant (*Ribes aureum*): This shrub produces fragrant, yellow flowers followed by small, edible black or yellow fruits. Buffalo Currant is highly drought-tolerant and can thrive in various soil types, making it ideal for low-maintenance food forests. The fruit is often used in jams, jellies, or eaten fresh.

4. HERBACEOUS LAYER:

Scorzonera (*Scorzonera hispanica*): Also known as black salsify, Scorzonera is a root vegetable with long, edible roots that have a sweet, nutty flavor. The plant grows best in well-drained soils with full sun exposure. In addition to its culinary uses, Scorzonera is drought-tolerant and can help improve soil structure with its deep taproots.

Mallow (*Malva sylvestris*): Commonly known as common mallow, this plant has edible leaves, flowers, and seeds, which are used in salads, teas, and as a thickener in soups. Mallow grows well in full sun or partial shade and prefers well-drained soils. It is also known for its soothing medicinal properties and its ability to attract pollinators.

5. ROOT LAYER:

Ground Plum (*Astragalus crassicarpus*): A native plant to North America, the ground plum produces sweet, edible pods that resemble small plums, though it's part of the legume family. The plant is drought-tolerant, grows in open prairies, and contributes to soil fertility through nitrogen fixation. It offers a unique and rare edible fruit in a Zone 5 food forest setting.

Mashua (*Tropaeolum tuberosum*): This Andean tuberous plant produces edible tubers that are rich in flavor and can be eaten roasted or boiled. It is related to nasturtiums, with similar leaves and flowers, and brings a vibrant, ornamental touch to the food forest. The plant thrives in cooler zones like 5, and its tubers add an exotic, nutrient-rich food source.

6. GROUNDCOVER LAYER:

Creeping Thyme (*Thymus serpyllum*): Creeping Thyme is a fragrant herb that forms a dense mat of tiny, edible leaves. The plant produces small pink flowers in summer, attracting bees and other beneficial insects. It works well as a low-maintenance groundcover in well-drained soils and full sun.

Pineapple Sage (*Salvia elegans*): This aromatic perennial produces edible, sweet-smelling leaves that are often used in teas or as a culinary herb. Pineapple Sage also has bright red flowers that attract hummingbirds. It's an excellent groundcover for sunny areas and is drought-tolerant once established.

7. VERTICAL LAYER:

Hardy Grape (*Vitis riparia*): Known as riverbank grape, this hardy variety thrives in USDA Zone 5 and produces small, flavorful grapes that are ideal for making jelly, wine, and fresh eating. It's a native vine that supports local wildlife and grows rapidly, offering vertical coverage and contributing to the biodiversity of the food forest.

Silver Lace Vine (*Fallopia baldschuanica*): Silver Lace Vine is a fast-growing climber that provides a stunning vertical element in the garden. It produces small, fragrant white flowers in late summer and fall, attracting pollinators. While its berries are not widely consumed, the plant's ornamental value and vigorous growth make it a valuable addition to the vertical layer.

8. FUNGAL LAYER:

Blewit Mushroom (*Clitocybe nuda*): Known for its fragrant aroma and blue to lilac color, Blewit is an edible mushroom that grows in wooded areas on decaying plant material. Its firm flesh makes it a popular choice for soups and sauces. As a decomposer, it plays a vital role in breaking down organic matter and improving the soil's nutrient content in the food forest.

Honey Mushroom (*Armillaria mellea*): This mushroom is edible when cooked and has a sweet, mild flavor. It grows in clusters on the roots of trees, particularly oaks and other hardwoods. Honey Mushroom plays a role in the natural decomposition process, breaking down dead and decaying trees, and helps return nutrients to the soil, supporting the overall health of the food forest ecosystem.

Zone 5 Plan Four: Mild Marvel Method

Celebrating the marvels of a temperate climate, this method layers two key species per level for a balanced ecosystem. It's focused on achieving delicious, sustainable produce in a nurturing environment.

1. CANOPY LAYER:

Hican (*Carya illinoinensis* × *Carya ovata*): Hican trees are a hybrid between pecan and hickory, suitable for USDA Zone 5. The nuts are edible, combining the best qualities of both parent species, and are rich in flavor and nutrition. Hican trees contribute to soil improvement and provide valuable habitat for wildlife.

Asian Pear (*Pyrus pyrifolia*): Asian pear trees produce crisp, juicy fruits that are excellent for fresh eating or cooking. These trees thrive in USDA Zone 5, adding diversity to the canopy layer and providing high-yield fruit crops. Asian pears attract pollinators, enhance biodiversity, and contribute organic material to the soil.

2. UNDERSTORY LAYER:

Cornelian Cherry (*Cornus mas*): Cornelian cherry is a small tree or large shrub that thrives in USDA Zone 5, producing bright red edible fruits. The fruits can be eaten fresh or used to make jams, sauces, or wine. This tree attracts pollinators in early spring and provides year-round interest with its yellow flowers and vibrant fruits.

Russian Olive (*Elaeagnus angustifolia*): Russian olive is a nitrogen-fixing tree that produces small, edible fruits with a sweet and tart flavor, similar to a cranberry. Suitable for USDA Zone 5, this tree is hardy, drought-tolerant, and attracts pollinators. Its nitrogen-fixing abilities improve soil fertility, making it a valuable addition to a food forest.

3. SHRUB LAYER:

Western Sand Cherry (*Prunus pumila*): This small shrub produces edible black cherries, which can be eaten raw or used in preserves. Western Sand Cherry is drought-tolerant and thrives in poor, sandy soils, making it suitable for challenging growing conditions. It also provides habitat for pollinators and wildlife.

Sea Buckthorn (*Hippophae rhamnoides*): Sea Buckthorn is a nitrogen-fixing shrub that produces bright orange, edible berries high in vitamins C and E. The berries are often used in juices, oils, or jams. Sea Buckthorn thrives in poor soils and helps improve soil fertility, making it a valuable shrub in a permaculture setting.

4. HERBACEOUS LAYER:

Giant Hyssop (*Agastache foeniculum*): This herb produces edible leaves and flowers with a strong anise flavor, commonly used in teas and salads. Giant Hyssop thrives in well-drained soils and full sun to partial shade. It is highly attractive to bees and butterflies, making it a key pollinator-friendly plant in a food forest.

Elecampane (*Inula helenium*): Elecampane is a tall perennial herb with edible roots that are often used in herbal remedies and have a flavor similar to licorice. It grows well in moist, well-drained soils and full sun or partial shade. The plant's large flowers attract pollinators, and it is valued for its medicinal properties, particularly for respiratory health.

5. ROOT LAYER:

Sweet Cicely (*Myrrhis odorata*): Sweet Cicely is a perennial herb whose roots, seeds, and leaves are all edible. The plant has a sweet, anise-like flavor and is used in both culinary and medicinal applications. It grows well in partial shade and is highly attractive to beneficial insects, making it a valuable addition to the root layer of a Zone 5 food forest.

Earthnut (*Conopodium majus*): This plant, native to Europe, produces small edible tubers that are similar in texture to chestnuts when roasted. Earthnut grows in woodland settings, preferring partial shade and moist, well-drained soils. The tubers are a rare and delicious foraged food that adds biodiversity to the food forest.

6. GROUNDCOVER LAYER:

Wild Ginger (*Asarum canadense*): Wild Ginger is a native groundcover that grows well in the shade and moist, well-drained soils. The roots are edible and have a flavor similar to ginger, often used in traditional herbal medicine. Its heart-shaped leaves form a dense carpet, helping to suppress weeds.

Creeping Jenny (*Lysimachia nummularia*): Creeping Jenny is a vigorous groundcover with bright green leaves that turn golden in the fall. It spreads rapidly, making it an excellent option for filling gaps between other plants. The small yellow flowers in spring provide nectar for pollinators, and its leaves can be used for medicinal purposes.

7. VERTICAL LAYER:

Cucamelon (*Melothria scabra*): Known as "mouse melon" or "Mexican sour gherkin," Cucamelon is a small, cucumber-like fruit that has a tangy taste. The vine is easy to grow in full sun and produces an abundance of tiny fruits that are perfect for pickling or snacking. Its climbing habit adds height to the food forest while providing a unique edible.

Grape Tomato Vine (*Solanum pimpinellifolium*): This small, cherry-sized tomato variety is incredibly sweet and perfect for snacking or adding to salads. It grows well in vertical spaces, producing numerous tiny, round tomatoes throughout the growing season. The plant's compact habit makes it suitable for containers or trellises, providing both food and height in the food forest.

8. FUNGAL LAYER:

Turkey Tail Mushroom (*Trametes versicolor*): A very common yet medicinal fungus, Turkey Tail is a polypore mushroom that grows on wood, breaking it down and recycling nutrients back into the soil. Known for its immune-boosting properties, the fruiting body is edible, but often used in extracts or teas. It thrives in temperate environments and adds both medicinal and ecological benefits to the food forest.

Enoki Mushroom (*Flammulina velutipes*): Enoki is a delicate, edible mushroom often used in soups and salads. It grows on decayed wood, particularly on hardwood trees, and has a slightly sweet, crunchy texture. The mushroom is prized for its health benefits, including anti-inflammatory and antioxidant properties, and it contributes to forest health by aiding in the decomposition of organic materials.

Zone 5 Plan Five: Seasonal Synergy Scheme

This plan brings together the strengths of seasonal plantings in a synergistic design. Its layered approach ensures that each level contributes to a cycle of food production and lasting ecological support.

1. CANOPY LAYER:

Osage Orange (*Maclura pomifera*): Osage orange trees, though not commonly known for their fruit, produce edible seeds inside the inedible, large, green fruit. The tree is hardy in USDA Zone 5 and is often used as a windbreak or living fence. It supports the food forest by providing dense wood for wildlife habitat and soil stability.

Sweet Chestnut (*Castanea sativa*): Sweet chestnut trees are well-suited to USDA Zone 5 and produce edible chestnuts, which are high in carbohydrates and can be roasted or ground into flour. This tree brings multiple benefits to the food forest, including soil improvement, wildlife habitat, and windbreaks, making it a valuable addition.

2. UNDERSTORY LAYER:

Hawthorn (*Crataegus monogyna*): Hawthorn trees produce small, edible haws, which can be made into jams, jellies, or used medicinally. This tree grows well in USDA Zone 5 and brings ornamental value, wildlife habitat, and support for pollinators. Its dense branching also provides protection for small birds and adds structure to the understory.

Mountain Ash (*Sorbus americana*): Mountain ash produces small, bright orange berries that are high in vitamin C and can be used in jellies or fermented into beverages. Thriving in USDA Zone 5, this tree attracts birds, provides food for pollinators, and adds vibrant fall color. Its compact size makes it ideal for the understory layer of the food forest.

3. SHRUB LAYER:

Silver Buffaloberry (*Shepherdia canadensis*): This hardy shrub produces small red, edible berries that are tart and used in jams or syrups. It is highly drought-tolerant and can fix nitrogen, which improves soil quality. Silver Buffaloberry is an excellent choice for wildlife, providing food and shelter.

Cranberry Cotoneaster (*Cotoneaster apiculatus*): This low-growing shrub produces small red fruits that are edible but tart, often used in jellies. It provides excellent ground cover and erosion control while also offering habitat for pollinators. Cranberry Cotoneaster thrives in full sun to partial shade and well-drained soils.

4. HERBACEOUS LAYER:

Groundnut (*Apios americana*): Groundnut is a perennial vine with edible tubers and seeds, making it a valuable nitrogen-fixer and food source in a food forest. It grows best in moist, well-drained soils and partial shade. Groundnut's tubers are often compared to potatoes in flavor, and the plant is historically significant as a food crop.

Ramps (*Allium tricoccum*): Ramps, also known as Wild leeks, have edible leaves and bulbs that offer a garlicky-onion flavor. They thrive in moist, shaded woodland areas, making them ideal for the herbaceous layer of a food forest. Ramps are highly prized for their culinary uses and help support early-season pollinators.

5. ROOT LAYER:

Yam Daisy (*Microseris lanceolata*): Also known as Murnong, this perennial plant was traditionally cultivated by indigenous peoples of Australia for its edible roots. The roots have a sweet, nutty flavor and can be eaten raw or cooked. Yam daisy thrives in well-drained soils and sunny locations, making it a suitable addition to the root layer in Zone 5.

Chinese Yam (*Dioscorea polystachya*): This vining plant produces long, edible tubers that are highly valued in Chinese cuisine. The tubers can be cooked in various ways, including stir-frying or roasting. The plant is resilient to cold climates and can naturalize easily, making it a rare but hardy root crop for Zone 5 food forests.

6. GROUNDCOVER LAYER:

Lamb's Quarters (*Chenopodium album*): This fast-growing edible plant is rich in nutrients, and its young leaves can be eaten raw or cooked like spinach. Lamb's Quarters thrives in full sun and well-drained soils, acting as a low-maintenance groundcover while providing nutritious greens.

Dandelion (*Taraxacum officinale*): Dandelions are edible throughout, from the roots to the leaves and flowers. This common plant offers high nutritional value and provides a food source for pollinators. It also works well in the groundcover layer by helping to break up compacted soil with its deep taproot.

7. VERTICAL LAYER:

Passionflower (*Passiflora incarnata*): Known for its stunning flowers, Passionflower produces edible fruit called "maypops" that are tangy and flavorful. The vine is fast-growing and can cover fences or trellises, making it an ideal vertical plant. Additionally, the plant has medicinal properties and provides habitat for pollinators like bees and butterflies.

Hops (*Humulus lupulus*): While often associated with brewing, hops is a fast-growing climbing vine that also offers edible young shoots that are tender and flavorful. The plant thrives in full sun and moist soil, creating vertical coverage and helping to improve the aesthetic value of the food forest. The flowers (also called cones) are used to impart bitterness to beer, but they also attract pollinators.

8. FUNGAL LAYER:

Lobster Mushroom (*Hypomyces lactifluorum*): Lobster mushrooms are parasitic fungi that grow on other mushrooms, giving them a bright orange, lobster-like appearance. They are edible and have a seafood-like flavor, often used as a meat substitute in vegetarian dishes. As decomposers, these mushrooms aid in breaking down organic matter and recycling nutrients back into the food forest.

Puffball Mushroom (*Lycoperdon perlatum*): This edible mushroom is known for its round shape and the way it releases a cloud of spores when it matures. Puffballs grow on the ground in areas with rich, organic soil. The flesh of young puffballs is firm and edible, with a mild flavor, and it serves as a decomposer, adding nutrients to the surrounding soil and supporting the health of the food forest.

Zone 5 Plan Six: Vibrant Valley Vision

ZONE 5 PLAN SIX: VIBRANT VALLEY VISION

Envisioning a valley of vitality, this plan layers a vibrant mix of edible species for robust yields. It is dedicated to producing nutritional abundance while strengthening the landscape's natural support.

1. CANOPY LAYER:

Silver Lime (*Tilia tomentosa*): Silver lime trees, also known as silver linden, thrive in USDA Zone 5 and produce edible flowers and leaves. The flowers are often used to make tea, while the leaves are consumed in salads. This tree provides shade, supports pollinators, and contributes to biodiversity in the food forest.

Siberian Pea Tree (*Caragana arborescens*): The Siberian pea tree is a hardy, nitrogen-fixing tree suitable for USDA Zone 5, producing edible pea-like seeds. These seeds are high in protein and can be eaten raw or cooked. The tree is drought-tolerant and enhances soil fertility, making it a valuable addition to any food forest.

2. UNDERSTORY LAYER:

Autumn Olive (*Elaeagnus umbellata*): Autumn olive is a nitrogen-fixing shrub that produces small, edible red berries rich in antioxidants. These berries can be eaten fresh or made into jams and sauces. Suitable for USDA Zone 5, this plant enhances soil fertility, attracts pollinators, and provides an abundant fruit crop in late summer.

Mulberry (Red Mulberry) (*Morus rubra*): Red mulberry is a small tree that produces sweet, edible berries, often used in desserts, jams, and wines. Growing well in USDA Zone 5, this tree supports wildlife, attracts pollinators, and offers high fruit yields in the understory. Its fruit is a valuable food source for humans and animals alike.

3. SHRUB LAYER:

Serviceberry *(Amelanchier laevis)*: Serviceberry produces edible berries that taste like blueberries, and the fruit can be eaten raw or used in jams and pies. It thrives in both full sun and partial shade and is an early-season food source for pollinators. The shrub also offers vibrant fall foliage.

Snowberry (*Symphoricarpos albus*): Snowberry produces white, inedible berries but is valued for its dense growth, which provides excellent habitat for wildlife and pollinators. It thrives in poor soils and can tolerate full sun or partial shade, making it a resilient addition to a Zone 5 food forest.

4. HERBACEOUS LAYER:

Skirret (*Sium sisarum*): Skirret is a lesser-known perennial vegetable with sweet, edible roots that resemble carrots or parsnips. It thrives in moist, well-drained soils and can tolerate partial shade. Skirret is easy to grow and provides a long-lasting, reliable root crop in a permaculture system.

Sorrel (*Rumex acetosa*): Sorrel is a hardy perennial herb with edible, lemon-flavored leaves that are used in salads, soups, and sauces. It prefers full sun to partial shade and thrives in well-drained soils. Sorrel is also known for its early spring growth, providing a valuable source of food when other plants are still dormant.

5. ROOT LAYER:

Bunium Bulbocastanum (*Bunium bulbocastanum*): Known as Earth Chestnut, this plant produces small, flavorful tubers that can be eaten raw or cooked. It thrives in well-drained soils and is a low-maintenance, long-lived perennial. The tubers have a rich, nutty flavor and are considered a delicacy, making this an excellent choice for the root layer.

Ulluco (*Ullucus tuberosus*): Ulluco is a rare tuber crop from the Andes, producing vibrant, colorful roots that are both ornamental and edible. The tubers are rich in nutrients and are cooked similarly to potatoes. The plant prefers cooler climates and grows well in Zone 5, adding a unique and nutritious root vegetable to the forest garden.

6. GROUNDCOVER LAYER:

Cress (*Lepidium sativum*): This fast-growing herb has edible leaves and seeds that are commonly used in salads, sandwiches, and soups. Cress thrives in full sun and moist soil, spreading quickly to form a dense, low-growing cover. It also attracts beneficial insects like pollinators and predatory insects that help control pests.

Moss Phlox (*Phlox subulata*): Moss Phlox is a perennial groundcover that produces beautiful pink, purple, or white flowers in spring. It is edible, with the flowers and leaves having medicinal uses, and it attracts bees and butterflies. This plant thrives in full sun and poor, well-drained soils, making it ideal for a Zone 5 food forest.

7. VERTICAL LAYER:

Winter Squash Vine (*Cucurbita maxima*): Winter squash, like Butternut or Hubbard, grows as a vine and can be trained up a trellis or fence to save space. It produces large, nutritious squash fruits that are high in vitamins and perfect for fall cooking. The vine's large leaves provide shade and shelter for other plants in the food forest.

Trio of Climbing Beans (*Phaseolus vulgaris*): These beans offer vertical growth and produce edible legumes that are a good source of protein and fiber. They grow rapidly on trellises and add a colorful element with their purple, red, or yellow beans. The plants also fix nitrogen in the soil, improving the soil health for other plants in the food forest.

8. FUNGAL LAYER:

Chanterelle Mushroom (*Cantharellus cibarius*): Chanterelles are highly prized, aromatic mushrooms with a yellow to golden color. They grow in association with certain tree species, forming symbiotic relationships with their roots. The mushroom's fruity, slightly peppery flavor adds a rich depth to dishes, and it plays a role in breaking down organic matter and enhancing soil fertility in the food forest.

Birch Polypore (*Piptoporus betulinus*): This mushroom grows on birch trees and has a tough, cork-like texture. Although not typically eaten raw, it has been used traditionally in medicinal preparations. As a decomposer, it helps break down birch wood, releasing essential nutrients into the surrounding soil, and contributes to the forest ecosystem's overall health.

Zone 5 Plan Seven: Balanced Bounty

I designed this plan to achieve a perfect balance between lush harvests and ecological integrity, this blueprint pairs species across layers. It aims for a system that's as sustainable as it is bountiful.

1. CANOPY LAYER:

Chinese Chestnut (*Castanea mollissima*): Chinese chestnut trees are cold-hardy and produce edible chestnuts that are sweet and nutritious. Well-suited to USDA Zone 5, these trees provide wildlife habitat, stabilize soil, and add organic matter to the forest ecosystem. The nuts are excellent for roasting or making flour.

Manchurian Walnut (*Juglans mandshurica*): Manchurian walnut trees, native to northeastern Asia, grow well in USDA Zone 5 and produce edible nuts with a rich flavor. These trees improve the food forest by fixing nitrogen, providing valuable shade, and attracting wildlife. The nuts are ideal for baking and snacks.

2. UNDERSTORY LAYER:

Yellowhorn (*Xanthoceras sorbifolium*): Yellowhorn, also known as Chinese flowering chestnut, produces edible seeds that can be roasted or pressed for oil. This small tree is suitable for USDA Zone 5 and adds ornamental value with its beautiful flowers. Yellowhorn enhances the food forest by supporting pollinators and providing a unique edible crop.

Black Haw (*Viburnum prunifolium*): Black haw is a native shrub or small tree that produces small, edible berries in the fall. The fruits are often used in jellies, jams, or eaten fresh. Suitable for USDA Zone 5, this plant provides shade, supports wildlife, and brings seasonal interest with its clusters of white flowers and colorful fruits.

3. SHRUB LAYER:

Thimbleberry (*Rubus parviflorus*): This deciduous shrub produces sweet, raspberry-like fruits that are delicious fresh or used in jams. Thimbleberry grows well in partial shade and prefers moist, well-drained soils. It's also a valuable plant for erosion control due to its spreading nature.

Ninebark (*Physocarpus opulifolius*): Ninebark is a hardy shrub with attractive peeling bark and small clusters of white or pink flowers. While its fruit is not typically eaten, the shrub provides important habitat for pollinators and wildlife. Ninebark is adaptable to a range of soils and light conditions.

4. HERBACEOUS LAYER:

Violets (*Viola odorata*): The leaves and flowers of violets are edible and often used in salads, teas, or as decorative garnishes. Violets thrive in moist, partially shaded areas and serve as a host plant for various pollinators. Their beautiful purple flowers make them an aesthetic and edible addition to the food forest.

Crosne (*Stachys affinis*): Also known as Chinese artichoke, Crosne produces edible tubers with a nutty flavor and crunchy texture. The plant grows well in full sun to partial shade and well-drained soils. Crosne is a rare and productive tuber crop, making it a unique addition to the herbaceous layer of a food forest.

5. ROOT LAYER:

Scorzonera (*Scorzonera hispanica*): This plant, often called black salsify, produces long, edible roots that are prized for their mild, oyster-like flavor when cooked. It thrives in cool, well-drained soils and is tolerant of poor soil conditions. The plant also produces yellow flowers that attract pollinators, adding both food and ecological benefits to the forest.

Tuberous Vetchling (*Lathyrus tuberosus*): This

perennial legume produces small, starchy tubers that can be harvested and eaten. It fixes nitrogen into the soil, promoting soil health while offering a rare and delicious root crop. The plant's pinkish flowers also make it an attractive addition to the forest.

6. GROUNDCOVER LAYER:

Strawberry Spinach (*Blitum capitatum*): A rare edible plant, Strawberry Spinach produces edible leaves and bright red berries. The plant thrives in full sun and moist soil, making it an excellent groundcover while offering an unusual and nutritious addition to the food forest.

Wild Strawberry (*Fragaria vesca*): This native ground cover produces small, sweet strawberries that attract pollinators and small mammals, while its dense foliage helps prevent weed growth and improve soil structure.

7. VERTICAL LAYER:

Climbing Okra (*Abelmoschus esculentus*): This vertical-growing okra plant produces edible pods that are rich in vitamins A, C, and K. The plant thrives in warm conditions and can climb up vertical structures, providing a striking green backdrop in the food forest. The leaves and young pods are often used in cooking, and the plant's rapid growth adds vertical layers to the food forest.

Purslane Vine (*Portulaca oleracea*): Known for its succulent leaves and edible seeds, Purslane is a nutritious groundcover that can also climb when trained. The leaves are packed with omega-3 fatty acids and are perfect for adding to salads or cooking. Its sprawling nature makes it a great addition to vertical spaces as well as ground-level coverage.

8. FUNGAL LAYER:

Red Reishi Mushroom (*Ganoderma tsugae*): Reishi mushrooms are known for their medicinal properties, particularly for enhancing the immune system. They grow on dead and decaying trees, often forming shelf-like structures. While not typically eaten fresh, their dried form is used in teas, tinctures, and powders, contributing both medicinal value and ecological benefits to the food forest.

Giant Puffball (*Calvatia gigantea*): This large, round mushroom is edible when young and has a mild, earthy flavor. It grows on the forest floor and can reach impressive sizes, making it easy to identify. It serves an important ecological role by decomposing organic material and improving soil health, and it can be used in a variety of dishes, including soups, stews, and stir-fries.

Zone 5 Plan Eight: Green Growth Grid

Zone 5 Plan Eight: Green Growth Grid

Structured like a dynamic grid, this plan coordinates growth in every layer to maximize food production. It's built on the idea that every plant contributes to a sustainable and resilient whole.

1. CANOPY LAYER:

Shellbark Hickory (*Carya laciniosa*): Shellbark hickory trees thrive in USDA Zone 5 and produce large, flavorful edible nuts. The nuts are a valuable food source for humans and wildlife alike. Shellbark hickories are long-lived, drought-tolerant trees that improve soil fertility and contribute to ecosystem stability.

Mulberry (White Mulberry) (*Morus alba*): White mulberry trees produce sweet, edible berries that are used fresh, dried, or in preserves. This fast-growing tree is suitable for USDA Zone 5 and attracts pollinators, provides shade, and improves the biodiversity of the food forest. Its fruits are also beneficial for wildlife.

2. UNDERSTORY LAYER:

Spicebush (*Lindera benzoin*): Spicebush produces aromatic berries and leaves that can be used in teas or as a spice. Thriving in USDA Zone 5, this shrub is an excellent source of nectar for pollinators and adds a fragrant, edible layer to the forest understory. Its early blooms attract butterflies, while its fruits feed birds in the fall.

Hoptree (*Ptelea trifoliata*): The hoptree produces small, citrus-scented fruits that are used as a substitute for hops in brewing. Suitable for USDA Zone 5, this small tree attracts pollinators, supports wildlife, and adds an aromatic touch to the food forest. Its drought tolerance and ornamental appeal make it a valuable understory plant.

3. SHRUB LAYER:

Mountain Currant (*Ribes montigenum*): This native shrub produces small, edible berries that are commonly used in preserves or eaten fresh. Mountain Currant thrives in rocky, well-drained soils and tolerates partial shade. The plant also attracts pollinators and is a good companion in food forests.

Chokeberry (*Photinia melanocarpa*): Chokeberry produces small, black edible fruits that are used in jams, jellies, or juices. The shrub is highly adaptable to different soils and moisture conditions, making it useful in erosion control and wetland areas. It also supports pollinators and wildlife.

4. HERBACEOUS LAYER:

Sweet Woodruff (*Galium odoratum*): Sweet Woodruff is a low-growing, shade-tolerant herb with edible leaves that are used to flavor drinks and desserts. It thrives in moist, shady areas and is often planted as ground cover. The plant also has medicinal properties and is valued for its sweet fragrance, especially when dried.

Jerusalem Artichoke (*Helianthus tuberosus*): Jerusalem Artichoke, also known as sunchoke, produces edible tubers that are often roasted or used in soups. It thrives in full sun and can tolerate poor soils, making it a great addition to a food forest. The plant also provides a tall, attractive presence in the landscape, supporting pollinators with its sunflower-like blooms.

5. ROOT LAYER:

Camas (*Camassia quamash*): A native North American plant, Camas produces edible bulbs that were an important food source for indigenous peoples. The bulbs have a sweet flavor when baked or roasted. The plant thrives in moist, well-drained soils and produces striking blue flowers, adding beauty and food value to the forest.

Yarrow-Leaf Sundew (*Drosera anglica*): This rare carnivorous plant grows in boggy soils and produces edible tubers that are traditionally used in folk medicine. Though it requires very specific growing conditions, it adds an unusual and functional element to a food forest. The plant also aids in pest control by trapping insects.

6. GROUNDCOVER LAYER:

Violets (*Viola spp.*): Violets are edible, with both the leaves and flowers offering a mild, slightly sweet flavor. The flowers are often used in salads or as garnishes, while the leaves are rich in vitamins. Violets are shade-tolerant and spread easily, providing groundcover while supporting pollinators.

Purslane (*Portulaca oleracea*): Purslane is a succulent groundcover that thrives in full sun and poor soils. Its fleshy leaves are highly nutritious, rich in omega-3 fatty acids, and can be eaten raw in salads or cooked. This plant is drought-tolerant, making it a reliable groundcover for Zone 5 food forests.

7. VERTICAL LAYER:

Climbing Okra (*Abelmoschus esculentus*): This vertical-growing okra plant produces edible pods that are rich in vitamins A, C, and K. The plant thrives in warm conditions and can climb up vertical structures, providing a striking green backdrop in the food forest. The leaves and young pods are often used in cooking, and the plant's rapid growth adds vertical layers to the food forest.

Hops (*Humulus lupulus*): A climbing vine that thrives in vertical spaces, Hops produces flowers used for brewing and medicinal purposes, while supporting pollinators and beneficial insects.

8. FUNGAL LAYER:

Red Reishi Mushroom (*Ganoderma tsugae*): Reishi mushrooms are known for their medicinal properties, particularly for enhancing the immune system. They grow on dead and decaying trees, often forming shelf-like structures. While not typically eaten fresh, their dried form is used in teas, tinctures, and powders, contributing both medicinal value and ecological benefits to the food forest.

Giant Puffball (*Calvatia gigantea*): This large, round mushroom is edible when young and has a mild, earthy flavor. It grows on the forest floor and can reach impressive sizes, making it easy to identify. It serves an important ecological role by decomposing organic material and improving soil health, and it can be used in a variety of dishes, including soups, stews, and stir-fries.

Zone 5 Plan Nine: Lively Layered Legacy

ZONE 5 PLAN NINE: LIVELY LAYERED LEGACY

This design plan is all about creating a living legacy through intelligent layering. It pairs species to provide both immediate nourishment and a foundation for future ecological stability.

1. CANOPY LAYER:

Yellow Buckeye (*Aesculus flava*): Yellow buckeye trees produce large seeds that are toxic when raw but can be leached and processed to become edible. They are suitable for USDA Zone 5 and provide important wildlife habitat and dense shade in the canopy layer. Yellow buckeyes contribute to nutrient cycling in the soil.

Japanese Raisin Tree (*Hovenia dulcis*): The Japanese raisin tree is a rare species that produces sweet, raisin-like fruit from its flower stalks, which are edible and can be used fresh or dried. This tree thrives in USDA Zone 5 and contributes to biodiversity, offering a unique fruit crop while providing habitat for wildlife.

2. UNDERSTORY LAYER:

Silver Buffaloberry (*Shepherdia argentea*): Silver buffaloberry is a nitrogen-fixing shrub that produces tart, edible berries. These berries are rich in vitamins and can be made into jellies or used in baking. The shrub thrives in USDA Zone 5 and improves soil fertility, making it a valuable addition to the understory layer of a food forest.

Siberian Pea Tree (*Caragana arborescens*): The Siberian pea tree is a hardy, nitrogen-fixing shrub that produces edible seeds resembling peas. Thriving in USDA Zone 5, this shrub enhances soil health, provides food for humans and animals, and adds diversity to the understory. Its drought resistance and resilience make it highly valuable.

3. SHRUB LAYER:

Flowering Quince (*Chaenomeles speciosa*): This shrub produces tart, edible fruits that are often used in jellies and preserves. Flowering Quince is drought-tolerant and thrives in well-drained soils, offering early spring flowers that attract pollinators. It also adds aesthetic value to a food forest.

Leatherwood (*Dirca palustris*): Leatherwood is a rare understory shrub that produces small, yellow flowers and provides early-season nectar for pollinators. While its berries are not typically eaten by humans, it plays an important role in supporting wildlife. Leatherwood prefers moist, well-drained soils and partial shade.

4. HERBACEOUS LAYER:

Wild Mint (*Mentha arvensis*): Wild Mint is an aromatic herb with edible leaves used in teas, salads, and as flavoring for various dishes. It grows well in moist soils and partial shade and spreads rapidly, making it an excellent ground cover plant. Wild Mint also attracts beneficial insects and offers medicinal uses.

Salad Burnet (*Sanguisorba minor*): Salad Burnet is a perennial herb with edible, cucumber-flavored leaves that are commonly used in salads and herbal drinks. It thrives in well-drained soils and full sun, and its drought tolerance makes it a resilient plant in food forests. The plant's flowers also attract pollinators.

5. ROOT LAYER:

Camas (*Camassia quamash*): A native North American plant, Camas produces edible bulbs that were an important food source for indigenous peoples. The bulbs have a sweet flavor when baked or roasted. The plant thrives in moist, well-drained soils and produces striking blue flowers, adding beauty and food value to the forest.

Yarrow-Leaf Sundew (*Drosera anglica*): This rare carnivorous plant grows in boggy soils and produces edible tubers that are traditionally used in folk medicine. Though it requires very specific growing conditions, it adds an unusual and functional element to a food forest. The plant also aids in pest control by trapping insects.

6. GROUNDCOVER LAYER:

Basil Thyme (*Acinos alpinus*): This low-growing herb provides aromatic leaves that can be used in cooking, making it an excellent addition to the groundcover layer. Basil Thyme thrives in well-drained soil and full sun, and its small pink flowers in late spring attract pollinators like bees.

Carpet Bugleweed (*Ajuga reptans*): This low-growing perennial offers edible leaves that can be used in salads. It thrives in full sun to partial shade and spreads quickly, making it an excellent groundcover option. Its blue-purple flowers also attract pollinators, adding color and biodiversity to the food forest.

7. VERTICAL LAYER:

Trout Lily Vine (*Erythronium americanum*): Known for its stunning early spring flowers, this plant produces edible bulbs that have a mild flavor similar to garlic. Trout Lily prefers woodland conditions, making it ideal for vertical growth in shaded or partial-shaded areas of the food forest. The plant provides an early-season food source and adds visual interest with its delicate blooms.

Climbing Chayote (*Sechium edule*): Chayote, also known as vegetable pear, grows on climbing vines that can be trained up trellises. It produces edible fruits that are often eaten cooked, raw, or in soups and stews. This perennial plant thrives in Zone 5, offering both vertical coverage and edible fruits that are high in fiber and antioxidants.

8. FUNGAL LAYER:

Wood Blewit Mushroom (*Clitocybe nuda*): Known for its sweet, savory flavor, the Wood Blewit is an edible mushroom that grows in forested areas on decaying wood. It's a choice mushroom for many chefs and offers a nice addition to a food forest's fungal layer. Its role as a decomposer enriches the soil while adding valuable edible fungi to the ecosystem.

Chicken Fat Mushroom (*Pseudohydnum gelatinosum*): This unique, gelatinous fungus is often found growing on decaying wood and has a mildly gelatinous texture when cooked. It has medicinal properties and is used in traditional medicine for its immune-supportive effects. It thrives in temperate zones, aiding in the decomposition of organic material while providing a unique addition to the fungal layer.

Zone 5 Plan Ten: Ecological Essence Edict

Zone 5 Plan Ten: Ecological Essence Edict

With a focus on the very essence of ecological balance, this plan layers food-producing plants for maximum benefit. It serves as a guiding edict for sustainable living while delivering a plentiful harvest.

1. CANOPY LAYER:

European Larch (*Larix decidua*): European larch trees are coniferous but lose their needles in winter, making them a deciduous conifer. The inner bark is edible and has been used as a food source historically. These trees are well-suited to USDA Zone 5, improving soil stability and providing valuable habitat in the food forest.

Hawthorn (*Crataegus monogyna*): Hawthorn trees produce small, edible fruits known as haws, which are often used in jams, jellies, and herbal remedies. This hardy tree grows well in USDA Zone 5 and is valued for its medicinal properties, ability to attract pollinators, and role in stabilizing soil in the forest ecosystem.

2. UNDERSTORY LAYER:

Ninebark (*Physocarpus opulifolius*): Ninebark is a deciduous shrub that can grow to a small tree size and is known for its peeling bark and attractive, edible seed pods. The plant is hardy and tolerant of poor soil conditions, making it suitable for USDA Zone 5. Ninebark is often used in permaculture designs for its drought tolerance and ability to provide cover for wildlife.

Dwarf Serviceberry (*Amelanchier spicata*): This small tree produces sweet, edible berries in early summer that can be used in pies, preserves, and eaten fresh. It grows well in USDA Zone 5 and can tolerate a variety of soil conditions, including poor, rocky soils. The tree is also beneficial for attracting pollinators and providing food for birds and other wildlife.

3. SHRUB LAYER:

Redtwig Dogwood (*Cornus sericea*): Redtwig Dogwood produces small, white berries that are technically edible but not commonly consumed due to their bitterness. The shrub is known for its striking red stems, which add winter interest to a food forest. It thrives in wet soils and provides habitat for wildlife.

Fragrant Sumac (*Rhus aromatica*): Fragrant Sumac produces small red berries that can be made into a tart beverage similar to lemonade. This drought-tolerant shrub thrives in poor soils and offers vibrant fall foliage. It's also a good plant for stabilizing slopes and preventing erosion.

4. HERBACEOUS LAYER:

Lovage (*Levisticum officinale*): Lovage produces edible leaves, stems, seeds, and roots, with a strong celery-like flavor. It grows well in full sun to partial shade and improves soil health through deep root growth. Lovage attracts beneficial insects and adds both culinary and ecological value to the food forest.

Alpine Bistort (*Bistorta vivipara*): Alpine Bistort is a rare, edible herb that produces small, nutty tubers and leafy greens that are often foraged in high-altitude regions. It thrives in moist, well-drained soils and partial shade, making it suitable for cooler climates like Zone 5. The plant also provides ground cover and supports pollinators.

5. ROOT LAYER:

Turnip-Rooted Chervil (*Chaerophyllum bulbosum*): This biennial herb produces edible, turnip-like roots that can be harvested in the second year. The roots have a delicate, sweet flavor and are often roasted or used in soups. It thrives in partial shade and is relatively low-maintenance, making it an excellent choice for the root layer in Zone 5.

Wild Sarsaparilla (*Aralia nudicaulis*): This low-growing perennial produces edible roots that have a licorice-like flavor. The roots are often used to make teas or flavor beverages. Wild sarsaparilla is shade-tolerant and grows well in woodland environments, making it an excellent choice for the root layer in Zone 5.

6. GROUNDCOVER LAYER:

Golden Mock Strawberry (*Duchesnea indica*): Golden Mock Strawberry produces small, edible yellow berries that resemble strawberries. The plant spreads rapidly, making it an ideal groundcover for shaded or partial-shaded areas. It also has attractive green foliage and can help suppress weeds.

Golden Oregano (*Origanum vulgare*): This oregano variety has edible, aromatic leaves that can be used in cooking. Golden Oregano spreads to form a dense groundcover that thrives in full sun and well-drained soils. Its attractive yellow-green foliage also provides a pop of color in the food forest.

7. VERTICAL LAYER:

Fuzzy Kiwi Vine (*Actinidia arguta*): A close relative of the Arctic Kiwi, this variety of kiwi is smaller and fuzz-free, providing sweet, nutrient-packed fruits. It climbs easily up trellises and is an excellent addition to the vertical layer, providing fruit in late summer. This kiwi is hardy enough to thrive in Zone 5 and adds a unique vertical element with edible, health-boosting fruits.

Pole Lima Bean (*Phaseolus lunatus*): This climbing variety of lima bean produces large, edible beans that are rich in protein and fiber. The vines are perfect for growing vertically and provide a beautiful backdrop of lush greenery and white flowers. The beans are an excellent source of nutrients and add vertical height while improving soil nitrogen levels.

8. FUNGAL LAYER:

Sulphur Shelf Mushroom (*Laetiporus sulphureus*): Commonly found on the trunks of hardwood trees, Sulphur Shelf is an edible mushroom known for its vibrant yellow-orange color. The tender caps are used in cooking and have a meaty texture that can be a substitute for chicken or other proteins. Its role as a decomposer contributes to the breakdown of dead wood and improves nutrient cycling within the forest.

Jelly Ear Fungus (*Auricularia auricula-judae*): Jelly Ear, also known as Wood Ear, is a medicinal mushroom that grows on decaying wood. It has a soft, jelly-like texture and is used in various culinary dishes, particularly in Asian cuisines. It's valued for its antioxidant properties and plays an important ecological role in breaking down organic matter and recycling nutrients in the food forest.

ZONE 6

Plant Hardiness Zone 6

Zone 6 represents a moderate climate with cold winters and diverse growing opportunities. It experiences minimum temperatures ranging from -10°F to 0°F (-23°C to -18°C), found in parts of the northeastern, central, and Pacific Northwest regions of the U.S., as well as parts of southern Canada and the U.K. This climate supports a wide variety of plants due to its relatively long growing season and moderate rainfall.
Locations in Zone 6.

Locations In Zone 6

Connecticut: Most of the state, including cities like Hartford, New Haven, and Bridgeport.

Delaware: Much of the state, including Wilmington.

Illinois: Parts of the state, including Chicago, Springfield, and Urbana.

Indiana: Most of the state, including Indianapolis, Lafayette, and Bloomington.

Iowa: Western and central areas, including Des Moines and Cedar Rapids.

Kansas: The eastern half, including Topeka and Lawrence.

Kentucky: Most of the state, including Louisville, Lexington, and Bowling Green.

Maryland: Northern and western parts, including Baltimore and Frederick.

Massachusetts: Western and central areas, including Worcester and Springfield.

Michigan: Southern and central regions, including Detroit and Kalamazoo.

Missouri: Central and southern areas, including

St. Louis, Columbia, and Jefferson City.

New Jersey: Northern and central parts, including Trenton and Newark.

New York: Central and eastern regions, including Albany, Syracuse, and Rochester.

Ohio: Most of the state, including Cleveland, Columbus, and Cincinnati.

Pennsylvania: Much of the state, including Pittsburgh, Harrisburg, and Scranton.

Tennessee: Central and eastern areas, including Nashville and Knoxville.

Virginia: Northern and western parts, including Roanoke and Winchester.

West Virginia: Most of the state, including Charleston and Morgantown.

Wisconsin: Southern and central regions, including Madison and Milwaukee.

Other notable locations in Zone 6

Arkansas: Northern parts, including Little Rock and Fayetteville.

Colorado: Areas at higher elevations, including Denver and Colorado Springs.

Nevada: Regions like Reno.

New Mexico: Higher elevations like Santa Fe.

Oregon: Higher elevation areas in the east.

California: Some parts of the Sierra Nevada mountains and northernmost areas.

Zone 6 Climate

Zone 6 winters range from -10°F to 0°F, with an average growing season lasting 140 to 180 days, generally beginning in mid-to-late March and ending with the first frost in October or November. Precipitation is moderate, ranging from 30 to 50 inches (75 to 127 cm) annually, with fairly even distribution throughout the year. However, some areas experience summer droughts, making soil management crucial. Microclimates also exist, particularly in urban areas or areas with varying elevations.

Zone 6 Soil Conditions

Soil types in Zone 6 range from clay to sandy to loamy, with pH levels generally between 6.0 and 7.0, allowing for diverse plant growth. However, soil conditions can vary, so testing and amending soil is recommended for optimal fertility and drainage. Proper drainage is especially crucial during winter months to avoid waterlogging in clay-heavy soils.

Zone 6 Growing Conditions

Zone 6 has 140 to 180 frost-free days, with an average last frost date in mid-April and a first frost date around mid-October. Windy conditions are common, particularly in plains and mountainous areas, which can affect plant health. Adequate sunlight is also a key factor for fruiting plants, with summer months providing ample sunshine. However, during peak heat, it is important to maintain soil moisture through watering and mulching.

Winter winds and cold temperatures can be harsh in Zone 6, necessitating protective measures such as mulch and cold frames for sensitive plants. Snowfall, while insulating for roots, may need to be managed to avoid excessive accumulation on fruit trees and bushes.

Zone 6 Planting Considerations

Plants grown in Zone 6 need to be cold-hardy

and frost-tolerant. Perennials such as apples, pears, cherries, berries, root vegetables, lilacs, and daylilies are well-suited to this zone. Frost-sensitive plants should be carefully timed with the growing season. Season-extension tools like greenhouses or row covers can help extend the growing season by several weeks.

Zone 6 Pest and Disease Management

Common pests in Zone 6 include aphids, Japanese beetles, squash bugs, and caterpillars. Integrated pest management strategies, such as companion planting and encouraging beneficial insects, can help control populations. Diseases like powdery mildew and blight are also common due to humid conditions. Proper plant spacing and resistant varieties are essential for disease control.

Zone 6 Wildlife Considerations

Zone 6 is home to various wildlife, including deer, rabbits, and groundhogs, which can damage crops. Protective measures such as fencing and selecting deer-resistant plants can help manage these threats. Additionally, pollinators like bees, butterflies, and hummingbirds thrive in Zone 6, enhancing crop yields.

Zone 6 Regional Tips

Midwest: The Great Lakes moderate temperatures, creating milder winters and early springs, though increased moisture can present challenges.

Northeast: The Appalachian region is ideal for growing fruit trees, berries, and root crops.

Rocky Mountains: Higher elevations and wind

exposure require windbreaks and irrigation systems for successful gardening.

Zone 6 offers diverse growing opportunities with a balance of moderate winters, sufficient rainfall, and a long growing season. With careful management of planting schedules, soil, and protection from winter extremes, this zone can support productive food forests, ornamental gardens, and homesteads.

Zone 6 Plan One: Meadow Mosaic Mission

A mission to transform open meadows into productive food forests, this plan uses layered pairings for dynamic results. It emphasizes both the nutritional bounty and the inherent sustainability of each level.

1. CANOPY LAYER:

Chinquapin (*Castanea pumila*): A smaller cousin of the American Chestnut, Chinquapin produces sweet, edible nuts that are often compared to chestnuts. The tree itself grows as a large shrub or small tree, thriving in well-drained soils with full sun. It can help improve soil health through its nutrient cycling and provides a valuable food source for both humans and wildlife, supporting biodiversity within the food forest.

Shagbark Hickory (*Carya ovata*): While similar to the Butternut, the Shagbark Hickory offers edible nuts that are sweet and high in protein. The tree provides excellent shade and grows in well-drained soils. Its bark, which peels off in strips, also adds aesthetic value and contributes to the wildlife habitat, making it a great addition to the canopy layer of a food forest.

2. UNDERSTORY LAYER:

American Beautyberry (*Callicarpa americana*): American Beautyberry produces striking purple berries that are edible and can be made into jellies, sauces, or wines. This plant thrives in partial shade and moist, well-drained soils. It attracts birds and pollinators while providing a vibrant pop of color in the understory, and the berries offer a valuable food source in the fall.

Blackberry (*Rubus fruticosus*): Blackberry bushes offer sweet, edible berries that are delicious when eaten fresh or used in jams and desserts. Growing well in partial shade and moist, well-drained soils, blackberries provide an abundance of fruit, attract pollinators, and serve as protective cover for wildlife.

3. SHRUB LAYER:

Aronia (*Aronia prunifolia*): Aronia, or purple chokecherry, produces dark purple to black berries that are highly nutritious and rich in antioxidants. Though the berries are quite tart, they are great for juicing or making preserves. This shrub is also highly resilient and grows in a variety of soil types.

New Jersey Tea (*Ceanothus americanus*): A nitrogen-fixing shrub that supports butterflies and other beneficial insects while improving soil fertility.

4. HERBACEOUS LAYER:

Horseradish (*Armoracia rusticana*): Known for its pungent, spicy flavor, horseradish is the root of this herbaceous plant that is commonly used to flavor sauces and condiments. It grows in full sun to partial shade and thrives in well-drained soils. Horseradish also helps to break up compacted soils, making it a good addition to a food forest for soil health.

Lovage (*Levisticum officinale*): Lovage is a perennial herb that resembles celery and has a similar flavor. The leaves, stems, and seeds are edible, and they can be used in soups, salads, or herbal teas. It grows best in moist, well-drained soils and can tolerate both full sun and partial shade, offering a strong aromatic addition to the food forest.

5. ROOT LAYER:

Dandelion (*Taraxacum officinale*): Dandelion roots are edible and can be used for medicinal purposes or roasted to make a coffee substitute. This plant thrives in full sun or partial shade and is highly beneficial for its deep taproot, which aerates and improves soil health. It also attracts pollinators and adds biodiversity to the food forest.

Wild Yam (*Dioscorea villosa*): A native root crop

with medicinal properties, it supports soil health and acts as a ground cover in shaded areas.

6. GROUNDCOVER LAYER:

Wood Sorrel (*Oxalis acetosella*): Wood Sorrel is a low-growing groundcover that thrives in shaded areas. Its clover-like leaves and small, delicate flowers provide both aesthetic appeal and ecological benefits. The leaves, flowers, and seed pods are edible, with a tangy, lemon-like flavor due to their oxalic acid content. It is a great nitrogen fixer and adds biodiversity to the food forest.

Dwarf Mondo Grass (*Ophiopogon japonicus*): This evergreen groundcover forms dense mats and is often used to fill gaps between larger plants. It's not typically edible, but it plays an important role in stabilizing soil and preventing erosion. Its slow growth rate makes it ideal for low-maintenance ground coverage in zone 6 forests.

7. VERTICAL LAYER:

Kiwi Berry (*Actinidia arguta*): A hardy kiwi vine that produces grape-sized fruits with smooth skin, edible raw or cooked. It grows well in full sun to partial shade and can climb trellises, providing both food and habitat for wildlife.

Chocolate Vine (*Akebia quinata*): Known for its purple-brown flowers and edible sausage-shaped fruit, this vigorous climber thrives in partial shade. The fruit is mildly sweet, while the vine provides dense cover, aiding biodiversity.

8. FUNGAL LAYER:

Elm Oyster Mushroom (*Hypsizygus ulmarius*): A choice edible mushroom, the Elm Oyster grows on dead or dying hardwoods, making it excellent for recycling nutrients in your food forest. The cap is thick and meaty, providing a flavorful addition to meals. It also promotes the breakdown of organic material, enriching the soil with nutrients while providing food for both you and beneficial insects.

Blewit Mushroom (*Clitocybe nuda*): This edible mushroom is prized for its slightly sweet and earthy flavor. The Blewit thrives in leaf litter and decaying plant matter, making it an ideal choice for maintaining soil health in a forest garden. It also assists in breaking down organic material, contributing to soil fertility and nutrient cycling.

Zone 6 Plan Two: Flourish Framework

This framework is built to help plants flourish by organizing them into supportive layers. It champions robust food production while ensuring the longevity and health of the ecosystem.

1. CANOPY LAYER:

Heartnut (*Juglans ailantifolia var. cordiformis*): A relative of the Black Walnut, the Heartnut produces heart-shaped edible nuts that are easier to crack and have a sweet, mild flavor. This tree grows best in full sun to partial shade and thrives in fertile, well-drained soils. Besides its edible nuts, it adds beauty to the forest canopy with its unique foliage and spreading habit.

Black Cherry (*Prunus serotina*): Black Cherry produces edible cherries, although they must be harvested when fully ripe to avoid bitterness. The tree is fast-growing and provides valuable shade, while its fruits support a variety of wildlife. It thrives in full sun and well-drained soils and is an excellent choice for adding diversity to the canopy.

2. UNDERSTORY LAYER:

Serviceberry (*Amelanchier laevis*): Serviceberry trees produce sweet, edible berries in early spring, perfect for fresh consumption or used in cooking. They thrive in partial shade and moist, fertile soils. In addition to providing food for humans, the flowers attract pollinators, and the branches offer nesting sites for birds.

Wild Plum (*Prunus americana*): Wild Plum produces small, tart plums that are edible and can be eaten fresh, dried, or used in preserves. This deciduous shrub grows well in partial shade and moist, well-drained soils, providing a valuable food source for both humans and wildlife, while its blossoms attract bees and butterflies.

3. SHRUB LAYER:

Goji Berry (*Lycium barbarum*): Goji berries are nutrient-dense and contain a variety of vitamins and antioxidants, making them a powerful addition to the diet. They are easy to grow in a range of soils and climates and can provide a rich harvest of small, red berries that have a slightly sweet flavor. This shrub also works well as a hedge or screening plant.

Currant (*Ribes aureum*): The golden currant produces small, sweet-tart berries that are used in jams, jellies, or fresh eating. It thrives in a variety of soils and light conditions, making it an adaptable choice for the understory layer. Its yellow flowers in the spring attract beneficial pollinators.

4. HERBACEOUS LAYER:

Bergamot (*Monarda fistulosa*): This herb produces fragrant leaves and purple flowers that are edible and can be used in teas. It also attracts pollinators and is highly beneficial to a food forest for its biodiversity and ability to attract bees and butterflies. Bergamot thrives in well-drained soils and prefers partial to full sun.

Sweet Cicely (*Myrrhis odorata*): Sweet Cicely is a perennial herb with an aromatic flavor that combines hints of anise and licorice. The young leaves and seeds are edible and can be used in cooking or as flavorings in desserts. This herb prefers rich, moist, and well-drained soils, making it an excellent addition to a shaded food forest understory.

5. ROOT LAYER:

Ginseng (*Panax quinquefolius*): Ginseng roots are prized for their medicinal properties and can be used to boost energy and support overall health. This plant prefers rich, moist, well-drained soils in partial to full shade, making it ideal for a shaded understory in a food forest. Ginseng also adds medicinal diversity to your garden.

Oca (*Oxalis tuberosa*): Oca is a tuberous root vegetable native to the Andes and provides a starchy, edible root with a tangy flavor. This perennial plant thrives in well-drained, slightly acidic soils and full sun, and the tubers are harvested in the fall. It is an excellent choice for adding a unique crop to the root layer of a food forest.

6. GROUNDCOVER LAYER:

Alpine Strawberry (*Fragaria vesca*): Alpine Strawberry is a compact, perennial groundcover producing small, intensely flavored strawberries. The berries, as well as the leaves, are edible, and it offers pollinator support. This plant spreads easily and thrives in partial shade, making it an excellent addition to diversify ground layers.

Sweet Woodruff (*Galium odoratum*): Sweet Woodruff is an aromatic herb that spreads quickly to cover the ground, particularly in shady areas. It is edible, though typically used as a flavoring in beverages and desserts. It's a great companion for other shade-loving plants and attracts pollinators while acting as a living mulch.

7. VERTICAL LAYER:

Climbing Hydrangea (*Hydrangea anomala petiolaris*): This shade-tolerant climber is mainly ornamental, but its roots have been used in traditional medicine. It creates beautiful vertical structure with its white lacecap flowers, attracting pollinators.

Chinese Yam (*Dioscorea polystachya*): A fast-growing, twining vine with edible tubers that can be cooked like potatoes. It is drought-tolerant and adds valuable root biomass to the system, supporting soil health.

8. FUNGAL LAYER:

Elm Bracket Mushroom (*Polyporus umbellatus*):

Also known as Umbrella Polypore, this rare mushroom is edible and grows on the roots of hardwood trees, providing a significant mycelial network that supports tree health. It is mildly flavored, best when cooked, and helps improve soil structure by breaking down wood into organic material.

Shaggy Parasol Mushroom (*Chlorophyllum rhacodes*): An edible mushroom that prefers growing in shaded, damp areas, the Shaggy Parasol is a good candidate for the fungal layer. It is known for its mild, nutty flavor when cooked. Additionally, this mushroom breaks down organic material and contributes to the formation of rich, loamy soil.

Zone 6 Plan Three: Sustainable Sprout Scheme

Focused on early growth and lasting sustainability, this scheme pairs sprouting edibles across every layer. It's designed to provide a consistent yield while reinforcing the system's natural resilience.

1. CANOPY LAYER:

American Beech (*Fagus grandifolia*): Known for its smooth, gray bark and dense canopy, the American Beech produces edible nuts, known as beechnuts. These nuts are sweet and nutritious and attract wildlife. The tree is slow-growing but durable, often living for centuries, and it thrives in moist, fertile soils, making it a long-term staple in a food forest.

Serviceberry (*Amelanchier canadensis*): Serviceberry, or Juneberry, offers sweet, edible berries that are enjoyed by both people and wildlife. The tree blooms early in the spring, providing early-season food for pollinators, and its berries are rich in antioxidants. It's adaptable to a range of soil types and provides year-round interest with its beautiful flowers, fruits, and vibrant fall foliage.

2. UNDERSTORY LAYER:

Elderberry (*Sambucus canadensis*): Elderberry shrubs produce clusters of edible berries that can be made into syrups, jellies, or wine. Thriving in partial shade and moist soils, they support pollinators with their late-spring flowers and provide abundant fruit for birds and wildlife in the fall.

Almond (*Prunus dulcis*): Almond trees produce edible nuts that are a valuable source of protein, fat, and vitamins. These small trees grow well in well-drained, slightly acidic soils and partial shade. The fragrant flowers in spring attract pollinators, and the nuts provide a food source for both humans and wildlife.

3. SHRUB LAYER:

Buffalo Berry (*Shepherdia argentea*): The tart red berries of the buffalo berry are edible and have been used traditionally in preserves, juices, and baked goods. This shrub is a hardy choice for a food forest, as it thrives in a variety of soil conditions and is drought-tolerant once established. Its silver foliage also adds a unique texture to the landscape.

Autumn Olive (*Elaeagnus umbellata*): This nitrogen-fixing shrub produces small, reddish-orange edible berries that are high in antioxidants. Autumn olive is hardy and can thrive in a variety of soils, making it a great choice for reclamation projects or as a companion plant in a food forest. The shrub also attracts pollinators with its fragrant flowers.

4. HERBACEOUS LAYER:

Chives (*Allium schoenoprasum*): Chives are an easy-to-grow herbaceous perennial that provides edible leaves used for garnishing or flavoring dishes. They grow well in full sun and well-drained soils, and their purple flowers attract pollinators. Chives offer both culinary benefits and support for beneficial insects in the food forest.

Dandelion (*Taraxacum officinale*): Though often considered a weed, dandelions are highly nutritious with their leaves, flowers, and roots all being edible. The leaves can be used in salads, and the root can be roasted for use in herbal teas. Dandelions are resilient and grow in almost any soil, making them a valuable addition to a food forest for both edible use and soil health.

5. ROOT LAYER:

Mache (*Valerianella locusta*): Mache, also known as lamb's lettuce, produces small, edible leaves and grows a long, slender taproot. The root can be eaten raw or cooked, and the plant is highly nutritious. It grows well in cool, moist soils and can provide a fresh, early-season crop in the food forest.

Sweet Cicely (*Myrrhis odorata*): Sweet Cicely's edible root has a licorice-like flavor and can be used in a variety of culinary applications. It prefers moist, well-drained soils and grows well in shaded or semi-shaded areas. Sweet Cicely is beneficial in a food forest by providing an early-season root harvest.

6. GROUNDCOVER LAYER:

Strawberry Clover (*Trifolium fragiferum*): A nitrogen-fixing groundcover, Strawberry Clover is excellent for building soil health. Its small pink flowers attract pollinators, and while its leaves and flowers are edible, they are more commonly used in herbal teas. It's also resilient and can tolerate a variety of soil conditions.

Cranberry (*Vaccinium macrocarpon*): Cranberry is a low-growing, evergreen groundcover plant that thrives in acidic, well-drained soils. The tart, edible berries are packed with antioxidants, making it both a productive and nutritious addition to the food forest. It also supports wildlife with its dense foliage and fruit.

7. VERTICAL LAYER:

Kiwi Vine (*Actinidia kolomikta*): A cold-hardy kiwi producing small, sweet fruit, it climbs structures well and grows best in partial shade. This plant provides both food and a vertical element to the food forest.

Five-leaf Akebia (*Akebia trifoliata*): Similar to the chocolate vine, it bears small, cucumber-like fruit with an edible, mildly sweet pulp. The vine is hardy and can grow in various light conditions.

8. FUNGAL LAYER:

Dryad's Saddle (*Cerioporus squamosus*): This edible polypore mushroom thrives on decaying hardwood trees and can be found in abundance in USDA Zone 6. With its large, fan-shaped caps, it breaks down wood fibers into compostable material. Its flavor is best when the mushroom is young and tender, often compared to mild chicken.

Cauliflower Mushroom (*Sparassis crispa*): Known for its unique, frilly appearance, this edible mushroom has a mild flavor and a crunchy texture. Cauliflower mushrooms grow on the roots of coniferous trees, helping decompose organic matter and benefiting the soil. This species adds a rare and ornamental touch to the fungal layer.

Zone 6 Plan Four: Edible Ecology Ensemble

Like a well-rehearsed orchestra, this ensemble coordinates each plant layer to produce a symphony of edibles. Its design ensures that every species plays a role in both nutrition and ecological support.

1. CANOPY LAYER:

Black Locust (*Robinia pseudoacacia*): Black Locust is known for its edible flowers, which are sweet and can be eaten raw or used to make syrups. The tree is fast-growing and nitrogen-fixing, improving soil fertility for other plants in the food forest. It is drought-tolerant and can thrive in a variety of soil types, making it a versatile and functional addition to the canopy layer.

Sweet Birch (*Betula lenta*): The leaves, bark, and sap of Sweet Birch are all edible. The sap is commonly harvested in early spring to make birch syrup, similar to maple syrup. The tree grows best in moist, well-drained soils and provides an attractive canopy with its aromatic bark and foliage, supporting both human use and wildlife needs.

2. UNDERSTORY LAYER:

Black Chokeberry (*Aronia melanocarpa*): Black Chokeberry produces small, tart berries that are rich in antioxidants and can be used in jams or smoothies. It thrives in partial shade and moist, well-drained soils. The shrub adds biodiversity to the food forest by attracting pollinators and providing food for birds and small mammals.

Wild Crabapple (*Malus pumila*): Wild Crabapple produces small, tart apples that can be used in sauces, jellies, or ciders. This tree grows well in partial shade and well-drained soils. It supports pollinators with its spring blossoms and offers shelter and food for wildlife, including birds and mammals.

3. SHRUB LAYER:

Mulberry (*Morus nigra*): The black mulberry is a rare, edible shrub that produces sweet, juicy berries that are perfect for eating fresh, making jams, or baking. It prefers full sun to partial shade and thrives in well-drained soils. This plant is also known for attracting pollinators and providing shade and shelter to smaller plants in the food forest.

Gooseberry (*Ribes uva-crispa*): The gooseberry is a hardy, edible shrub that produces tart, green to red berries that are used in cooking, preserves, and pies. It thrives in well-drained, slightly acidic soils and partial shade. Gooseberries also have spiny branches that can provide additional security in your food forest by deterring larger animals from browsing.

4. HERBACEOUS LAYER:

Mullein (*Verbascum thapsus*): Mullein is known for its tall, spiky flowers, but its leaves and flowers are also edible when cooked and can be used for medicinal purposes. It thrives in poor soils and full sun, making it an excellent choice for disturbed areas in a food forest, where it can contribute to soil improvement.

Fennel (*Foeniculum vulgare*): Fennel is a fragrant, herbaceous plant with edible leaves, seeds, and bulbs. The seeds are often used as a spice, while the bulbs and leaves can be added to salads and cooked dishes. Fennel prefers full sun and well-drained, fertile soils, and it attracts beneficial insects, which helps support biodiversity in the food forest.

5. ROOT LAYER:

Mizuna (*Brassica rapa*): Mizuna is an edible root vegetable that produces tender, peppery-flavored leaves and a small, edible root. It is highly adaptable, thriving in well-drained, fertile soils and full sun. Mizuna can be used for salads and stir-fries and adds an early season harvest to the food forest.

Cress (*Lepidium sativum*): Cress is a fast-growing, cold-hardy herb with edible roots and peppery-tasting leaves. It prefers moist, well-drained soils and is a great plant for adding flavor to salads and sandwiches. Its roots are used in herbal medicine and can also be consumed for their mild, spicy flavor.

6. GROUNDCOVER LAYER:

Bearberry Cotoneaster (*Cotoneaster dammeri*): This low-growing shrub is excellent for erosion control and provides a dense groundcover with white spring flowers and small red berries. While the berries are edible, they are usually more attractive to birds. It offers year-round greenery and supports pollinators.

Sweet Violet (*Viola odorata*): Sweet Violet is a small, shade-loving groundcover with edible flowers and leaves often used in salads or desserts. Its dense growth helps to suppress weeds, and the plant attracts pollinators with its fragrant blooms, making it a beneficial addition to shaded areas of the forest.

7. VERTICAL LAYER:

Hopniss (*Apios priceana*): This rare climbing legume has edible tubers and beans, traditionally used as a food source by Indigenous peoples. It also fixes nitrogen, improving soil fertility.

Winged Bean (*Psophocarpus tetragonolobus*): A tropical climbing bean that thrives in warm climates. The entire plant is edible—leaves, flowers, pods, and tubers. It enriches the soil with nitrogen and offers diverse edible parts.

8. FUNGAL LAYER:

Amethyst Deceiver (*Laccaria amethystina*): This edible mushroom, with its striking purple color, is a great addition to a food forest. It grows well in leaf litter and contributes to nutrient cycling in the soil. Its cap is small, and the mushroom is best consumed when young. Though not commonly cultivated, it can thrive in woodland environments.

Lobster Mushroom (*Hypomyces lactifluorum*): This unique mushroom is a parasitic fungus that colonizes other mushrooms, turning them into vibrant red, edible "lobster" mushrooms with a seafood-like flavor. It helps maintain a balance in the fungal ecosystem and can be found on decaying wood, assisting with nutrient breakdown in forest soils.

Zone 6 Plan Five: Layered Lush Landscape

This plan transforms a patch of land into a lush, multi-layered mosaic of food-producing plants. It's all about creating a sustainable landscape where each tier contributes to a thriving whole.

1. CANOPY LAYER:

Eastern Black Pine (*Pinus thunbergii*): This unique pine offers edible seeds (pine nuts), which are rich in protein and healthy fats. The tree is drought-tolerant and grows well in a range of soil types, including sandy soils. It provides excellent shelter and shade for other plants, making it a valuable component in a food forest's canopy layer.

Sugar Maple (*Acer saccharum*): Known for its syrup, the Sugar Maple's sap is collected in the early spring to make maple syrup. This tree also produces beautiful fall foliage and grows best in fertile, moist soils. It adds both economic value through syrup production and aesthetic beauty to the food forest while providing shade and shelter to other plants.

2. UNDERSTORY LAYER:

Spicebush (*Lindera benzoin*): Spicebush produces aromatic berries that are edible when dried and can be used to make teas or flavor food. Thriving in partial shade and moist, well-drained soils, it attracts pollinators with its early spring flowers, while its dense growth provides cover for wildlife.

American Black Currant (*Ribes americanum*): American Black Currant produces small, tart berries that are high in vitamin C and can be used in jellies, syrups, or baked goods. It grows well in partial shade and moist soils, and its blossoms attract pollinators while providing a valuable early-season food source for birds.

3. SHRUB LAYER:

Chilean Guava (*Ugni molinae*): A rare shrub that produces small, aromatic berries with a unique sweet-tart flavor, often used in jams and desserts. It is ideal for low-maintenance food forest design due to its resilience and adaptability to various soil conditions. This shrub also offers evergreen foliage, which helps retain moisture in the understory.

Red-berried Elder (*Sambucus racemosa*): This elder species is not only a beautiful addition to any food forest but also provides small, red berries that are edible when cooked. These berries are great for jellies and syrups. Red-berried elder thrives in partial shade and moist soils and is often used as a wildlife-friendly shrub.

4. HERBACEOUS LAYER:

Lovage (*Levisticum officinale*): Lovage is a perennial herb with a flavor similar to celery, and the leaves, stems, and seeds are all edible. It prefers moist, fertile soils and full to partial sun, offering a high-yield perennial herb that can be used for soups, salads, and sauces, while also improving soil with its deep root system.

Comfrey (*Symphytum officinale*): Comfrey has large, broad leaves that are edible when young and can be used in salads or as an herbal remedy. Comfrey's deep roots also help improve soil structure by pulling up nutrients from deep in the soil. It thrives in full sun to partial shade and works well in food forests for its dynamic root system and ability to improve soil health.

5. ROOT LAYER:

Jerusalem Artichoke (*Helianthus tuberosus*): Jerusalem Artichoke, also known as sunchoke, is a tuberous plant that produces edible, knobby roots. The roots have a nutty, sweet flavor and are used in soups, stir-fries, and salads. It is a hardy plant that thrives in full sun and can help improve soil health.

Fennel (*Foeniculum vulgare*): While often grown for its bulbs, fennel also has edible roots that can be used in culinary dishes, adding a mild licorice flavor. It grows well in full sun and well-drained soils. Fennel's deep roots help to break up compacted soil, making it a good choice for the root layer.

6. GROUNDCOVER LAYER:

Corsican Mint (*Mentha requienii*): Corsican Mint is a tiny, creeping groundcover with a strong mint aroma. While the leaves are edible, typically used in teas or desserts, it is prized more for its role as a fragrant and attractive groundcover that thrives in moist, shaded areas. It helps retain moisture in the soil.

Lingonberry (*Vaccinium vitis-idaea*): Lingonberry is a hardy evergreen groundcover that produces tart, red berries similar to cranberries. The berries and leaves are edible and commonly used in sauces, jams, and teas. It thrives in acidic soils and offers year-round ground coverage while supporting wildlife.

7. VERTICAL LAYER:

Malabar Spinach (*Basella alba*): Although technically not a true spinach, its fleshy leaves are edible and high in vitamins. This fast-growing vine thrives in warm weather and adds greenery to trellises while providing nutritious leaves.

Chayote (*Sechium edule*): This climbing vine produces edible fruits, shoots, and roots. It is highly productive and works well on trellises, offering a multipurpose food source in Zone 6.

8. FUNGAL LAYER:

King Alfred's Cake (*Daldinia concentrica*): Though not typically considered edible in the culinary sense, this mushroom is used as tinder for fires. It grows on dead or dying trees, helping with decomposition and nutrient cycling. Its unique, black, spherical appearance adds diversity to the fungal layer of a food forest.

Morel Mushroom (*Morchella esculenta*): A prized edible species, morels are known for their honeycomb-like appearance and rich, earthy flavor. These mushrooms grow in symbiosis with certain trees, benefiting from forest debris. Morels help break down organic matter and contribute to the forest's nutrient cycling, making them a valuable part of the ecosystem.

Zone 6 Plan Six: Nourishing Nexus Network

At the nexus of nourishment and natural design, this network pairs species in a strategic, layered fashion. It aims to boost food output and build an interdependent system that supports itself over time.

1. CANOPY LAYER:

Yellow Birch (*Betula alleghaniensis*): Yellow Birch produces edible sap, which can be tapped in the spring to make syrup. It is a slow-growing tree that thrives in cool, moist conditions and can grow to impressive heights. The aromatic bark adds interest to the forest canopy, and it supports a range of wildlife, making it a valuable addition to a food forest.

American Hazelnut (*Corylus americana*): The American Hazelnut produces small, edible nuts that are sweet and nutritious. This tree is an excellent choice for providing food for both humans and wildlife and grows well in fertile, well-drained soils. Its dense, spreading canopy provides shelter for other plants and animals within the food forest.

2. UNDERSTORY LAYER:

Chokecherry (*Prunus virginiana*): Chokecherry trees produce clusters of small, tart cherries that are edible when cooked and can be used in sauces or syrups. This tree thrives in partial shade and moist, well-drained soils. The flowers attract pollinators, and the cherries provide food for birds and other wildlife.

Elderberry (*Sambucus nigra*): European Elderberry produces edible berries when cooked, perfect for making syrups, jellies, or wines. It grows well in moist, well-drained soils and partial shade. The flowers attract pollinators, and the berries provide a valuable food source for both humans and birds.

3. SHRUB LAYER:

Nanking Cherry (*Prunus tomentosa*): This hardy shrub produces small, tart cherries that can be eaten fresh, used in preserves, or made into sauces. Nanking cherries thrive in well-drained, fertile soils and are valued for their ornamental qualities, especially their fragrant spring flowers and red berries. The plant is also a great pollinator attractor.

New Jersey Tea (*Ceanothus americanus*): This small, nitrogen-fixing shrub produces clusters of white flowers that are highly attractive to pollinators. Its roots improve soil fertility, and its leaves can be used to make tea.

4. HERBACEOUS LAYER:

Calendula (*Calendula officinalis*): Known for its vibrant orange and yellow flowers, calendula is edible and often used in teas, salads, or as a garnish. This herbaceous plant grows well in full sun and moist, well-drained soils, and its flowers attract pollinators, making it a valuable addition to the food forest.

Red Clover (*Trifolium pratense*): Red clover is a nitrogen-fixing herbaceous plant that provides edible flowers and leaves. It grows in full sun and well-drained soils and is a valuable plant in a food forest for improving soil fertility and providing forage for bees and other pollinators.

5. ROOT LAYER:

Taro (*Colocasia esculenta*): Taro is a tropical plant known for its starchy, edible root, which can be used in a variety of dishes. It requires warm, moist soil to grow well and can be cultivated in food forests with access to water. Taro's deep roots are also valuable for soil aeration.

Scorzonera (*Scorzonera hispanica*): Scorzonera, also known as black salsify, has edible roots that taste similar to artichokes. This plant thrives in well-drained, fertile soils and full sun. The roots are harvested in late fall and are a nutritious addition to the root layer of a food forest.

6. GROUNDCOVER LAYER:

Creeping Jenny (*Lysimachia nummularia*): Creeping Jenny forms a dense mat of bright green foliage that thrives in moist areas. It's not commonly edible but plays a significant role in erosion control and soil moisture retention. It is ideal for covering large areas in damp parts of the food forest.

Partridgeberry (*Mitchella repens*): Partridgeberry is a low-growing evergreen groundcover that produces small red berries, which are edible but not particularly flavorful. It thrives in shaded, acidic soils and provides excellent ground cover while supporting wildlife with its berries and dense foliage.

7. VERTICAL LAYER:

Arctic Kiwi (*Actinidia kolomikta*): A cold-hardy kiwi species producing small, smooth-skinned fruits that are rich in Vitamin C. It thrives in partial shade and helps attract pollinators with its fragrant flowers.

Purple Bell Vine (*Rhodochiton atrosanguineus*): While mainly ornamental with its unique, bell-shaped flowers, the plant offers nectar for pollinators. It creates vertical interest and thrives in sunny, sheltered spots.

8. FUNGAL LAYER:

Black Trumpet Mushroom (*Craterellus cornucopioides*): This edible, funnel-shaped mushroom is highly sought after for its smoky, rich flavor. Black Trumpets grow in symbiosis with hardwood trees, supporting nutrient exchange in the soil. They thrive in shaded, moist environments and contribute to the decomposition of organic matter, enhancing soil health.

Mica Cap Mushroom (*Coprinellus micaceus*): An edible mushroom that grows in dense clusters on decaying wood, the Mica Cap helps decompose dead organic material, releasing nutrients back into the soil. It has a delicate flavor and should be eaten quickly after harvesting, as it deteriorates rapidly.

Zone 6 Plan Seven: Green Grail Guide

Embarking on a quest for the ultimate green harvest, this guide maps out a layered approach to sustainable edibles. It's designed to yield a "grail" of nutritious food while fortifying ecosystem dynamics.

1. CANOPY LAYER:

Chestnut Oak (*Quercus montana*): Chestnut Oak produces edible acorns, which can be processed and used in cooking. The tree grows well in dry, rocky soils and is drought-tolerant once established. Its large, spreading canopy provides excellent shade and helps improve soil moisture retention, benefiting other plants in the food forest.

Pecan (*Carya illinoinensis*): Known for its edible nuts, the Pecan tree produces large, sweet nuts that are commonly used in pies and other recipes. This tree grows best in well-drained, deep soils and is a hardy addition to the canopy, offering shade and habitat for wildlife. Its nuts are a significant food source for both people and animals.

2. UNDERSTORY LAYER:

Goji Berry (*Lycium barbarum*): Goji Berry is a small shrub that produces edible, tangy berries packed with antioxidants. It thrives in partial shade and well-drained soils. This shrub provides fruit throughout the growing season, supports pollinators, and offers a nutritional food source for both humans and wildlife.

Chilean Guava (*Ugni molinae*): Chilean Guava is a low-growing shrub that produces small, edible, sweet berries that can be eaten fresh or used in preserves. It thrives in partial shade and moist, well-drained soils. This rare shrub adds unique flavor and diversity to the understory, while also providing food for birds and small mammals.

3. SHRUB LAYER:

Tart Cherry (*Prunus cerasus*): Tart cherries are well-known for their use in pies and baking. This shrub provides a heavy yield of small, sour cherries and thrives in well-drained, slightly acidic soils. Its ability to grow in both full sun and partial shade makes it a versatile choice for a food forest.

Wild Gooseberry (*Ribes missouriense*): Wild gooseberries are a rare edible shrub that produces small, tart berries. They are used in cooking or can be eaten raw. They thrive in partial to full sun and are adaptable to a wide range of soil types. This shrub is a great wildlife plant and serves as a natural deterrent due to its thorny branches.

4. HERBACEOUS LAYER:

Lemon Balm (*Melissa officinalis*): Lemon balm is a fragrant herb that produces edible leaves, often used in teas or as a culinary herb. It prefers well-drained, slightly moist soils and partial sun, adding both culinary value and an aromatic touch to the food forest, while attracting beneficial insects like bees.

Oregano (*Origanum vulgare*): Oregano is a perennial herb known for its strong, savory flavor. The leaves are edible and can be used in a variety of dishes, especially in Mediterranean cuisine. It thrives in full sun and well-drained, slightly alkaline soils, and is an excellent plant to attract pollinators in the food forest.

5. ROOT LAYER:

Chicory (*Cichorium intybus*): Chicory's roots can be roasted and ground into a coffee substitute, offering a unique flavor. The plant prefers well-drained soils and full sun, and its deep roots help with soil aeration. Chicory is also beneficial for attracting pollinators and adding diversity to the food forest.

Celeriac (*Apium graveolens var. rapaceum*): Celeriac,

also known as celery root, has an edible, knobby root that can be used in soups, stews, or mashed. This plant thrives in well-drained, fertile soils and full sun. It is a valuable root crop for food forests, providing a nutritious harvest and benefiting the soil with its deep roots.

6. GROUNDCOVER LAYER:

Wintergreen (*Gaultheria procumbens*): Wintergreen is a low, creeping groundcover with edible red berries and aromatic leaves used in teas and flavorings. It prefers acidic soils and shaded conditions, making it a great choice for the forest floor. Its leaves and berries provide both culinary and medicinal uses.

Wild Blue Phlox (*Phlox divaricata*): Wild Blue Phlox is a shade-tolerant groundcover that produces clusters of fragrant blue or lavender flowers. While not typically considered edible, it attracts a variety of pollinators and adds beauty to the understory. Its spreading habit helps to fill in shaded areas.

7. VERTICAL LAYER:

Japanese Honeysuckle (*Lonicera japonica*): Though invasive in some areas, it can be managed to produce edible flowers and leaves. The fragrant blooms attract pollinators, and it provides dense cover for wildlife.

Chilean Potato Vine (*Solanum crispum*): This climbing vine bears potato-like flowers and small berries, though the berries are toxic. It offers vertical growth and habitat for beneficial insects in the food forest.

8. FUNGAL LAYER:

Brick Cap Mushroom (*Hypholoma lateritium*): This edible mushroom grows in clusters on decaying wood and tree stumps, helping to decompose organic matter and improve soil fertility. It is known for its reddish-brown cap and mild flavor. Brick Caps are a good addition to the fungal layer, contributing to nutrient cycling in the ecosystem.

Deer Mushroom (*Pluteus cervinus*): Found on decaying hardwood, the Deer Mushroom is an edible species with a mild taste. It helps break down wood and leaves, returning nutrients to the soil. The mushroom's cap is a smooth brown, and it thrives in woodland environments, making it an ideal addition to a food forest.

Zone 6 Plan Eight: Abundant Arbour

This array of plantings is arranged to mimic a natural arbour, where each layer supports abundant growth. Its purpose is to deliver a constant supply of food while reinforcing a self-sustaining system.

1. CANOPY LAYER:

Ironwood (*Carpinus caroliniana*): The Ironwood tree produces small, edible nuts that are rich in fats and nutrients. This tree is slow-growing and prefers moist, well-drained soils but can tolerate a range of conditions. Its tough, durable wood and unique leaves make it a beautiful addition to the food forest's canopy, and it provides shade for other layers.

American Hornbeam (*Carpinus caroliniana*): Known for its edible seeds, the American Hornbeam is a small to medium-sized tree that thrives in moist, well-drained soils. Its dense canopy provides shade and shelter, making it a good companion to other food forest plants. The tree's beautiful blue-gray bark and striking fall color provide additional aesthetic value.

2. UNDERSTORY LAYER:

Saskatoon Berry (*Amelanchier alnifolia*): Saskatoon Berry shrubs produce small, sweet berries that can be eaten fresh or used in pies, jams, and other treats. This plant grows well in partial shade and moist, well-drained soils. It provides food for wildlife, attracts pollinators, and offers a bountiful harvest in the summer.

Fuzzy Kiwi (*Actinidia chinensis*): Fuzzy Kiwi vines produce sweet, edible kiwis that grow on vines, providing a delicious fruit harvest. They thrive in well-drained soils and partial shade. Kiwi vines offer a vertical component to the understory and attract pollinators while producing highly nutritious fruit.

3. SHRUB LAYER:

Black Chokeberry (*Aronia melanocarpa*): This shrub produces dark purple berries that are extremely high in antioxidants. The berries are edible but are typically consumed in jams, juices, or syrups. Black chokeberry is a hardy plant that thrives in various soil types and is beneficial to pollinators.

Red Huckleberry (*Vaccinium parvifolium*): A rare shrub that produces small, bright red berries that are slightly tart and often used in pies, jams, and syrups. This shrub thrives in well-drained, acidic soils and partial to full shade. It is a great addition to a food forest due to its ability to grow in difficult, shaded conditions.

4. HERBACEOUS LAYER:

Lemon Verbena (*Aloysia citrodora*): This herb is known for its aromatic, lemon-scented leaves, which are edible and often used to flavor teas and desserts. Lemon verbena thrives in full sun and well-drained soils, making it a fragrant and flavorful addition to a food forest, while also attracting beneficial insects.

Tarragon (*Artemisia dracunculus*): Tarragon is a perennial herb with a distinct, slightly anise-like flavor. The leaves are edible and used in cooking, particularly in French cuisine. It prefers well-drained soils and full sun and can provide an aromatic contribution to the food forest while supporting biodiversity through its fragrant foliage.

5. ROOT LAYER:

Japanese Knotweed (*Fallopia japonica*): Japanese Knotweed is an invasive species in some areas, but its young shoots and roots are edible, with a tart, rhubarb-like flavor. It grows in moist, well-drained soils and is highly resilient. If managed carefully, it can be a unique and sustainable crop for food forests.

Burdock (*Arctium lappa*): Burdock's large, edible

root is often used in Japanese cuisine and herbal medicine. The plant grows well in moist, well-drained soils and full sun. Burdock's deep taproot helps improve soil structure by breaking up compacted soil layers.

6. GROUNDCOVER LAYER:

Mazus (*Mazus reptans*): Mazus is a fast-growing groundcover that produces small purple or white flowers in spring. It's not commonly edible but is highly effective for erosion control and covering bare soil. Its creeping habit helps retain soil moisture and suppress weeds.

Creeping Raspberry (*Rubus calycinoides*): Creeping Raspberry is a ground-hugging, drought-tolerant groundcover with small, tart edible berries. It forms a dense mat that prevents soil erosion and provides excellent ground coverage in well-drained soils. Its berries are similar to traditional raspberries but smaller.

7. VERTICAL LAYER:

Perennial Pea (*Lathyrus latifolius*): This nitrogen-fixing vine adds vertical growth with beautiful pink or white flowers. While not edible for humans, it enriches the soil and provides nectar for bees and butterflies.

Snake Gourd (*Trichosanthes cucumerina*): An exotic climbing plant that produces long, edible fruits. The young gourds can be eaten like zucchini, and the vine adds vertical interest with its sprawling growth.

8. FUNGAL LAYER:

Velvet Foot Mushroom (*Flammulina velutipes*): Known for its soft, velvety stems and smooth golden caps, this mushroom grows on decaying wood, especially in colder seasons. It has a mild, slightly sweet flavor and contributes to the decomposition of organic material in the forest. It can also help maintain soil moisture.

Hen of the Woods (*Grifola umbellata*): This rare mushroom is an edible polypore that grows at the base of hardwood trees. Its dense clusters of fronds contribute to decomposing wood and creating rich organic material. Known for its slightly earthy and savory flavor, it provides soil health benefits while also offering edible yields.

Zone 6 Plan Nine: Plant Power Plan

Emphasizing the strength that comes from thoughtful layering, this plan pairs edibles to create maximum plant power. It supports immediate nutritional needs and builds a resilient ecological network.

1.CANOPY LAYER:

American Hornbeam (*Carpinus caroliniana*): Known for its edible seeds, the American Hornbeam is a small to medium-sized tree that thrives in moist, well-drained soils. Its dense canopy provides shade and shelter, making it a good companion to other food forest plants. The tree's beautiful blue-gray bark and striking fall color provide additional aesthetic value.

Hackberry (*Celtis occidentalis*): The Hackberry tree produces small, sweet, edible drupes that are enjoyed by both humans and wildlife. The tree is adaptable to various soil types, including dry, rocky soils, and it grows well in full sun to partial shade. Its fast growth and dense canopy provide valuable shade and shelter for other plants in the food forest.

2. UNDERSTORY LAYER:

Sea Buckthorn (*Hippophae rhamnoides*): Sea Buckthorn produces small, orange berries that are edible and high in vitamin C. The plant grows well in partial shade and well-drained soils. Its berries can be used to make juices or jams, and it provides food for birds and wildlife while improving soil health with its nitrogen-fixing properties.

Haskap Berry (*Lonicera caerulea*): Haskap Berry bushes produce elongated, sweet, and tart berries that are rich in antioxidants. They grow well in partial shade and moist, well-drained soils. Haskap berries provide a unique early-season food source for both humans and wildlife, with its blossoms attracting pollinators.

3. SHRUB LAYER:

Japanese Wineberry (*Rubus phoenicolasius*): The Japanese wineberry produces small, red to purple berries with a tart-sweet flavor. These berries are edible and perfect for jams, pies, or fresh eating. The shrub is hardy and fast-growing, making it an excellent ground cover or berry producer in a food forest.

Fuchsia (*Fuchsia spp.*): Known for its attractive, pendulous flowers, the fuchsia shrub produces small, edible berries. These berries are sweet and can be eaten raw or used in cooking. Fuchsia grows well in partial to full shade and thrives in slightly acidic, well-drained soils. It's a perfect choice for the understory layer.

4. HERBACEOUS LAYER:

Yarrow (*Achillea millefolium*): Yarrow is a hardy herb with edible leaves and flowers that can be used in teas or salads. Known for its medicinal properties, it thrives in well-drained soils and full sun, making it a valuable addition to a food forest both for its medicinal use and its ability to attract pollinators.

Borage (*Borago officinalis*): Borage is a herbaceous plant that produces beautiful blue flowers, which are edible and often used in salads or drinks. The leaves and flowers have a mild cucumber taste and are known to attract pollinators. It grows best in full sun and moist, well-drained soils, making it an excellent companion plant in a food forest.

5. ROOT LAYER:

Shallots (*Allium cepa var. aggregatum*): Shallots produce small, edible bulbs with a mild, onion-like flavor. They thrive in well-drained soils and full sun, and their roots help improve soil structure. Shallots are a great addition to a food forest for their culinary value and their ability to enhance the soil.

Ginger (*Zingiber officinale*): Ginger's rhizomes (roots) are widely used in cooking and medicinal applications. It thrives in warm, moist soils and partial shade, making it suitable for the food forest's understory. Ginger contributes both culinary and medicinal value to the food forest.

6. GROUNDCOVER LAYER:

Ground Plum (*Astragalus crassicarpus*): Ground Plum is a low-growing legume that fixes nitrogen and produces edible seed pods resembling small plums. It is a rare plant that supports soil health while providing forage for wildlife. The pods are edible when young and tender.

Sea Thrift (*Armeria maritima*): Sea Thrift is a small, tufted groundcover that grows well in poor, sandy soils. While it is not typically considered edible, it adds color and structure to the ground layer with its pink, globe-like flowers. It is drought-tolerant and helps prevent soil erosion in dry areas.

7. VERTICAL LAYER:

Sweet Autumn Clematis (*Clematis terniflora*): An aggressive climber with sweetly scented, small white flowers that bloom in late summer. While primarily ornamental, it provides nectar for pollinators and is hardy in Zone 6.

Hyacinth Bean (*Lablab purpureus*): This fast-growing vine produces edible beans when cooked properly. It offers stunning purple flowers and pods, adding vertical beauty and supporting nitrogen fixation.

8. FUNGAL LAYER:

Pear-shaped Puffball (*Lycoperdon pyriforme*): This edible mushroom is found in clusters on decaying logs, helping to decompose wood into rich humus. The Pear-shaped Puffball is best eaten when young, with a mild, nutty flavor. It enhances soil fertility and contributes to maintaining the forest's nutrient balance.

Shrimp of the Woods (*Entoloma abortivum*): This fascinating edible mushroom is created through a parasitic relationship between fungi. It helps maintain a healthy fungal ecosystem and grows well in decaying wood and leaf litter. The mushroom has a firm, shrimp-like texture, with a mild, pleasant flavor when cooked.

Zone 6 Plan Ten: Eco-Edible Expedition

Embark on an expedition into sustainable gardening with this eco-edible plan that layers life at every level. It is designed to yield hearty harvests while charting a course for long-term environmental success.

1. CANOPY LAYER:

American Elm (*Ulmus americana*): The American Elm produces small, edible seeds, though they are not as commonly consumed. This tree grows best in moist, well-drained soils and is valued for its large, spreading canopy, which provides excellent shade. It supports a range of wildlife and is an important part of the ecosystem in temperate zones.

Black Ash (*Fraxinus nigra*): The Black Ash tree produces small seeds that are edible, though they must be processed before eating. It thrives in wet, poorly drained soils and provides dense shade. The tree's strong wood and ability to grow in wetlands make it a great addition to the food forest's canopy layer.

2. UNDERSTORY LAYER:

Mulberry (*Morus nigra*): Morus Nigra, or Black Mulberry, produces dark purple, sweet, edible berries that can be eaten fresh or used in jams and baked goods. The tree thrives in partial shade and well-drained soils. It provides valuable fruit for wildlife and offers a food source for both humans and birds.

Creeping Raspberry (*Rubus idaeus*): Creeping Raspberry is a low-growing shrub that produces small, sweet berries that are perfect for snacking or making preserves. It thrives in partial shade and moist soils. This shrub is an excellent groundcover in the understory and provides food for pollinators and small wildlife.

3. SHRUB LAYER:

Elderflower (*Sambucus nigra*): Known for its large white flowers, the elderflower shrub produces small, dark berries that are edible when cooked. The flowers are used to make elderflower syrup and can also be used in teas. This shrub thrives in moist soils and partial sun, and is an excellent choice for a food forest.

Mountain Holly (*Ilex mucronata*): Mountain holly is a small shrub that produces bright red berries that are edible, though they should be eaten in moderation. This shrub thrives in moist, well-drained soils and partial shade, and it provides shelter and food for wildlife.

4. HERBACEOUS LAYER:

Arugula (*Eruca vesicaria*): Arugula is an edible herb known for its peppery leaves, which can be used fresh in salads, sandwiches, or pestos. This herb grows well in cool weather and can thrive in both full sun and partial shade, making it a great choice for early-season harvests in a food forest.

Chervil (*Anthriscus cerefolium*): Chervil is a delicate herb often used in culinary dishes, especially in French cuisine. Its leaves are edible and have a mild, parsley-like flavor with a touch of anise. Chervil prefers moist, well-drained soils and partial sun, offering both flavor and texture to the herbaceous layer in the food forest.

5. ROOT LAYER:

Black Garlic (*Allium sativum*): Black garlic is made by fermenting garlic bulbs, creating a rich, sweet flavor. The bulbs and roots of garlic are edible, and this plant thrives in full sun and well-drained soils. Black garlic's garlic relatives, such as elephant garlic, can also be included for their large, edible roots.

Fiddlehead Fern (*Matteuccia struthiopteris*): Fiddlehead ferns are edible in the early spring stage, with tightly coiled fronds that can be sautéed or pickled. This perennial plant thrives in shaded, moist environments, making it ideal for the shaded understory in a food forest. The roots help keep the soil moist and support biodiversity.

6. GROUNDCOVER LAYER:

Ajuga (*Ajuga reptans*): Ajuga, also known as Bugleweed, is a low-growing groundcover with blue flowers. Although not widely edible, it plays an important role in controlling soil erosion and filling in bare patches in shaded or semi-shaded areas. Its rapid growth provides excellent coverage in forest settings.

Wood Betony (*Stachys officinalis*): Wood Betony is a clumping perennial groundcover with edible leaves used in teas and salads. It thrives in woodland settings and adds value as a pollinator attractor while offering herbal benefits. It also supports soil health by improving organic matter in the forest floor.

7. VERTICAL LAYER:

Scarlet Bean (*Phaseolus coccineus*): Known for its striking red flowers, this climbing bean produces edible pods and seeds. It is a heavy nitrogen-fixer and attracts hummingbirds and other pollinators.

Jícama (*Pachyrhizus erosus*): This climbing vine produces edible tubers that are crisp and sweet. It thrives in warm climates, and both the roots and the plant structure provide benefits to the ecosystem, improving soil health.

8. FUNGAL LAYER:

Rosy Veincap (*Rhodotus palmatus*): While not typically grown for consumption, this rare and beautiful mushroom adds aesthetic value to the forest. It grows on decaying wood and aids in the decomposition process, enriching soil health. The unique, pink-veined caps make it a striking addition to the fungal layer.

Parasol Mushroom (*Macrolepiota procera*): Known for its large, umbrella-like cap, the Parasol Mushroom is an edible species with a delicate, nutty flavor. It grows well in open, grassy areas of the forest, where it breaks down organic material and contributes to nutrient cycling in the soil.

ZONE 7

Plant Hardiness Zone 7

Zone 7 represents a temperate climate with moderate winters and warm summers, offering an ideal environment for a wide range of perennial plants, trees, shrubs, and vegetables. The flexibility of this zone makes it suitable for creating diverse food forests, as the average annual minimum temperature ranges from 0°F to 10°F (-17.8°C to -12.2°C). This climate provides the opportunity to grow a variety of crops while accommodating both cool and warm season plants.

Locations within Zone 7

Zone 7 is commonly found across coastal and inland regions, encompassing elevations from sea level to mountain foothills. It covers areas from the Pacific Northwest to the Appalachian Mountains, and includes parts of the Southeastern U.S., where milder winters and extended growing seasons are typical.

Alabama: Northern areas, such as Huntsville and Muscle Shoals.

Arkansas: Central and northern regions, including Little Rock and Fayetteville.

Arizona: Higher elevations, like Prescott and Sedona.

California: Inland parts of the Central Valley and mountain regions like Fresno.

Georgia: Northern areas, including Atlanta and Athens.

Kentucky: Southern regions, including Bowling Green and Somerset.

Mississippi: Northern parts, such as Tupelo.

New Mexico: Higher elevations, like Albuquerque and Taos.

North Carolina: Central and western regions, such as Raleigh, Charlotte, and Asheville.

South Carolina: Northwest regions, such as Greenville and Spartanburg.

Texas: Northern and central areas, including Dallas, Austin, and parts of Houston.

Additional areas in other states, such as high elevations in Utah, Washington, and Nevada, also fall into Zone 7.

Zone 7 Climate

Winter Temperatures: Winter minimum temperatures range from 0°F to 10°F (-17.8°C to -12.2°C). Although Zone 7 experiences freezing temperatures, they are not severe enough to significantly hinder plant growth.

Summer Conditions: Summers in Zone 7 are typically warm, with temperatures ranging from 70°F to 90°F (21°C to 32°C), supporting a wide range of plants.

Frost-Free Period: The frost-free growing season typically lasts from 180 to 210 days, allowing ample time for both cool-season and warm-season crops.

Rainfall: Annual rainfall varies from 30 to 50 inches (76 to 127 cm), with more concentrated rain in spring and summer months.

Soil Types in Zone 7

The soils in Zone 7 vary greatly due to the geographic diversity, with common types including:

Loamy Soil: Ideal for most plants, offering both good drainage and moisture retention.

Clay Soil: While challenging due to drainage

issues, clay soil can be improved with compost to support a range of plants.

Sandy Soil: Found in coastal and foothill areas, this soil type drains quickly and may require frequent watering and amendments.

Growing Season in Zone 7

The growing season in Zone 7 spans 180 to 210 days, with frost dates between mid-April and mid-October. This long season supports a mix of cool-season crops like lettuce and kale, and warm-season crops like tomatoes and peppers.

Microclimates and Wind

Regions within Zone 7 may experience variations in climate due to wind exposure and topography:

Wind Exposure: Open or elevated areas may be subject to strong winds, which can be mitigated with windbreaks.

Microclimates: Certain areas, such as south-facing slopes or sheltered spots, may provide warmer conditions or extend the growing season, allowing for the cultivation of crops suited to warmer zones.

Challenges in Zone 7

Although Zone 7 provides excellent growing conditions, there are some challenges:

Winter Protection: Occasional frosts can harm tender plants, so mulching and row covers are helpful for protection.

Pests: Warmer temperatures may attract pests like aphids and beetles, requiring organic pest control strategies.

Drought: Some areas, especially those with sandy soils or in the Southwest, may experience summer droughts. Implementing drip irrigation and mulching helps conserve moisture.

Maximizing Potential

With its long growing season, mild winters, and sufficient rainfall, Zone 7 offers excellent opportunities for food forests. Managing microclimates, addressing occasional frosts and droughts, and implementing water-efficient practices will ensure a productive and resilient ecosystem.

Zone 7 Plan One: Sunny Sustain Strategy

Bathed in abundant sunshine, this plan leverages warm conditions to create a productive, layered food forest. It balances immediate edible output with strategies for ongoing sustainability.

1. CANOPY LAYER:

Japanese Raisin Tree (*Hovenia dulcis*): This tree produces raisin-like fruit that can be eaten fresh or dried. It is a fast-growing tree that also provides shade, attracting beneficial insects and improving the soil with its leaf litter.

Pawpaw (*Asimina triloba*): Known for its creamy, custard-like fruit with a tropical flavor, the pawpaw thrives in shaded understories but can also grow in full sun. Its large leaves help capture and retain moisture in the soil.

2. UNDERSTORY LAYER:

American Persimmon (*Diospyros virginiana*): This native understory tree produces small, orange fruit that is edible when fully ripened and has a rich, sweet flavor. Persimmons support wildlife, as birds and mammals are attracted to the fruit, and they improve biodiversity in food forests. The tree grows well in USDA Zone 7 and prefers well-drained soils, thriving in partial shade. It also offers great adaptability to different soil types, making it a resilient choice.

Strawberry Tree (*Arbutus unedo*): Also known as the "Irish Strawberry Tree," this evergreen shrub or small tree produces small, round, red berries that are edible but mild in flavor, often used in jams or fermented into alcohol. The tree thrives in well-drained soils and can tolerate drought conditions once established. It also has stunning white or pink flowers that attract pollinators, contributing to ecosystem diversity in the food forest.

3. SHRUB LAYER:

Silver Buffaloberry (*Shepherdia argentea*): Silver Buffaloberry produces bright red edible berries high in vitamin C, making them excellent for jams or jellies. It thrives in poor, dry soils and helps fix nitrogen, improving soil fertility for surrounding plants. This tough, drought-tolerant shrub is well-suited for erosion control and wildlife habitat.

Beach Plum (*Prunus maritima*): The small, tart fruits of the Beach Plum are used in jams, jellies, and even wines. It grows in coastal or sandy soils and is tolerant of poor, dry conditions. Its dense, low-growing habit makes it excellent for hedgerows or windbreaks.

4. HERBACEOUS LAYER:

Meadow Rue (*Thalictrum dasycarpum*): Meadow Rue is a tall, graceful herbaceous plant that adds structure to the food forest with its feathery foliage and airy blooms. Though not commonly eaten, the young shoots are sometimes foraged as a vegetable. It thrives in moist, shaded areas and attracts pollinators, offering habitat support.

Ramps (*Allium tricoccum*): Ramps, also known as wild leeks, are prized for their edible bulbs and leaves, which have a strong onion-garlic flavor. They grow in rich, moist soils in the understory and are one of the first edible greens to emerge in spring. Their strong flavor and seasonal availability make them a rare and sought-after delicacy.

5. ROOT LAYER:

Chinese Yam (*Dioscorea polystachya*): Chinese Yam produces starchy, edible tubers that can be used similarly to potatoes or sweet potatoes. It grows vigorously in Zone 7, preferring well-drained soils and partial shade. This vine can also support vertical layers by climbing, and its tubers are prized for their high nutritional content and medicinal uses.

Yam Daisy (*Microseris lanceolata*): Yam Daisy, also known as Murnong, is a rare edible root plant traditionally used by Indigenous Australians. It thrives in well-drained soils and full sun. The roots are sweet and nutty, making them a unique and nutritious addition to a food forest while also benefiting pollinators with its yellow, daisy-like flowers.

6.GROUNDCOVER LAYER:

Mossy Stonecrop (*Sedum acre*): Mossy Stonecrop is a low-growing succulent with tiny, star-shaped yellow flowers and fleshy, edible leaves that are best consumed in small quantities. It thrives in well-drained soils and full sun, making it perfect for dry, rocky areas of a food forest. Its ability to tolerate drought and poor soils adds resilience to the groundcover layer while supporting pollinators with its flowers.

Sea Kale (*Crambe maritima*): Sea Kale is a rare, perennial groundcover with edible leaves, stems, and flowers. It prefers well-drained, sandy soils and full sun, making it ideal for coastal or dry environments in Zone 7. The young shoots are harvested like asparagus, and its attractive blue-green foliage adds beauty to the landscape while providing an edible crop.

7. VERTICAL LAYER:

American Bittersweet (*Celastrus scandens*): A vigorous vine that thrives in Zone 7, providing vertical interest and attracting pollinators with its clusters of white flowers.

Hardy Kiwi (*Actinidia arguta*): Hardy Kiwi is a smaller, smooth-skinned relative of the traditional kiwifruit, producing sweet, bite-sized fruits. It grows well in full sun to partial shade, thriving in well-drained soils. The entire plant is hardy to colder climates, making it suitable for Zone 7, and its small, grape-like fruits offer a unique edible crop for the vertical layer. Hardy Kiwi is also beneficial for supporting pollinators with its fragrant flowers.

8. FUNGAL LAYER:

Pine Mushrooms (*Tricholoma matsutake*): Pine Mushrooms are a rare and highly prized edible fungi found growing near pine trees in forests. They are known for their distinctive spicy aroma and rich, earthy flavor, which makes them a delicacy. These mushrooms thrive in well-drained soils under specific conditions, usually in symbiotic relationships with pine trees. Their mycorrhizal association with the trees helps improve soil health, while the mushrooms themselves provide a unique, flavorful addition to the food forest's fungal layer.

Hen of the Woods (*Grifola frondosa*): Hen of the Woods, also known as Maitake, is a medicinal and edible mushroom that grows at the base of hardwood trees like oaks. It produces a large cluster of ruffled, edible caps that have a rich, earthy flavor. This fungus helps break down organic matter in the soil and supports overall soil health. It is known for its immune-boosting properties and provides a nutritious, sustainable food source for the food forest.

Zone 7 Plan Two: Warmth & Wellness Way

This design capitalizes on the warmth of its locale to foster both food production and ecosystem wellness. Its layered approach ensures that every level supports health and vitality.

1. CANOPY LAYER:

Butternut (*Juglans cinerea*): This nut tree produces edible butternuts, which have a rich, buttery flavor. Butternuts are a source of valuable food for wildlife and humans alike, and the tree has a broad canopy that provides shelter and cooling in the summer.

Hardy Pecan (*Carya illinoinensis*): This tree produces edible pecans, which are high in healthy fats and proteins. It can help with soil stabilization due to its deep root system and provides significant habitat for wildlife.

2. UNDERSTORY LAYER:

Fragrant Snowbell (*Styrax obassia*): A rare understory tree that produces fragrant white flowers in late spring. While not widely known for its edibility, the flowers can be used to infuse flavor in syrups or teas. This plant grows best in partial shade and moist, well-drained soils. It brings beauty and a light fragrance to the food forest, attracting pollinators with its sweet-smelling blooms.

Japanese Raisin Tree (*Hovenia dulcis*): The "raisins" of this tree are actually edible, sweet fruit stalks that form after the flowers. It prefers full sun to partial shade and thrives in well-drained soils in Zone 7. Its unique fruiting habit and drought tolerance make it a rare but intriguing addition to the understory of a food forest, offering an alternative source of sweetness and novelty.

3. SHRUB LAYER:

Indigo Bush (*Amorpha fruticosa*): Indigo Bush is

a nitrogen-fixing shrub that supports soil health and fertility. It produces purple flowers that attract pollinators, and while not commonly eaten, its seeds are edible when cooked. It thrives in moist soils and can be planted along water edges for erosion control.

Sweetshrub (*Calycanthus floridus*): Sweetshrub, also called Carolina Allspice, has aromatic, deep red flowers that emit a fruity scent. The seeds are considered edible in small amounts, but the plant is mainly grown for its fragrance and as a pollinator attractor. It prefers partial shade and moist soils.

4. HERBACEOUS LAYER:

Sweet Cicely (*Myrrhis odorata*): Sweet Cicely has sweet, anise-flavored leaves, seeds, and roots that can be used in cooking, particularly for flavoring desserts. It thrives in partial shade and moist soils, helping to stabilize soil and support pollinators with its delicate white flowers.

Stinging Nettle (*Urtica dioica*): Stinging Nettle is a nutrient-dense plant used for edible greens when cooked, neutralizing its sting. The leaves are rich in vitamins and minerals and can be used in soups, teas, or smoothies. It grows in moist, nitrogen-rich soils and benefits the food forest by attracting beneficial insects.

5. ROOT LAYER:

Earthnut Pea (*Lathyrus tuberosus*): Earthnut Pea produces small, edible tubers with a nutty flavor. It is a nitrogen-fixing legume, benefiting the soil by improving fertility, and it grows well in a variety of soils, including poorer ones. This plant also produces delicate pink flowers, adding aesthetic value to the food forest.

Groundnut (*Apios americana*): Groundnut produces edible tubers that have been an important food source for centuries. It grows in moist areas and can spread vigorously, providing nitrogen to the soil and supporting biodiversity with its legume-like properties. The plant also has attractive, fragrant flowers that appeal to pollinators.

6. GROUNDCOVER LAYER:

Salad Burnet (*Sanguisorba minor*): Salad Burnet is a hardy, drought-tolerant groundcover with edible leaves that have a cucumber-like flavor. It thrives in well-drained soils and full sun to partial shade, making it an excellent choice for erosion control on slopes. The leaves can be harvested year-round, providing a fresh, green addition to salads and herb mixtures.

Strawberry Spinach (*Blitum capitatum*): Strawberry Spinach is a unique edible groundcover that produces both leafy greens and small red berries. It thrives in moist, well-drained soils and partial shade, making it versatile for different areas of the food forest. The leaves can be used like spinach, and the berries, while not as sweet as cultivated strawberries, add a colorful, edible element to the ground layer.

7. VERTICAL LAYER:

Grapes (*Vitis vinifera*): Grapes are well-known for their edible fruit, which thrives in full sun and well-drained soils. The vines can be trained along trellises or arbors, creating both aesthetic appeal and high yields. In addition to fruit, the leaves can be used in cooking, such as for wrapping stuffed dishes. Grapes also improve air circulation in the food forest and attract pollinators.

Climbing Fleeceflower (*Fallopia aubertii*): Climbing Fleeceflower is a fast-growing, edible climbing vine that produces small, edible berries that are rich in antioxidants. It thrives in moist, well-drained soils and partial shade to full sun. The plant's climbing nature makes it an excellent candidate for covering trellises or fences, and its edible berries can be harvested in late summer. It's an excellent choice for providing both food and habitat for beneficial insects.

8. FUNGAL LAYER:

Reishi Mushroom (*Ganoderma lucidum*): Reishi mushrooms are known for their medicinal properties and are highly regarded in traditional Chinese medicine for their potential to boost immune health and reduce stress. These mushrooms grow on decaying hardwood, often near the base of trees, and are not typically eaten raw but are used in teas or supplements. While it is not typically consumed for its taste, its contribution to health and the food forest ecosystem through its role in decomposing organic matter and enriching the soil makes it a valuable fungal species.

Maitake Mushroom (*Grifola frondosa*): Maitake, also called Hen of the Woods, is an edible mushroom that grows in clusters at the base of oak trees, among other hardwoods. It offers both culinary and medicinal benefits, with its rich, savory flavor making it a favorite in cooking, while also having immune-boosting properties. This fungus breaks down organic matter, contributing to soil health and supporting the surrounding plant life. Maitake thrives in nutrient-rich, moist environments, making it suitable for food forest applications.

Zone 7 Plan Three: Solar Synergy

Harnessing the power of the sun, this plan organizes plant layers in a way that each contributes to a synergistic whole. It's all about blending nutrition with a naturally sustainable design.

1. CANOPY LAYER:

Heartnut (*Juglans ailantifolia var. cordiformis*): This Japanese walnut variety produces heart-shaped nuts that are edible and delicious. The tree is hardy and provides shade, contributing to the microclimate of the food forest.

Ginkgo (*Ginkgo biloba*): The seeds of the female ginkgo tree are edible when cooked. This ancient species is known for its resilience to pests and diseases and helps in creating a beneficial ecosystem.

2. UNDERSTORY LAYER:

Silverbell (*Halesia tetraptera*): Known for its showy white bell-shaped flowers, the Silverbell also produces seeds that can be eaten when roasted, though they are not commonly harvested for food. This tree grows well in partial shade and provides excellent springtime visual interest, contributing to the aesthetic and ecological diversity of the food forest.

Mexican Plum (*Prunus mexicana*): This understory tree produces small, tart plums that are great for making jams, jellies, and preserves. It thrives in full sun to partial shade and tolerates a range of soil conditions, making it a versatile option for Zone 7 food forests. The tree also provides early spring flowers that attract pollinators, enhancing ecosystem health.

3. SHRUB LAYER:

Buffalo Currant (*Ribes aureum*): This hardy shrub produces clusters of small, edible black or yellow berries that are great for fresh eating, preserves, or baking. The plant is highly adaptable to different soil types and can tolerate drought, making it ideal for low-maintenance gardens.

Hog Peanut (*Amphicarpaea bracteata*): Hog Peanut is an unusual vining shrub that produces small, edible underground tubers and seeds. It thrives in shady, woodland environments and can help stabilize soil. This rare plant adds unique edibility and nitrogen-fixing benefits to the forest garden.

4. HERBACEOUS LAYER:

Blue Flag Iris (*Iris versicolor*): Blue Flag Iris is an attractive plant for wet areas, producing striking blue-purple flowers. Its rhizomes are sometimes used medicinally, though caution is needed due to toxicity. It thrives in wet, boggy areas and can help with water filtration in a food forest setting.

Wild Sarsaparilla (*Aralia nudicaulis*): Wild Sarsaparilla produces edible, small black berries and a root used as a flavoring similar to root beer. It prefers shady, forested areas and spreads via underground rhizomes, making it a useful ground cover plant for preventing soil erosion.

5. ROOT LAYER:

Golden Root (*Rhodiola rosea*): Golden Root, also known as Arctic Root or Rose Root, produces edible roots known for their adaptogenic properties, used to reduce stress and fatigue. It thrives in cold, rocky environments and is suited for well-drained soils in Zone 7. Its bright yellow flowers attract beneficial insects, making it a valuable addition to the food forest.

Camas (*Camassia quamash*): Camas produces edible bulbs that were historically a staple food for Native Americans. It prefers moist, well-drained soils and full to partial sunlight. The plant produces striking blue flowers that attract pollinators, and its nutritious bulbs are harvested after flowering for roasting or baking

6. GROUNDCOVER LAYER:

Cranberry (*Vaccinium macrocarpon*): Cranberry is a creeping groundcover known for its tart, edible berries. It thrives in acidic, moist soils and full sun, making it an excellent addition to wet areas of the food forest. Cranberries provide a nutritious, antioxidant-rich crop, while the plant's dense growth habit helps prevent soil erosion and adds visual interest.

Prostrate Rosemary (*Rosmarinus officinalis 'Prostratus'*): Prostrate Rosemary is a low-growing, aromatic herb with edible leaves used in cooking. It thrives in well-drained soils and full sun, adding both culinary value and drought tolerance to the groundcover layer. Its fragrant, needle-like leaves are used to flavor dishes, and its small blue flowers attract pollinators.

7. VERTICAL LAYER:

Chilean Guava (*Ugni molinae*): Chilean Guava is a rare evergreen vine that produces aromatic, sweet-tasting berries. It thrives in well-drained soils and partial sun, making it ideal for the vertical layer of a food forest. The berries are edible and can be eaten fresh or used in preserves, providing a flavorful addition to the landscape. The plant's dense, evergreen foliage also offers year-round cover and support for pollinators.

Passionfruit (*Passiflora edulis*): Passionfruit is a tropical vine that produces large, fragrant, edible fruits. It requires full sun and well-drained, fertile soils, making it ideal for vertical growth on a trellis. The fruits are filled with aromatic pulp that is rich in vitamins and antioxidants. In addition to its edible fruits, the plant's striking flowers attract pollinators and add visual interest to the food forest.

8. FUNGAL LAYER:

Chanterelle (*Cantharellus cibarius*): Chanterelles are a highly sought-after edible mushroom known for their distinctive golden color and delicate, fruity flavor. They grow symbiotically with trees, particularly oak and conifers, in well-drained, acidic soils. Their mycorrhizal relationship with plants enhances nutrient cycling in the food forest while providing an exceptional culinary ingredient. Chanterelles are not only a delicious and valuable addition but also help improve soil health by decomposing organic material.

Boletus edulis (King Bolete) (*Boletus edulis*): Also known as the Porcini mushroom, the King Bolete is an edible and prized species in culinary circles. It grows in association with trees such as pines, spruces, and oaks, and thrives in well-drained, acidic soils. The large, meaty cap of the King Bolete has a rich, nutty flavor and is perfect for cooking. By aiding in the decomposition of organic matter and enhancing soil quality, it supports the growth of other food forest plants.

Zone 7 Plan Four: Bright Bounty Blueprint

A blueprint for a radiant harvest, this plan layers sun-loving plants to produce a bright bounty. It's designed to generate plentiful edibles while reinforcing the system's eco-friendly backbone.

1. CANOPY LAYER:

European Chestnut (*Castanea sativa*): This chestnut species produces edible nuts that are rich in carbohydrates. The tree is a great addition for wildlife, offering both food and shelter, while its leaves improve soil fertility.

Empress Tree (*Paulownia tomentosa*): This fast-growing tree produces large, fragrant flowers that attract pollinators. Though its wood is valuable, the tree itself also provides shade and nitrogen-rich leaf litter to the forest floor.

2. UNDERSTORY LAYER:

Dwarf Pomegranate (*Punica granatum var. nana*): This smaller version of the pomegranate tree produces edible fruit, though smaller than regular pomegranates. It grows well in partial shade to full sun and prefers well-drained soils. The bright red flowers attract pollinators, and its compact size makes it ideal for food forests where space is a premium.

Buffalo Berry (*Shepherdia argentea*): This nitrogen-fixing shrub or small tree produces edible berries that are rich in vitamins and antioxidants, commonly used for jams or medicinal purposes. It thrives in USDA Zone 7, tolerating poor soils and drought conditions, making it a hardy addition to the understory layer. It also supports biodiversity by attracting birds and beneficial insects.

3. SHRUB LAYER:

American Bladdernut (*Staphylea trifolia*): American Bladdernut produces small edible seeds inside papery pods that can be eaten raw or cooked. It grows in moist, rich soils and provides shelter and food for wildlife. Its rapid growth and dense foliage make it useful for creating natural screens.

Silky Dogwood (*Cornus amomum*): Silky Dogwood produces small, blue fruits that are edible and attract a wide variety of birds. It thrives in wet conditions, making it a great choice for rain gardens or riparian plantings. The shrub also helps with erosion control along water banks.

4. HERBACEOUS LAYER:

Solomon's Seal (*Polygonatum biflorum*): Solomon's Seal produces edible shoots that are often eaten like asparagus when young, though the berries are toxic. This plant thrives in shaded areas with rich soil, adding texture with its graceful arching stems. It also provides wildlife habitat and stabilizes soil.

Woodland Lettuce (*Lactuca floridana*): Woodland Lettuce is an uncommon edible green that thrives in shaded woodland environments. Its leaves can be used in salads or cooked like spinach. It adds diversity to a food forest by providing a unique, early-season crop and supporting local wildlife.

5. ROOT LAYER:

Skirret (*Sium sisarum*): Skirret is a rare perennial root vegetable with sweet, white roots that can be roasted or boiled. It grows best in moist, well-drained soils and partial shade. Skirret is easy to grow and low maintenance, making it a great addition to the root layer while supporting pollinators with its umbrella-shaped flowers.

Salsify (*Tragopogon porrifolius*): Salsify, also known as Oyster Plant due to its oyster-like flavor, produces long, edible roots that are commonly used in soups and stews. It prefers well-drained soils and sunny conditions. This biennial plant also adds value to the food forest with its striking purple flowers and pollinator appeal.

6. GROUNDCOVER LAYER:

Lamb's Lettuce (*Valerianella locusta*): Lamb's Lettuce, also known as Mâche or Corn Salad, is a hardy, cold-tolerant groundcover with tender, edible leaves. It thrives in moist, well-drained soils and partial shade, providing a fresh salad green during cooler seasons. Its rosette-like growth habit makes it a great low-maintenance addition to the forest floor.

Good King Henry (*Blitum bonus-henricus*): Good King Henry is a rare, perennial groundcover with edible leaves, shoots, and seeds. It grows well in well-drained soils and full sun to partial shade, producing early spring shoots that can be harvested like asparagus. The leaves can be used like spinach, and its ability to fix nitrogen helps improve soil fertility.

7. VERTICAL LAYER:

Yam Bean (*Pachyrhizus erosus*): Also known as Jicama, Yam Bean is a climbing legume that produces large, edible tubers that are crisp and mildly sweet. It grows well in full sun and well-drained soils, with vines that can reach up to 10 feet in length. The tubers are harvested in the fall, providing a valuable and unique addition to the food forest. The vines also help fix nitrogen in the soil, benefiting other plants.

Arctic Kiwi (*Actinidia kolomikta*): Arctic Kiwi is a hardy kiwi variety that produces small, smooth-skinned, sweet fruits similar to the larger kiwifruit but more tolerant of cold climates. It thrives in full sun and well-drained soils, making it perfect for Zone 7 vertical growth. The fruit is edible, providing a refreshing snack or addition to dishes, and the plant's attractive foliage adds visual interest throughout the year.

8. FUNGAL LAYER:

Lion's Mane Mushroom (*Hericium erinaceus*):

Lion's Mane is a unique edible mushroom known for its white, hair-like appearance, resembling a lion's mane. It grows on hardwood trees, primarily on dead or decaying wood, and is prized for its medicinal benefits, particularly its neuroprotective properties. It has a mild, seafood-like flavor, making it an interesting culinary addition. This mushroom also contributes to the food forest by aiding in decomposing wood and enriching soil health.

Turkey Tail Mushroom (*Trametes versicolor*): Turkey Tail is a medicinal and edible mushroom that often grows on decaying wood and fallen logs. It is known for its colorful, fan-shaped caps and its potential health benefits, particularly for immune support. While it is not commonly consumed for taste, its role in breaking down organic matter and supporting plant health in the food forest makes it a valuable fungal species. The mushroom's high antioxidant content also adds an extra dimension to the forest's ecosystem.

Zone 7 Plan Five: Radiant Roots Regimen

Centered on deep-rooted sustainability, this regimen ensures that every plant layer draws strength from the sun's energy. It produces a steady supply of food and builds resilience from the ground up.

1. CANOPY LAYER:

Korean Pine (*Pinus koraiensis*): This pine tree produces large, edible pine nuts. It's a hardy species that offers both food and habitat for wildlife while stabilizing soil with its strong root system.

Sweet Chestnut (*Castanea mollissima*): The sweet chestnut tree produces flavorful nuts that are a popular food source. It's a great tree for adding biodiversity and creating a balanced ecosystem in the canopy layer.

2. UNDERSTORY LAYER:

Mayhaw (*Crataegus aestivalis*): The Mayhaw tree produces small, tart fruits that are traditionally used in jellies and sauces. This tree thrives in moist soils and partial shade, making it a good candidate for a riparian food forest environment. Its blossoms also attract pollinators, while the fruit feeds birds and other wildlife.

Cornelian Cherry (*Cornus mas*): A hardy, small tree producing edible red fruits that are often used for jams, syrups, and wines. Cornelian cherry thrives in partial shade and well-drained soils, making it an excellent fit for Zone 7. Its yellow early-spring flowers provide one of the first sources of pollen for bees, supporting early-season pollinators.

3. SHRUB LAYER:

Gooseberry (*Ribes uva-crispa*): Gooseberry is a hardy shrub that produces tart, vitamin C-rich berries used in pies, jellies, and wines. It grows in a variety of soil conditions, including poor soils, and prefers partial shade. Its thorny branches make it useful for forming natural barriers.

Japanese Bayberry (*Morella rubra*): Japanese Bayberry, also known as Chinese Bayberry, produces sweet, edible berries that can be consumed fresh or processed into jams. This plant prefers moist, well-drained soils and has a unique ornamental value with its dark green, glossy leaves.

4. HERBACEOUS LAYER:

Golden Alexanders (*Zizia aurea*): Golden Alexanders are a native herbaceous plant with edible leaves and seeds, traditionally used in salads or as a spice. It thrives in moist soils and partial shade, producing clusters of bright yellow flowers that attract pollinators and beneficial insects.

Ostrich Fern (*Matteuccia struthiopteris*): Ostrich Fern is best known for its edible fiddleheads, which are harvested in early spring and have a mild, nutty flavor. It grows in moist, shady environments and is perfect for adding a lush, ferny texture to the food forest. The plant also helps stabilize soils along streams.

5. ROOT LAYER:

Alexanders (*Smyrnium olusatrum*): Alexanders is a rare edible plant with a history of use in ancient Roman cuisine. Its roots, stems, and leaves are all edible, though the roots are most often used in cooking. Alexanders grows in a wide range of soils, including coastal and chalky environments, and supports pollinators with its umbrella-like yellow flowers.

Chufa (*Cyperus esculentus*): Chufa, also known as Tiger Nut, produces small, edible tubers that are rich in fiber and healthy fats. It thrives in moist, well-drained soils and full sun, growing rapidly once established. The plant also supports biodiversity, providing habitat and food for beneficial insects.

6. GROUNDCOVER LAYER:

Creeping Jenny (*Lysimachia nummularia*): This versatile ground cover thrives in moist, shaded areas, providing a lush, green carpet that helps suppress weeds and protect soil.

Wild Strawberry (*Fragaria vesca*): A low-growing ground cover that produces sweet, tiny strawberries, which attract pollinators and birds. It's an excellent option for filling spaces while providing food.

7. VERTICAL LAYER:

Clematis Virgin's Bower (*Clematis virginiana*): A native climbing vine that produces white, fragrant flowers and is a favorite of pollinators like bees and butterflies. It adds vertical interest and works well as a ground cover or trellis plant.

American Climbing Bittersweet (*Celastrus scandens*): A vigorous vine with bright orange berries that attract birds. It climbs well and provides a stunning vertical presence with its colorful autumn foliage.

8. FUNGAL LAYER:

Lion's Mane Mushroom (*Hericium erinaceus*): This medicinal mushroom improves soil health by breaking down organic matter and enhancing soil fertility. It also has neuroprotective properties, making it a beneficial addition to the food forest ecosystem.

Shiitake Mushroom (*Lentinula edodes*): A popular edible mushroom that thrives on logs, adding both culinary value and soil enrichment by breaking down wood and organic material.

Zone 7 Plan Six: Luminous Layer Layout

This plan features a luminous layout where each tier is carefully arranged to capture sunlight and support growth. It delivers a layered system of edibles that shine with both flavor and sustainability.

1. CANOPY LAYER:

Japanese Pagoda Tree (*Styphnolobium japonicum*): Known for its nitrogen-fixing ability, this tree improves soil fertility and supports nearby plants. The flowers are highly attractive to pollinators, and its seeds are sometimes used in traditional medicine.

Monkey Puzzle Tree (*Araucaria araucana*): This rare tree produces edible seeds similar to pine nuts. It is slow-growing and can live for hundreds of years, adding a unique, ancient presence to the food forest.

2. UNDERSTORY LAYER:

Dwarf Mulberry (*Morus nigra*): This compact version of the mulberry tree produces dark, sweet berries that are edible and highly nutritious, rich in vitamins and antioxidants. It thrives in partial shade, tolerating a range of soil conditions in Zone 7. The tree attracts a variety of wildlife, from birds to pollinators, contributing to a diverse food forest ecosystem.

Hop Tree (*Ptelea trifoliata*): Also known as the wafer ash, this small tree produces seeds that can be used as a hop substitute in brewing. It grows well in partial shade and dry to moist soils, making it adaptable to a variety of conditions in USDA Zone 7. The tree's fragrant flowers attract pollinators, particularly bees, supporting the ecological balance of the food forest.

3. SHRUB LAYER:

Sand Cherry (*Prunus pumila*): Sand Cherry is a hardy, drought-tolerant shrub that produces dark purple edible fruits, often used in pies, preserves, or fresh. It grows well in sandy, poor soils, making it ideal for erosion control or dry areas. Its showy white flowers also attract pollinators.

Farkleberry (*Vaccinium arboreum*): Farkleberry is a close relative of the blueberry and produces small, edible berries with a sweet, tart flavor. It tolerates poor soils and drought, making it suitable for dry, sandy areas. Its glossy leaves and delicate white flowers add beauty to any food forest.

4. HERBACEOUS LAYER:

Valerian (*Valeriana officinalis*): Valerian is known for its medicinal root, which is used as a sleep aid, though its leaves can be used as an herb. It grows well in moist soils with partial shade and produces fragrant, white flowers that attract pollinators. Its deep roots also help break up compacted soils.

Cutleaf Coneflower (*Rudbeckia laciniata*): Cutleaf Coneflower produces tall, striking yellow flowers and is sometimes referred to as Green-headed Coneflower. While the young shoots are occasionally eaten, this plant is mostly grown for its pollinator-friendly flowers and its ability to thrive in moist, shaded areas.

5. ROOT LAYER:

Perennial Leek (*Allium ampeloprasum var. porrum*): Perennial Leek produces edible bulbs and leaves similar to cultivated leeks but with a more robust growth habit. It thrives in well-drained soils and sunny areas, providing a steady, reliable crop. The plant is also beneficial for soil health and helps deter pests from other crops.

Roselle (*Hibiscus sabdariffa*): Roselle is primarily grown for its edible calyces used to make tea, but its roots are also edible and have medicinal uses. It grows well in warm climates, preferring full sun and well-drained soils. This plant also produces stunning red flowers that attract pollinators, adding ornamental value to the food forest.

6. GROUNDCOVER LAYER:

Houttuynia (*Houttuynia cordata*): Houttuynia, also known as Fish Mint, is a creeping groundcover with edible leaves that have a strong, unique flavor used in Southeast Asian cuisine. It thrives in moist, partially shaded areas and spreads rapidly, making it useful for covering large areas. Its leaves are commonly used fresh in salads or cooked as a vegetable.

Creeping Raspberry (*Rubus pentalobus*): Creeping Raspberry is a hardy groundcover with edible, tangy berries. It thrives in well-drained soils and full sun to partial shade, forming dense mats that help control erosion. In addition to its fruit, it offers attractive foliage that turns a bronze color in the fall, adding aesthetic value to the forest floor.

7. VERTICAL LAYER:

Kiwi Berry (*Actinidia arguta*): Kiwi Berry is a small, smooth-skinned kiwi fruit that grows on a hardy, climbing vine. It thrives in full sun to partial shade and requires well-drained, fertile soils. The small fruits are rich in vitamin C and fiber, making them a nutritious addition to the food forest. Its compact size and vertical growth habit make it suitable for smaller spaces while providing a high yield of edible fruit.

Cucamelon (*Melothria scabra*): Cucamelon, also known as Mexican Sour Gherkin, is a small, climbing vine that produces tiny, cucumber-like fruits with a refreshing, sour taste. It thrives in full sun and well-drained soils, growing well on trellises or fences. The fruits are edible and make an excellent snack or addition to salads. Its compact size makes it a good fit for smaller food forests or vertical layers.

8. FUNGAL LAYER:

Shaggy Mane Mushroom (*Coprinus comatus*): Shaggy Mane is an edible mushroom known for its long, shaggy white cap and delicate flavor. It typically grows in grassy areas and on decaying organic matter, including decomposing wood. Shaggy Mane's rapid growth and ability to break down organic material contribute to soil health in the food forest. The mushroom is edible when young and tender, but should be consumed immediately after harvesting due to its quick spoilage.

Coral Mushroom (*Ramaria formosa*): Coral Mushrooms are a unique, brightly colored edible fungus that grows in clusters resembling underwater coral. These mushrooms thrive in moist, well-drained soil and often appear at the base of conifers or hardwoods. Their delicate, nutty flavor makes them a choice edible, and they help decompose organic matter, enriching the forest's soil. Coral mushrooms also have potential medicinal uses, particularly for their antioxidant content.

Zone 7 Plan Seven: Sun-Kissed Cultivation

Celebrating the gentle caress of sunbeams, this cultivation method pairs plants that thrive under warm conditions. Its layered design nurtures both abundant food production and a supportive ecosystem.

1. CANOPY LAYER:

English Walnut (*Juglans regia*): Producing large, edible walnuts, this tree is valued for its food production and timber. It's a hardy species that provides plenty of shade and habitat for wildlife.

Kentucky Coffeetree (*Gymnocladus dioicus*): Though the seeds require roasting before they can be consumed as a coffee substitute, this tree offers both an edible product and habitat for birds and insects. It also tolerates a range of soil conditions.

2. UNDERSTORY LAYER:

Carolina Silverbell (*Halesia carolina*): This small understory tree produces white, bell-shaped flowers in spring, and while not typically harvested for food, the seeds are said to be edible when roasted. The tree thrives in partial shade and moist, well-drained soils. Its attractive flowers and adaptability make it a great choice for enhancing the beauty and biodiversity of a food forest.

Clove Currant (*Ribes odoratum*): A unique shrub that produces small, edible black currants with a clove-like fragrance, often used in jams and pies. The plant thrives in partial shade and well-drained soils, making it a versatile addition to the understory layer of Zone 7 food forests. It is also highly attractive to pollinators and wildlife.

3. SHRUB LAYER:

Buffaloberry (*Shepherdia canadensis*): Buffaloberry is a nitrogen-fixing shrub that produces small, tart red berries that can be eaten fresh or made into preserves. The plant is highly tolerant of poor, dry soils and can grow in rocky or sandy areas, making it excellent for difficult planting sites.

New Jersey Tea (*Ceanothus americanus*): New Jersey Tea is a nitrogen-fixing shrub with fragrant white flowers that attract pollinators. The leaves can be dried and used as a tea substitute, making this a unique edible addition to the garden. It thrives in well-drained soils and can tolerate drought once established.

4. HERBACEOUS LAYER:

Nodding Onion (*Allium cernuum*): Nodding Onion has edible bulbs and leaves with a mild onion flavor, making it a great addition to a food forest. It grows in clumps and thrives in well-drained soils, attracting pollinators with its pinkish-purple, nodding flower heads. This plant is drought-tolerant once established.

Sweet Woodruff (*Galium odoratum*): Sweet Woodruff is an herbaceous ground cover with fragrant, vanilla-scented leaves, which are sometimes used to flavor beverages like May wine. It thrives in shady, moist areas and helps suppress weeds in the food forest with its dense growth habit.

5. ROOT LAYER:

Parsnip (*Pastinaca sativa*): Parsnip produces edible roots that are sweet and nutty after exposure to frost. It thrives in well-drained soils and full sun to partial shade. Parsnip adds deep root structure to the food forest, helping to aerate the soil and improve water infiltration while providing a nutritious vegetable for winter harvests.

Burdock (*Arctium lappa*): Burdock produces large, edible taproots that are commonly used in soups, stews, and medicinal remedies. It grows well in disturbed soils and full sun to partial shade. The plant also has burrs that attach to animals, aiding seed dispersal, and is valued for its deep roots, which improve soil health.

6. GROUNDCOVER LAYER:

Cape Gooseberry (*Physalis peruviana*): Cape Gooseberry, also known as Goldenberry, is a creeping plant that produces small, edible fruits encased in papery husks. It grows well in well-drained soils and full sun, adding both a unique fruit crop and ornamental interest to the groundcover layer. The berries are rich in vitamins and can be eaten fresh or used in jams and desserts.

Alpine Strawberry (*Fragaria vesca var. semperflorens*): Alpine Strawberry is a low-growing, perennial groundcover that produces small, sweet, edible berries. It thrives in well-drained soils and partial shade, making it a great addition to the forest floor. Unlike traditional strawberries, Alpine Strawberries are everbearing, producing fruit throughout the growing season.

7. VERTICAL LAYER:

Malabar Spinach (*Basella alba*): Malabar Spinach is a fast-growing, heat-loving vine that produces edible, fleshy leaves used in cooking. It thrives in warm climates with full sun and moist, well-drained soils. The leaves have a mild, spinach-like flavor and are rich in vitamins A and C, making them a nutritious and versatile food source. The vine's rapid growth and ability to shade other plants make it valuable in the vertical layer.

Red-Fleshed Kiwi (*Actinidia melanandra*): Red-Fleshed Kiwi is a rare kiwi variety known for its striking red flesh and sweet, tropical flavor. The vines thrive in well-drained soils and full sun, producing fruit that is both visually appealing and rich in nutrients. The fruit is edible, offering a unique flavor and texture, while the vine's attractive foliage provides year-round beauty and support for pollinators.

8. FUNGAL LAYER:

Sparassis crispa (Cauliflower Mushroom) (*Sparassis crispa*): The Cauliflower Mushroom is an edible fungus that grows in clumps resembling a head of cauliflower. It thrives on decaying wood and is particularly common under conifer trees. The mushroom has a firm texture and a mild, nutty flavor, making it a delicious addition to meals. Its role in breaking down organic material in the food forest contributes to nutrient cycling and soil health.

Black Trumpet Mushroom (*Craterellus cornucopioides*): The Black Trumpet is an edible, trumpet-shaped mushroom that thrives in moist, shaded areas under oak, beech, or conifer trees. Known for its rich, smoky flavor, this mushroom is a delicacy in culinary dishes. It plays an essential role in breaking down organic matter, promoting soil health, and enhancing the nutrient content of the food forest. Black Trumpet mushrooms are also a valuable source of antioxidants.

Zone 7 Plan Eight: Vivid Verdure Venture

Embarking on a venture into vibrant greenery, this plan layers species to create a vivid tapestry of edibles. It focuses on harnessing warmth to provide nutritious yields while maintaining ecological balance.

1. CANOPY LAYER:

Persian Walnut (*Juglans regia*): This species produces edible walnuts with a thinner shell compared to other varieties. It's excellent for food production and provides a wide canopy that fosters biodiversity.

Silk Tree (*Albizia julibrissin*): Known for its beautiful pink flowers, the silk tree attracts pollinators like bees and butterflies. Its flowers are sometimes used in traditional herbal medicine, and it helps with nitrogen fixation.

2. UNDERSTORY LAYER:

Russian Olive (*Elaeagnus angustifolia*): This nitrogen-fixing shrub or small tree produces small, olive-like fruits that are edible and often used in preserves. It grows well in dry, sandy soils and tolerates drought, making it highly resilient. The tree also supports wildlife and soil fertility through nitrogen fixation, enhancing the overall health of the food forest.

Black Haw (*Viburnum prunifolium*): This small tree produces edible, sweet fruits that can be eaten raw or used in preserves. It thrives in partial shade and tolerates a variety of soil conditions, making it an adaptable choice for Zone 7. Its flowers attract pollinators, and the fruits are a valuable food source for birds and other wildlife.

3. SHRUB LAYER:

Wild Raisin (*Viburnum nudum*): Wild Raisin produces clusters of edible, dark blue berries that can be eaten fresh or dried like raisins. It thrives in wet soils and is perfect for planting in rain gardens or areas prone to flooding. The shrub also provides habitat for birds and beneficial insects.

Northern Spicebush (*Lindera benzoin*): Northern Spicebush produces aromatic red berries used as a spice in cooking. This shade-loving shrub thrives in moist, rich soils and is a valuable food source for pollinators. Its leaves and berries are fragrant, adding sensory appeal to the food forest.

4. HERBACEOUS LAYER:

Pale Indian Plantain (*Arnoglossum atriplicifolium*): Pale Indian Plantain is an uncommon herbaceous plant that produces large, edible roots traditionally foraged by Native Americans. It grows in sunny to partially shaded areas with moist, well-drained soils and helps attract butterflies and other pollinators.

Toothwort (*Cardamine concatenata*): Toothwort is a rare woodland herbaceous plant with edible, spicy-tasting roots, often foraged as a horseradish substitute. It prefers rich, moist soils in shaded environments and supports early spring pollinators with its white to pinkish flowers.

5. ROOT LAYER:

Japanese Knotweed (*Fallopia japonica*): Though considered invasive in some areas, Japanese Knotweed has edible shoots and roots, which can be used in cooking similar to rhubarb. It grows vigorously in a variety of soils and conditions, stabilizing stream banks and preventing soil erosion. Careful management is required to keep it in check in a food forest.

Daylily (*Hemerocallis fulva*): Daylily produces edible tubers, flowers, and shoots, making it a versatile addition to the root layer. It thrives in a range of soils and sun conditions and spreads quickly. Daylilies also attract pollinators with their vibrant, showy blooms, enhancing the aesthetic and ecological diversity of the food forest.

6. GROUNDCOVER LAYER:

Turkish Rocket (*Bunias orientalis*): Turkish Rocket is a rare, perennial groundcover with edible leaves, shoots, and flowers. It thrives in well-drained soils and full sun, producing early spring shoots that can be harvested like broccoli raab. The plant's deep roots help improve soil structure, while its ability to tolerate drought makes it a resilient addition to the food forest.

Wood Sorrel (*Oxalis acetosella*): Wood Sorrel is a shade-tolerant groundcover with edible leaves that have a tangy, lemon-like flavor. It thrives in moist, well-drained soils and is often found in forested areas. The leaves can be eaten fresh in salads or used as a garnish, while its delicate flowers add beauty to the forest floor.

7. VERTICAL LAYER:

Climbing Sorrel (*Oxalis tuberosa*): Climbing Sorrel is a vine that produces edible tubers, which can be harvested for their tangy, lemon-flavored taste. It thrives in full sun and well-drained soils, and the plant's vertical growth makes it an excellent choice for smaller spaces. The tubers are eaten boiled or roasted, providing a unique culinary addition to the food forest. The plant also adds nitrogen to the soil, benefiting surrounding plants.

Wineberry (*Rubus phoenicolasius*): Wineberry is a vigorous, edible climbing vine known for its sweet, tart berries. It thrives in well-drained, fertile soils and full sun, producing large quantities of fruit throughout the summer. The berries can be eaten fresh or made into preserves, and the plant's dense growth offers good cover for wildlife and pollinators. Wineberry is also valuable for attracting birds to the food forest.

8. FUNGAL LAYER:

Lobster Mushroom (*Hypomyces lactifluorum*): The Lobster Mushroom is a parasitic fungus that transforms other mushrooms into a brightly colored, orange-red, edible species with a texture resembling lobster meat. It thrives in rich, moist, well-drained soils, particularly in coniferous forests. This mushroom offers a savory, slightly seafood-like flavor and helps decompose other organic matter, improving soil health. It provides a unique culinary experience and enhances the diversity of edible fungi in the food forest.

Slippery Jack (*Suillus luteus*): Slippery Jack is an edible mushroom that grows in association with pine trees, producing a slimy cap when young. It is known for its smooth texture and mild flavor when cooked. The mushroom thrives in nutrient-rich, acidic soils and contributes to the nutrient cycling process by breaking down organic matter. While not as well-known as other edible mushrooms, it provides valuable nutrition and supports soil health in the food forest.

Zone 7 Plan Nine: Dappled Delight Design

Inspired by the dappled light filtering through canopies, this design pairs plants to maximize both shade and sun. It aims to create an environment where food production and sustainability delight in equal measure.

1. CANOPY LAYER:

Italian Stone Pine (*Pinus pinea*): This Mediterranean pine produces large, edible pine nuts, which are a gourmet delicacy. The tree is tolerant of dry conditions and supports soil stability.

Hazelnut (*Corylus colurna*): This nut tree produces edible hazelnuts that are high in protein and healthy fats. The tree also provides excellent cover for wildlife and contributes to soil fertility with its leaf litter.

2. UNDERSTORY LAYER:

Sweetbay Magnolia (*Magnolia virginiana*): Known for its fragrant white flowers, this tree also produces small red seeds that can be eaten by wildlife. While not commonly consumed by humans, its leaves and flowers are sometimes used for flavoring. It thrives in wet, acidic soils and partial shade, making it a good option for moist food forest sites in Zone 7.

Jujube (*Ziziphus jujuba*): This small tree produces edible fruits that are similar to dates in texture and flavor, often eaten fresh, dried, or used in teas. It grows well in USDA Zone 7, tolerating drought and a variety of soil types. Jujube is a rare but highly productive choice for the understory layer, offering a unique and nutritious fruit.

3. SHRUB LAYER:

Blue Elderberry (*Sambucus nigra subsp. cerulea*): Blue Elderberry produces clusters of small, blue edible berries that are rich in antioxidants and used in syrups, wines, and preserves. This drought-tolerant shrub can grow in poor soils and helps attract pollinators with its white flower clusters.

Silverberry (*Elaeagnus commutata*): Silverberry is a nitrogen-fixing shrub that produces edible, silver-colored berries, often used in jams or jellies. It tolerates a wide range of soil types, including poor and dry soils. Its silvery foliage adds ornamental interest to the food forest.

4. HERBACEOUS LAYER:

Caraway (*Carum carvi*): Caraway is an edible herb known for its flavorful seeds used in cooking and baking. It prefers sunny to partially shaded areas with well-drained soils. Caraway also attracts beneficial insects to the food forest and can be used as a medicinal herb for digestive health.

Lovage (*Levisticum officinale*): Lovage is a perennial herb with a celery-like flavor, and its leaves, stems, and seeds are all edible. It grows in rich, well-drained soils and can reach impressive heights, adding vertical interest to the herbaceous layer. Lovage is also a great plant for attracting pollinators.

5. ROOT LAYER:

Parsley Root (*Petroselinum crispum var. tuberosum*): Parsley Root is a rare variety of parsley that produces edible, carrot-like roots. It grows well in sunny to partially shaded areas with well-drained soil. The roots can be roasted, boiled, or used in soups, while the leaves offer a second edible component as a fresh herb.

Horseradish (*Armoracia rusticana*): Horseradish is known for its pungent, edible root, which is often grated and used as a condiment. It thrives in well-drained, fertile soils and full sun, spreading easily via underground runners. In addition to its culinary uses, horseradish is a hardy plant that can help deter pests in the food forest.

6. GROUND COVER LAYER:

Wintergreen (*Gaultheria procumbens*): Wintergreen is a low-growing, evergreen groundcover with edible leaves and berries. It thrives in acidic, well-drained soils and partial shade, making it an excellent choice for the forest floor. The berries have a minty flavor and can be eaten fresh or used to make tea, while the leaves contain medicinal properties.

Miner's Lettuce (*Claytonia perfoliata*): Miner's Lettuce is a shade-tolerant groundcover with tender, edible leaves that are high in vitamins and minerals. It thrives in moist, well-drained soils and is often one of the first greens to emerge in spring. The leaves can be eaten fresh in salads or cooked like spinach, providing a nutritious, early-season crop.

7. VERTICAL LAYER:

Climbing Fleeceflower (*Fallopia aubertii*): Climbing Fleeceflower is a fast-growing, edible climbing vine that produces small, edible berries that are rich in antioxidants. It thrives in moist, well-drained soils and partial shade to full sun. The plant's climbing nature makes it an excellent candidate for covering trellises or fences, and its edible berries can be harvested in late summer. It's an excellent choice for providing both food and habitat for beneficial insects.

Dutchman's Pipe (*Aristolochia macrophylla*): A fast-growing vine that thrives in shaded environments, this plant attracts pollinators with its unique pipe-shaped flowers and offers habitat for wildlife by providing dense cover.

8. FUNGAL LAYER:

Hen of the Woods (*Grifola frondosa*): A large, cluster-forming mushroom that grows at the base of oak and other hardwood trees. It is edible and prized for its meaty texture, commonly used in soups and stir-fries. Hen of the Woods also contributes to the forest's nutrient cycle by decomposing dead wood.

Bolete Mushroom (*Boletus edulis*): Known for its rich flavor and meaty texture, this edible mushroom is found in symbiosis with trees, especially pines and oaks. Bolete mushrooms help in nutrient cycling by breaking down organic material, and they are highly valued in culinary dishes for their robust taste.

Zone 7 Plan Ten: Harvest Horizon Hub

This hub is strategically designed to extend your harvest's horizon by layering sunlit edibles. It not only produces nutritious food but also establishes a foundation for long-term ecosystem resilience.

1. CANOPY LAYER:

Jujube (*Ziziphus jujuba*): Also called Chinese date, the jujube produces small, sweet fruits that can be eaten fresh or dried. The tree is drought-tolerant and supports pollinators with its nectar-rich flowers.

Arctic Kiwi Tree (*Actinidia kolomikta*): This cold-hardy species produces small, edible fruits similar to kiwis. Its unique variegated leaves add ornamental value while also attracting pollinators to the garden.

2. UNDERSTORY LAYER:

Blue Elderberry (*Sambucus cerulea*): Producing small, edible blue-black berries that are used in jams, wines, and medicinal syrups, the Blue Elderberry thrives in moist soils and partial shade. It is an important plant for supporting wildlife and attracts a range of pollinators. Its berries, rich in antioxidants and vitamins, are a valuable food source for both humans and animals.

Silk Tree (*Albizia julibrissin*): While known more for its beautiful pink flowers, the Silk Tree's seed pods are edible when young, though not widely used in modern cuisine. The tree grows well in USDA Zone 7, tolerating a range of soil types and thriving in full sun to partial shade. It provides pollinator support and enhances the aesthetic appeal of the food forest.

3. SHRUB LAYER:

Sweet Gale (*Myrica gale*): Sweet Gale is a nitrogen-fixing shrub that produces fragrant leaves and is sometimes used in brewing beer. While not commonly consumed, the plant provides habitat for wildlife and helps improve soil health. It thrives in wet soils, making it ideal for riparian zones.

Evergreen Huckleberry (*Vaccinium ovatum*): Evergreen Huckleberry produces small, black edible berries that are high in vitamin C and antioxidants. The plant grows well in partial shade and acidic soils, making it perfect for woodland areas. Its evergreen leaves add year-round beauty to the food forest.

4. HERBACEOUS LAYER:

Bitterroot (*Lewisia rediviva*): Bitterroot is a rare, low-growing herb with an edible, starchy root traditionally used as a food source. It thrives in dry, well-drained soils and adds diversity to the herbaceous layer of the food forest. Its showy pink or white flowers attract pollinators in spring.

Great Angelica (*Angelica atropurpurea*): Great Angelica is a tall, herbaceous plant with edible stems and roots often candied or used in herbal remedies. It prefers moist, shaded environments and is known for its large, umbrella-shaped flower clusters that attract pollinators and beneficial insects.

5. ROOT LAYER:

Celeriac (*Apium graveolens var. rapaceum*): Celeriac, or Celery Root, is grown for its knobby, edible root, which has a mild celery flavor. It prefers well-drained soils and full sun to partial shade, making it suitable for a variety of food forest environments. The roots can be roasted, mashed, or used in soups and stews.

Scorzonera (*Scorzonera hispanica*): Scorzonera, also known as Black Salsify, produces long, edible roots with a mild, nutty flavor. It grows well in well-drained soils and full sun, adding a unique root crop to the food forest. Its roots can be boiled, mashed, or used in soups, while its flowers attract pollinators.

6. GROUNDCOVER LAYER:

Clover Fern (*Marsilea quadrifolia*): Clover Fern, also

known as Water Clover, is a low-growing, edible fern that thrives in moist, well-drained soils and partial shade. Its leaves resemble clover and are commonly used as a vegetable in parts of Asia. Clover Fern adds a unique, aquatic element to the food forest and helps improve soil moisture retention.

Wild Leek (*Allium tricoccum var. canadense*): Wild Leek, also known as Canada Onion, is a native groundcover with edible leaves and bulbs. It thrives in moist, shaded areas, producing early spring greens that are often foraged for their onion-like flavor. This plant adds biodiversity to the forest floor and supports soil health through its allium family properties.

7. VERTICAL LAYER:

Cucurbita maxima (Winter Squash) (*Cucurbita maxima*): Winter Squash is a sprawling vine that produces edible fruits that are often harvested in the fall. It thrives in full sun and well-drained soils, with vines reaching long lengths. The fruits, including varieties such as butternut and hubbard squash, are nutrient-rich and can be stored for long periods. The vines' leaves also provide ground cover and shade for lower layers of the food forest.

Luffa Gourd (*Luffa aegyptiaca*): Luffa Gourd is a climbing vine that produces long, edible gourds when young, which can later be dried for use as sponges. It grows well in full sun and well-drained soils, making it an ideal candidate for the vertical layer of a food forest. The young fruit can be eaten when tender, providing a mild, cucumber-like flavor. The vine also helps shade the soil and other plants beneath it.

8. FUNGAL LAYER:

Lobster Mushroom (*Hypomyces lactifluorum*): An interesting fungus that parasitizes other mushrooms, turning them bright red-orange, it is edible and often used in cooking for its seafood-like flavor. Lobster mushrooms help with nutrient recycling in forest systems by transforming other fungi into edible forms.

Shiitake Mushroom (*Lentinula edodes*): Shiitake mushrooms are edible, widely known for their savory flavor and medicinal properties, often cultivated on logs or stumps. They play a significant role in forest health by breaking down decaying wood, enriching the soil, and boosting biodiversity in the ecosystem.

ZONE 8

Plant Hardiness Zone 8

Zone 8 is characterized by a relatively mild climate with moderate winters and warm summers, making it ideal for a wide variety of plants. The average annual minimum temperature in Zone 8 ranges from 10°F to 20°F (-12°C to -6°C). The growing season is typically long, with the first frost date around mid-November and the last frost date around early April, allowing ample opportunity for a diverse range of crops.

Locations within Zone 8

Zone 8 spans a broad area of the United States, offering a temperate climate that supports a variety of plants.

Locations within this zone include:

Alabama: Northern regions, including cities like Birmingham and Huntsville.

Arkansas: Central and southern parts, including Little Rock and Hot Springs.

California: Coastal areas such as parts of the San Francisco Bay Area, the Central Valley, and foothill regions.

Colorado: Areas along the Front Range, including Denver, Boulder, and Colorado Springs.

Florida: Northern parts, including Jacksonville and Tallahassee.

Georgia: Central and northern regions, such as Atlanta, Macon, and Augusta.

Idaho: Southwestern regions, including Boise.

Kansas: Eastern parts, particularly around Topeka and Wichita.

Kentucky: Central and eastern regions, including

Lexington and Louisville.

Louisiana: Northern and central areas, such as Baton Rouge and Shreveport.

Mississippi: Northern parts, including Jackson and Oxford.

Missouri: Southern and southwestern regions, such as Springfield and Joplin.

Montana: Western portions, including areas like Missoula.

New Mexico: Central and northern regions, including Santa Fe and Albuquerque.

Nevada: Western parts, including near Reno and Carson City.

North Carolina: Central and western regions, such as Raleigh, Charlotte, and Asheville.

Oklahoma: Central areas, including Oklahoma City and Norman.

Oregon: Western areas, including the Willamette Valley, Portland, and Salem.

South Carolina: Western regions, including Greenville, Spartanburg, and Anderson.

Tennessee: Central and northern parts, including Nashville, Chattanooga, and Knoxville.

Texas: Much of central Texas, including cities like Austin, San Antonio, and Dallas.

Utah: Lower elevations in the western and southern parts, including Salt Lake City and St. George.

Washington: Western regions, including Seattle, Olympia, and parts of the Puget Sound.

Wyoming: Southwestern parts, such as Jackson and nearby areas.

Climate Characteristics of Zone 8

Wind and Weather: Zone 8 can experience occasional droughts and heat stress, especially in inland areas during the summer. Coastal regions may encounter storms and high winds, particularly during hurricane season. Winter temperatures rarely dip below freezing, though occasional frosts may occur.

Temperature & Seasons: Winters in Zone 8 are cool, but freezing temperatures are rare. Summer temperatures can range from mild to hot, with daytime highs reaching into the 80s and 90s°F (27-35°C), while cooler evenings help promote plant growth.

Rainfall and Humidity: Zone 8 receives moderate rainfall, ranging from 30-50 inches (76-127 cm) per year, although this can vary by region. Areas closer to coastal regions may experience higher humidity, while inland areas may have drier conditions.

Frost-Free Period: With a frost-free period spanning 180-210 days, Zone 8 offers an extended growing season for both warm- and cool-season crops.

Soil Types in Zone 8

Soils in Zone 8 vary widely, from sandy loams to clay-rich types. Fertility is generally good, and gardeners often amend the soil with organic matter to improve structure and drainage. Raised beds or mounded gardens can help address drainage issues in areas with clay-heavy soils.

Pest and Disease Concerns

Pests like aphids, caterpillars, and slugs are common in Zone 8, especially in areas with high humidity. Fungal diseases and root rot can also be an issue if proper drainage isn't maintained. To combat pests, gardeners should focus on creating a balanced ecosystem with beneficial insects and good soil practices.

With its long growing season, mild winters, and abundant rainfall, Zone 8 provides an excellent environment for a wide variety of plants. With attention to soil care, drainage, and pest management, gardeners can grow a thriving food forest, benefiting from a diverse range of fruits, vegetables, herbs, and flowers.

Zone 8 Plan One: Tropical Tapestry

Weaving together a vibrant array of tropical edibles, this plan layers species to form a living tapestry. Its goal is to deliver flavorful yields while establishing a resilient, multi-layered ecosystem.

1. CANOPY LAYER:

American Beech (*Fagus grandifolia*): The American Beech produces edible nuts that are high in protein, fat, and nutrients. This deciduous tree is slow-growing but offers valuable habitat and shade, making it a beneficial canopy layer plant. It thrives in well-drained, slightly acidic soils and has a long lifespan, adding stability to the food forest.

American Chestnut (*Castanea dentata*): A historically important tree, American chestnut produces edible nuts that are rich in protein and carbohydrates. It grows quickly and thrives in well-drained soils, making it an excellent choice for food forests aiming to provide long-term yields. The tree has been making a comeback due to efforts in breeding blight-resistant varieties, adding a valuable nut-producing canopy layer.

2. UNDERSTORY LAYER:

Serviceberry (*Amelanchier arborea*): Serviceberry is a small, multi-stemmed tree known for its delicious, blueberry-like fruits, which are edible and high in antioxidants. It provides early spring blossoms, beneficial for pollinators, and its berries attract birds. This plant thrives in partial shade and adds biodiversity to the understory of the food forest.

Maypop (*Passiflora incarnata*): This native passionflower vine produces large, aromatic fruits that are edible and known as maypops. It's an excellent understory addition due to its ability to climb and fill vertical spaces without overtaking other plants. It also attracts pollinators with its striking flowers.

3. SHRUB LAYER:

Golden Currant (*Ribes aureum*): Golden Currant is a hardy shrub that produces edible, sweet-tart berries, which are excellent for jams, pies, and fresh eating. The plant thrives in partial sun and is drought-tolerant, making it well-suited for a food forest. Its fragrant yellow flowers attract pollinators, and the shrub provides valuable habitat for wildlife.

Silver Buffaloberry (*Shepherdia argentea*): This nitrogen-fixing shrub produces small, edible red berries that are tart and often used in jellies or as a vitamin C-rich supplement. It thrives in poor soils and dry conditions, making it a resilient addition to the shrub layer. Its silvery foliage adds visual interest to the food forest.

4. HERBACEOUS LAYER:

Good King Henry (*Blitum bonus-henricus*): This perennial herb produces edible leaves that are spinach-like and highly nutritious. Its shoots can also be eaten like asparagus in the spring. Good King Henry thrives in partial shade and tolerates poor soils, making it a low-maintenance addition to the herbaceous layer.

Turmeric (*Curcuma longa*): The rhizomes of Turmeric are widely known for their anti-inflammatory and medicinal properties, as well as for their use in cooking as a spice. This herbaceous plant prefers moist, well-drained soils and partial shade in Zone 8, and it adds vibrant color to the food forest.

5. ROOT LAYER:

Daylily (*Hemerocallis fulva*): The tubers of the daylily are edible and have a nutty flavor when cooked, making them a great addition to the root layer. This perennial plant also provides vibrant flowers that are attractive to pollinators. Daylilies thrive in various soils and can adapt to partial shade, contributing to soil health with their fibrous roots.

Earth Chestnut (*Bunium bulbocastanum*): Known for its small, flavorful tubers, the earth chestnut is a rare perennial that can serve as an excellent edible in the root layer. It offers benefits like drought tolerance and nitrogen fixation, making it a valuable addition to a food forest. The nutty-flavored tubers are typically roasted or boiled.

6. GROUNDCOVER LAYER:

Miner's Lettuce (*Claytonia perfoliata*): This edible, low-growing groundcover is known for its mild-tasting leaves and stems, often used in salads or cooked as a green. Miner's lettuce thrives in cool, shady areas, and its high vitamin C content adds a nutritious component to the food forest. Its small white flowers also attract pollinators.

Wintergreen (*Gaultheria procumbens*): Wintergreen is a creeping groundcover that produces edible berries and leaves with a distinct minty flavor. The berries can be used in jams or teas, while the leaves are often used medicinally. This evergreen plant prefers acidic soils and partial shade, adding year-round interest to the forest floor.

7. VERTICAL LAYER:

Chayote (*Sechium edule*): Chayote is a fast-growing vine that produces edible fruits, which are often used in salads, soups, and stews. The shoots and leaves are also edible and can be cooked like spinach. This plant thrives in warm climates, making it suitable for USDA Zone 8, and provides vertical coverage while also contributing to soil health with its extensive root system.

Akebia (*Akebia quinata*): Also known as chocolate vine, Akebia produces large clusters of edible fruits with a mild, sweet flavor. This rare vine is vigorous and can quickly cover trellises or fences, making it an excellent vertical layer plant. Its fragrant flowers attract pollinators, and its fruit is often used in jams or eaten fresh.

8. FUNGAL LAYER:

Cauliflower Mushroom (*Sparassis crispa*): The cauliflower mushroom is an edible and visually striking fungus that resembles a coral-like cluster. It thrives in USDA Zone 8 and is often found growing at the base of coniferous trees. This rare mushroom has a mild, nutty flavor and is prized for its high culinary value, making it a unique addition to the fungal layer. It also helps break down dead wood, contributing to nutrient cycling in the soil.

Wood Ear Mushroom (*Auricularia auricula-judae*): The wood ear mushroom, also known as jelly ear, is an edible fungus with a chewy texture and mild flavor. It grows well in USDA Zone 8, particularly on decaying hardwood trees. This mushroom is often used in Asian cuisine and is valued for its potential medicinal benefits, including improving circulation. Its ability to decompose deadwood makes it beneficial for soil health.

Zone 8 Plan Two: Lush Luminosity Layout

Zone 8 Plan Two: Lush Luminosity Layout

Designed for regions where moisture and warmth abound, this layout brings together layers of lush, tropical life. It emphasizes a radiant mix of produce and ecological support that thrives in humid conditions.

1. CANOPY LAYER:

American Hazelnut (*Corylus americana*): This native hazelnut produces small, edible nuts that are rich in fats, making them an excellent food source. The plant is relatively hardy, adapting well to various soil types and offering dense foliage that provides shelter for wildlife. It's a beneficial addition to a food forest, contributing to both the canopy and the understory layers.

American Linden (*Tilia americana*): The American linden produces edible flowers that are fragrant and can be used to make tea. It grows best in moist, fertile soils and provides dense, heart-shaped leaves that attract pollinators. Its large canopy offers shade while enhancing biodiversity in the food forest.

2. UNDERSTORY LAYER:

American Beautyberry (*Callicarpa americana*): This deciduous shrub produces bright purple berries that are edible when cooked and often made into jellies. It thrives in partial shade, making it a perfect understory plant, and its berries are a valuable food source for wildlife. Its leaves have also been traditionally used as a natural insect repellent.

Oregon Grape (*Mahonia aquifolium*): Oregon Grape is an evergreen shrub that produces clusters of edible, tart berries resembling grapes. The berries are often used in jams or wines. It tolerates shade well, making it ideal for the understory layer, and its early yellow flowers provide nectar for pollinators.

3. SHRUB LAYER:

Salmonberry (*Rubus spectabilis*): Salmonberry is a thorny shrub that produces edible, raspberry-like fruits that are rich in vitamins and antioxidants. It prefers moist, shaded areas and is an excellent plant for stabilizing soil and supporting wildlife. The bright pink flowers attract pollinators early in the season.

Red Elderberry (*Sambucus racemosa*): Red Elderberry produces clusters of bright red berries that are edible when cooked and used in jellies or syrups. The shrub grows well in moist soils and partial shade, making it ideal for the shrub layer. It also provides food for birds and beneficial insects.

4. HERBACEOUS LAYER:

Lovage (*Levisticum officinale*): Lovage is a hardy herb whose leaves, seeds, and roots are all edible, with a flavor similar to celery. It grows tall and thrives in partial to full sun. Lovage is great for attracting beneficial insects and providing a tall, structured element to the herbaceous layer.

Salsify (*Tragopogon porrifolius*): Salsify produces edible roots with a mild, oyster-like flavor, often used in soups or roasted. The plant grows well in full sun and well-drained soils and adds diversity to the food forest. It also has beautiful purple flowers that attract pollinators.

5. ROOT LAYER:

Mashua (*Tropaeolum tuberosum*): This Andean root crop produces spicy, edible tubers that are rich in nutrients. Mashua is a vigorous grower in USDA Zone 8, with beautiful trailing vines that add diversity to the root layer. The plant has pest-repellent properties, making it an excellent companion for other crops

.Purple Prairie Clover (*Dalea purpurea*): A nitrogen-fixing plant with a deep root system that helps aerate the soil and supports pollinators with its purple flowers.

6. GROUNDCOVER LAYER:

Sweet Woodruff (*Galium odoratum*): This aromatic perennial forms a dense mat of foliage and is known for its star-shaped leaves and small white flowers. The leaves are edible when dried and often used to flavor drinks or desserts. Sweet woodruff thrives in shady, moist areas and suppresses weeds, making it a beneficial addition to the groundcover layer.

Ground Plum (*Astragalus crassicarpus*): Ground plum produces small, plum-like fruits that are edible when young, offering a unique groundcover option for USDA Zone 8. This drought-tolerant perennial thrives in poor soils and helps improve nitrogen levels in the soil, benefiting other plants in the food forest.

7. VERTICAL LAYER:

Air Potato (*Dioscorea bulbifera*): Air potato is a climbing vine that produces small, aerial tubers that are edible when cooked. This plant can grow rapidly in USDA Zone 8 and is prized for its high yield of tubers, which can be harvested and eaten as a starchy vegetable. Air potato is ideal for covering arbors or trellises while also contributing to food production.

Hablitzia (*Hablitzia tamnoides*): Known as Caucasian spinach, Hablitzia is a cold-hardy vine that produces edible leaves, often used in salads or cooked like spinach. This rare perennial vine grows well in USDA Zone 8 and can quickly cover vertical structures. It is a valuable source of early-season greens and helps improve biodiversity in the food forest.

8. FUNGAL LAYER:

Shaggy Parasol Mushroom (*Chlorophyllum rhacodes*): This rare, edible mushroom is found in USDA Zone 8, often growing in wood chips or forested areas. The shaggy parasol has a nutty, rich flavor, and it can grow in clusters, making it a great choice for the fungal layer of a food forest. This mushroom contributes to the breakdown of organic matter, improving soil fertility and structure.

Hen of the Woods (*Grifola frondosa***):** This rare, edible mushroom grows at the base of oak trees and thrives in USDA Zone 8. Maitake is prized for its medicinal properties and savory flavor, often used in stir-fries or soups. This mushroom helps decompose fallen wood, enriching the soil with organic matter and aiding in nutrient recycling.

Zone 8 Plan Three: Exuberant Eden Expedition

Set on an expedition toward a modern Eden, this plan uses paired plantings to capture tropical exuberance. It is crafted to yield abundant food and cultivate a self-sustaining, vibrant ecosystem.

1. CANOPY LAYER:

Black Cherry (*Prunus serotina*): Black cherry produces edible cherries, although they are more bitter than sweet varieties. The tree's fast growth and ability to adapt to various soil types make it an ideal candidate for a diverse food forest. Additionally, the wood of black cherry is highly valued for furniture making.

Black Walnut (*Juglans nigra*): The black walnut is a large tree that produces highly nutritious, edible nuts. It has a distinctive canopy that casts dense shade, making it suitable for layering in food forests. The tree's wood is also valuable, while its leaves and bark are known for their medicinal properties.

2. UNDERSTORY LAYER:

Spicebush (*Lindera benzoin*): Spicebush is a deciduous shrub with aromatic leaves and stems, producing small red berries that are edible and can be used in cooking or as a spice. It supports biodiversity by attracting swallowtail butterfly larvae and other wildlife. It grows well in moist, shaded environments in the understory.

Gooseberry (*Ribes uva-crispa*): Gooseberries are small, shade-tolerant shrubs that produce edible, tart fruits, often used in pies, jams, or eaten fresh. Their thorny branches can create protective habitats for small animals, and their berries attract a variety of bird species.

3. SHRUB LAYER:

Goumi (*Elaeagnus multiflora*): Goumi is a nitrogen-fixing shrub that produces small, sweet-tart, edible berries. The berries are rich in vitamins and are often eaten fresh or made into preserves. This shrub thrives in poor soils, helping to improve fertility and create a sustainable food forest ecosystem.

Black Chokeberry (*Aronia melanocarpa*): Black Chokeberry is a small, resilient shrub that produces antioxidant-rich berries, which are commonly used in juices, jams, and wines. It is highly adaptable to various soil conditions and can tolerate partial shade. Its beautiful fall foliage adds seasonal interest to the food forest.

4. HERBACEOUS LAYER:

Daylily (*Hemerocallis fulva*): The flowers, buds, and tubers of Daylilies are edible and can be stir-fried, sautéed, or added to salads. These hardy perennials thrive in a variety of soil conditions and bring bright, cheerful blooms to the food forest. They are drought-tolerant and can spread to create a natural ground cover.

Miner's Lettuce (*Claytonia perfoliata*): This small, leafy herb produces edible leaves that are high in vitamin C and are perfect for salads. Miner's Lettuce thrives in moist, shady conditions and grows as a ground cover, adding diversity and providing early spring greens in the food forest.

5. ROOT LAYER:

Camas (*Camassia quamash*): The edible bulbs of camas have a sweet, chestnut-like flavor when roasted. Historically an important food source for Native Americans, this perennial plant grows well in moist, open areas. Camas also offers beautiful blue flowers that attract pollinators.

Skirret (*Sium sisarum*): Skirret produces long, white roots that have a sweet and slightly nutty flavor, typically boiled or roasted. This perennial is a hardy addition to the root layer, thriving in moist soils and offering edible roots year after year. Its attractive foliage and clusters of white flowers bring beauty and function to the garden.

6. GROUND COVER LAYER:

Wild Ginger (*Asarum canadense*): Wild ginger has heart-shaped leaves and spreads via rhizomes, making it an effective groundcover for shady areas. The roots are aromatic and can be used as a ginger substitute in cooking. This plant also attracts beneficial insects and improves soil health with its extensive root system.

Cape Gooseberry (*Physalis peruviana*): Also known as goldenberry, this low-growing perennial produces small, edible fruits encased in papery husks. The berries are rich in vitamins and antioxidants. Cape gooseberry thrives in full sun and well-drained soils, making it a unique groundcover that also attracts pollinators.

7. VERTICAL LAYER:

Scarlet Bean (*Phaseolus coccineus*): Also known as runner bean, Scarlet Bean is a vigorous climbing vine that produces edible pods and seeds. The bright red flowers attract pollinators, making it beneficial for the surrounding plants. This vine is ideal for covering vertical structures while also providing a reliable source of protein-rich beans.

Snake Gourd (*Trichosanthes cucumerina*): Snake gourd is a fast-growing vine that produces long, edible gourds used in various cuisines. The fruits are high in nutrients, and the plant thrives in warm climates, making it suitable for USDA Zone 8. Its vigorous growth habit allows it to cover trellises, fences, or arbors, contributing both food and visual interest.

8. FUNGAL LAYER:

King Oyster Mushroom (*Pleurotus eryngii*): The

king oyster mushroom is a robust, edible fungus that thrives in USDA Zone 8. It grows on dead wood and has a firm, meaty texture, making it ideal for grilling or sautéing. King oyster mushrooms contribute to soil health by breaking down organic matter and promoting fungal biodiversity.

Honey Mushroom (*Armillaria mellea*): The honey mushroom is a parasitic fungus that grows on the roots of trees and is edible when cooked. It thrives in USDA Zone 8 and has a sweet, mild flavor, often used in soups or stews. This mushroom can help decompose dead wood, although it can also parasitize living trees, making it a dual-edged addition to the food forest.

Zone 8 Plan Four: Vibrant Vista Venture

This plan creates a panoramic view of layered, tropical abundance. It is designed to showcase both diverse edibles and a robust support system that ensures sustainability over time.

1. CANOPY LAYER:

Chinese Chestnut (*Castanea mollissima*): Known for its high-quality edible nuts, Chinese chestnut is a fast-growing tree that thrives in fertile, well-drained soils. The tree's large canopy provides shade and shelter, while its nuts are a valuable food source in the late summer. It's also resistant to the chestnut blight that affects American chestnuts.

Chinkapin Oak (*Quercus muehlenbergii*): The small, sweet acorns of the Chinkapin Oak are edible after proper processing. This tree grows well in dry, rocky soils and offers excellent wildlife habitat. Its deep roots help improve soil structure and stability, making it a valuable canopy addition.

2. UNDERSTORY LAYER:

Nannyberry (*Viburnum lentago*): Nannyberry is a large shrub or small tree that produces edible, sweet fruits in the fall, which can be eaten raw or made into preserves. It thrives in partial shade and provides food for birds and pollinators. Its dense growth habit makes it a good understory plant for creating shelter in a food forest.

Elderberry (*Sambucus canadensis*): Elderberry is a fast-growing shrub that produces clusters of small, dark berries, which are edible when cooked and are used in syrups, wines, and jams. It prefers moist, partially shaded areas, making it suitable for the understory. The plant is also highly attractive to wildlife and pollinators.

3. SHRUB LAYER:

Blueberry (*Vaccinium corymbosum*): Highbush Blueberry is a classic edible shrub known for its sweet, juicy berries that are a favorite in pies and fresh eating. It prefers acidic, well-drained soils and full sun to partial shade. Blueberries support a wide range of pollinators and wildlife.

Sea Buckthorn (*Hippophae rhamnoides*): Sea Buckthorn is a nitrogen-fixing shrub that produces bright orange, edible berries rich in vitamin C. The berries are often used in juices and skincare products. This shrub is highly tolerant of poor soils and drought, making it a valuable addition to the food forest.

4. HERBACEOUS LAYER:

Wood Sorrel (*Oxalis acetosella*): Wood Sorrel has edible leaves with a tangy, lemony flavor, often used in salads or as a garnish. It prefers moist, shady areas and can thrive in the understory of a food forest. The plant also has attractive, delicate flowers that add to the aesthetic value of the forest floor.

Mallow (*Malva neglecta*): Common Mallow produces edible leaves and flowers that are nutritious and can be used in soups or salads. The plant grows well in disturbed soils and partial shade, making it a resilient addition to the herbaceous layer. It also attracts pollinators with its beautiful flowers.

5. ROOT LAYER:

Chinese Artichoke (*Stachys affinis*): This perennial root crop is prized for its small, crunchy tubers that are often eaten raw or pickled. Chinese artichoke spreads quickly, making it ideal for ground cover in the root layer, and its dense growth helps suppress weeds. It is easy to cultivate and thrives in partial shade.

Hopniss (*Apios priceana*): Hopniss is a rare leguminous vine that produces nutrient-dense, starchy tubers. The tubers are a traditional food source and are often boiled or roasted. In addition to providing edible roots, hopniss also fixes nitrogen in the soil, benefiting surrounding plants.

6. GROUNDCOVER LAYER:

Trailing Lantana (*Lantana montevidensis*): Trailing lantana is a low-growing, drought-tolerant plant that produces clusters of purple flowers. While not typically grown for food, its flowers and leaves are used in traditional medicine. It adds ornamental value to the groundcover layer and attracts pollinators, especially butterflies.

Beach Strawberry (*Fragaria chiloensis*): This rare species of wild strawberry spreads quickly and produces small, flavorful berries. It thrives in coastal and well-drained soils, making it an excellent groundcover for USDA Zone 8. Beach strawberry's dense growth helps control erosion and attract pollinators.

7. VERTICAL LAYER:

Malabar Spinach (*Basella alba*): Malabar spinach is a vining plant that produces edible, spinach-like leaves, which are rich in vitamins A and C. This heat-tolerant plant thrives in USDA Zone 8 and can quickly cover trellises or poles, providing vertical coverage. The leaves are often used in salads or cooked as a leafy green.

Jicama (*Pachyrhizus erosus*): Jicama is a vining plant that produces edible tubers with a sweet, crunchy texture. The tubers are often eaten raw in salads or cooked in various dishes. Jicama grows well in USDA Zone 8 and can cover vertical structures, adding a root crop to the food forest while also attracting beneficial insects.

8. FUNGAL LAYER:

Velvet Foot Mushroom (*Flammulina velutipes*):

Also known as enoki, the velvet foot mushroom is an edible fungus with a delicate, mild flavor. It grows well in USDA Zone 8, particularly on dead or decaying wood. Enoki mushrooms are often used in soups and salads, and they contribute to the decomposition process in a food forest, improving soil health.

Amethyst Deceiver (*Laccaria amethystina*): The amethyst deceiver is an edible mushroom with a vibrant purple color that grows well in USDA Zone 8. It can be found in forested areas, often in leaf litter, and plays an important role in breaking down organic matter. This mushroom is prized for its striking appearance and mild flavor, adding both aesthetic and ecological value to the fungal layer.

Zone 8 Plan Five: Sunlit Spice Spectrum

Focusing on the aromatic interplay of tropical spices and edibles, this plan layers plants to capture a full spectrum of flavor. Its purpose is to provide nutritional diversity while supporting the overall system with natural vigor.

1. CANOPY LAYER:

Cucumber Tree (*Magnolia acuminata*): The Cucumber Tree is known for its large, aromatic leaves and edible seeds that can be eaten after roasting. This tree prefers moist, well-drained soils and can grow rapidly. It provides shade and nectar for pollinators, enriching the food forest ecosystem.

Eastern Redbud (*Cercis canadensis*): The edible flowers of the Eastern redbud tree can be used in salads, teas, or as garnish. This small, understory tree thrives in a variety of soil types and is known for its beautiful pink blooms in early spring. Its heart-shaped leaves provide food and shelter for a range of creatures.

2. UNDERSTORY LAYER:

Wild Sarsaparilla (*Aralia nudicaulis*): Wild Sarsaparilla is a low-growing, shade-loving plant with edible roots that have been used historically as a flavoring agent for root beer. Its rhizomes spread, helping to stabilize soil, and it provides ground cover for the understory while supporting wildlife.

Coralberry (*Symphoricarpos orbiculatus*): This shrub produces small red berries that, while not typically eaten by humans, provide important food for birds. It thrives in the shade and is highly adaptable, making it a good ground cover in the understory. The plant also helps to prevent soil erosion.

3. SHRUB LAYER:

Blackcap Raspberry (*Rubus occidentalis*): This

native raspberry variety produces small, black, edible fruits that are sweet and rich in antioxidants. It thrives in partial sun and can tolerate a range of soil types. Its dense growth habit helps control erosion and provides shelter for wildlife.

Hawthorn (*Crataegus douglasii*): Also known as Black Hawthorn, this shrub produces small, edible berries that are often used in jellies or as a medicinal tonic. The shrub is tolerant of various soil types and can thrive in both sun and partial shade. Its thorny branches offer protection for birds and small animals.

4. HERBACEOUS LAYER:

Ramps (*Allium tricoccum*): Ramps are a wild onion species prized for their edible leaves and bulbs, which have a strong, garlicky flavor. They thrive in shaded, moist areas of the forest floor, making them ideal for the herbaceous layer. Ramps are also highly sought after in gourmet cuisine.

Sorrel (*Rumex acetosa*): Sorrel is a perennial herb known for its tangy, lemon-flavored leaves, which are great in soups, sauces, and salads. It thrives in partial shade and moist soil, making it an excellent choice for the herbaceous layer. Sorrel is also a rich source of vitamins and minerals.

5. ROOT LAYER:

Crosne (*Stachys sieboldii*): The crosne plant, also known as Chinese artichoke, offers edible tubers that are crunchy with a nutty flavor. Its vigorous growth makes it ideal for erosion control, while its shallow roots help improve soil aeration. The tubers are typically harvested in late fall and used in various culinary dishes.

Alexanders (*Smyrnium olusatrum*): Alexanders is a biennial plant with fleshy roots that can be used in soups or roasted. This ancient vegetable was once widely cultivated but is now considered rare. Alexanders grows in a wide range of soils and offers medicinal properties as well as edible leaves, seeds, and roots.

6. GROUNDCOVER LAYER:

Mazus (*Mazus reptans*): Mazus is a fast-growing, mat-forming perennial that produces small purple or white flowers. While not commonly used for food, it can be grown as an ornamental groundcover in moist areas. Its spreading habit helps prevent soil erosion, and it thrives in both sun and partial shade.

Corsican Mint (*Mentha requienii*): Corsican mint is a low-growing herb that forms a dense carpet of tiny, fragrant leaves. The leaves are edible and can be used to flavor drinks, sauces, or desserts. This plant thrives in partial shade and moist soils, and its creeping habit makes it ideal for covering bare ground.

7. VERTICAL LAYER:

Kiwi Vine (*Actinidia arguta*): Also known as hardy kiwi, this fast-growing vine produces small, smooth-skinned kiwis that are edible and rich in vitamin C. It thrives in USDA Zone 8 and can easily cover vertical structures, providing both fruit and habitat for pollinators. Kiwi vines are dioecious, so both male and female plants are needed for fruit production.

Chilean Glory Flower (*Eccremocarpus scaber*): This rare, fast-growing vine produces vibrant orange or red tubular flowers and edible pods. It is well-suited for USDA Zone 8, where it can thrive in warm conditions. The flowers attract pollinators, making it an excellent addition to the vertical layer for biodiversity and beauty.

8. FUNGAL LAYER:

Black Trumpet Mushroom (*Craterellus cornucopioides*): This rare, edible mushroom grows in USDA Zone 8 in association with hardwood trees. The black trumpet has a rich, smoky flavor and is often used in gourmet dishes. It is a mycorrhizal species, forming beneficial relationships with tree roots and enhancing nutrient uptake for the plants in the food forest.

Chicken Fat Mushroom (*Suillus americanus*): The chicken fat mushroom is a yellow, edible fungus that thrives in USDA Zone 8, often found growing under pine trees. It has a mild flavor and can be used in soups and sauces. This mushroom contributes to the fungal layer by decomposing organic material, aiding in soil enrichment.

Zone 8 Plan Six: Tropical Taste Triumph

A celebration of the unique flavors found in warm climates, this design plan pairs species to achieve a taste triumph. It focuses on producing succulent edibles while reinforcing the system's natural balance.

1. CANOPY LAYER:

Elderberry Tree (*Sambucus nigra*): The Elderberry Tree bears clusters of small berries that can be used in jams, syrups, and medicinal concoctions. It thrives in moist, fertile soil and can handle a range of environmental conditions. Its ability to attract pollinators and improve soil health makes it an excellent choice for the canopy.

Hackberry (*Celtis occidentalis*): The Hackberry produces small, sweet drupes that are edible and attract a variety of wildlife. This tree is hardy and tolerant of a variety of soil types. Its canopy provides shade and habitat, making it a versatile choice for the canopy layer.

2. UNDERSTORY LAYER:

Chokeberry (*Aronia melanocarpa*): Chokeberry is a small, shade-tolerant shrub that produces edible, antioxidant-rich berries that are used in juices, jams, and wines. Its berries are also an important food source for birds. The plant can thrive in a variety of soil conditions, adding resilience to the food forest.

Groundnut (*Apios americana*): Groundnut is a vining plant that produces edible tubers high in protein, traditionally consumed by Native Americans. It thrives in moist, shaded environments and can climb other plants without overshadowing them, making it a valuable addition to the understory.

3. SHRUB LAYER:

Autumn Olive (*Elaeagnus umbellata*): Autumn Olive is a nitrogen-fixing shrub that produces small, tart red berries that are high in lycopene. The berries are often used in jams or eaten fresh. This fast-growing shrub thrives in poor soils and helps improve soil fertility in the food forest.

Goji Berry (*Lycium barbarum*): Goji Berry is a drought-tolerant shrub that produces bright red, antioxidant-rich berries. The berries are often used in smoothies, teas, or eaten dried. This shrub can tolerate a wide range of soil conditions and adds valuable nutrition to the food forest ecosystem.

4. HERBACEOUS LAYER:

Skirret *(Sium sisarum)*: Skirret is an ancient root vegetable with sweet, edible tubers that can be roasted, boiled, or used in stews. The plant grows well in full sun or partial shade and thrives in moist soils. Skirret is highly nutritious and was once a staple food in Europe.

Cardoon (*Cynara cardunculus*): Cardoon is a tall, thistle-like plant related to the artichoke, with edible stalks that can be cooked and eaten like celery. It thrives in full sun and well-drained soils. The plant also provides a striking architectural element to the food forest while attracting pollinators with its large purple flowers.

5. ROOT LAYER:

Yellow Asphodel (*Asphodeline lutea*): The edible tubers of yellow asphodel have been used as a food source since ancient times. The roots can be boiled or roasted, and the plant produces stunning yellow flowers that attract pollinators. Yellow asphodel thrives in poor soils, contributing to soil health with its deep roots.

Black Salsify (*Scorzonera hispanica*): Black salsify is a perennial root vegetable known for its long, black-skinned roots with a delicate, slightly sweet flavor. The roots are harvested in late fall or winter, and the plant is drought-tolerant, making it a hardy choice for USDA Zone 8. Its flowers also attract pollinators, improving biodiversity.

6. GROUNDCOVER LAYER:

Houttuynia (*Houttuynia cordata*): Known for its heart-shaped leaves and strong flavor, houttuynia is a creeping groundcover that thrives in moist, shady areas. The leaves are edible and often used in Asian cuisine. This plant also has medicinal properties and helps suppress weeds in the food forest.

Alpine Strawberry (*Fragaria vesca*): Alpine strawberries are a compact, low-growing plant that produces small, intensely flavorful berries. The plant prefers well-drained soils and partial shade, making it a versatile groundcover. Its ability to produce fruit throughout the growing season adds value to any food forest.

7. VERTICAL LAYER:

Seven Year Bean (*Canavalia gladiata*): Also known as sword bean, this vigorous climbing vine produces edible seeds and pods, which are often cooked in stews and soups. The plant thrives in warm climates like USDA Zone 8, where it can cover trellises or arbors efficiently. Sword bean is nitrogen-fixing, helping to improve soil health.

Hyacinth Bean (*Lablab purpureus*): Hyacinth bean is a fast-growing vine that produces edible pods and seeds, although they must be cooked to remove toxins. The plant is highly ornamental with its purple flowers and deep green leaves, and it thrives in USDA Zone 8. It provides vertical coverage while also attracting pollinators and adding food value.

8. FUNGAL LAYER:

Morel Mushroom (*Morchella esculenta*): Morels are highly sought-after edible mushrooms known for their honeycomb-like caps and rich, earthy flavor. They grow well in USDA Zone 8, often in disturbed soils or near dying trees. Morels contribute to soil health by breaking down organic matter and are considered a gourmet delicacy.

Caesar's Mushroom (*Amanita caesarea*): This rare and edible mushroom, historically prized by Roman emperors, thrives in USDA Zone 8. Caesar's mushroom has a delicate, sweet flavor and is often used in salads or sautéed. It grows in symbiosis with oak and chestnut trees, contributing to the health of the surrounding ecosystem.

Zone 8 Plan Seven: Exotic Edible Ensemble

Bringing together an ensemble of exotic, food-producing plants, this plan layers species for mutual benefit. Its design celebrates diversity and aims to provide both delicious yields and ecosystem stability

1. CANOPY LAYER:

Heartnut (*Juglans ailantifolia var. cordiformis*): A rare variety of the Japanese walnut, heartnut produces unique, heart-shaped nuts that are sweet and edible. The tree is cold-hardy and grows well in well-drained soils. It adds beauty and a rare nut variety to the canopy layer of the food forest.

Persimmon (*Diospyros virginiana*): A native, drought-tolerant tree that produces sweet, orange fruits loved by humans and wildlife alike. Its deep roots help stabilize the soil, making it a resilient addition to the forest.

2. UNDERSTORY LAYER:

Huckleberry (*Gaylussacia baccata*): This low-growing shrub produces small, sweet, edible berries similar to blueberries. It prefers acidic soils and partial shade, making it ideal for the understory. Huckleberries are also a favorite food of birds and other wildlife.

Wintergreen (*Gaultheria procumbens*): Wintergreen is a creeping evergreen shrub with edible, mint-flavored berries and leaves that can be used for teas or flavoring. It thrives in the shady understory and provides year-round ground cover. Its berries also serve as a food source for wildlife.

3. SHRUB LAYER:

Lingonberry (*Vaccinium vitis-idaea*): Lingonberry is a low-growing, evergreen shrub that produces small, edible red berries with a tart flavor. The berries are often used in sauces, jams, and syrups.

Lingonberry thrives in acidic, well-drained soils and is an excellent ground cover in the shrub layer.

Thimbleberry (*Rubus parviflorus*): Thimbleberry is a fast-growing shrub that produces large, soft, edible berries similar to raspberries. It prefers partial shade and moist soils, making it ideal for the food forest shrub layer. Its dense growth provides habitat for wildlife and helps control erosion.

4. HERBACEOUS LAYER:

Chufa (*Cyperus esculentus*): Also known as Tiger Nut, Chufa is a grass-like plant that produces small, edible tubers used to make the traditional Spanish drink horchata. It thrives in moist soils and full sun, making it a valuable addition to the herbaceous layer. Chufa is also a great source of dietary fiber.

Ostrich Fern (*Matteuccia struthiopteris*): The young fiddleheads of Ostrich Ferns are edible and considered a delicacy in spring. This fern thrives in moist, shaded areas and adds a lush, tropical feel to the food forest. The plant also helps retain moisture in the soil, benefiting surrounding plants.

5. ROOT LAYER:

Canna Lily (*Canna indica*): The tubers of canna lilies are edible when cooked, providing a starchy, potato-like texture. This plant thrives in wet soils and is known for its vibrant flowers and large, broad leaves, which help improve soil structure. The tubers are a good source of carbohydrates.

Oca (*Oxalis tuberosa*): Oca is a perennial tuber crop that produces small, tangy tubers with a lemony flavor. Native to the Andes, this plant adapts well to USDA Zone 8 and provides an abundant root harvest. Its shallow roots help reduce soil erosion, and the plant thrives in partial shade.

6. GROUNDCOVER LAYER:

Rock Samphire (*Crithmum maritimum*): Rock

samphire is a salt-tolerant perennial that thrives in coastal regions, producing edible leaves with a salty, parsley-like flavor. It grows in rocky, well-drained soils and can help stabilize the ground. This rare herb adds diversity to the food forest, with its leaves often used in salads or pickled.

Salad Burnet (*Sanguisorba minor*): Salad burnet is a hardy, low-growing perennial with fern-like leaves that have a cucumber-like flavor. The leaves are often used in salads or as a garnish, and the plant is drought-tolerant, making it a good groundcover for dry areas. Its flowers attract pollinators, enhancing biodiversity.

7. VERTICAL LAYER:

Loofah (*Luffa cylindrica*): Loofah is a climbing vine that produces large, edible fruits when harvested young, often used in stir-fries or soups. The mature fruits are dried and used as natural sponges. Loofah grows well in USDA Zone 8 and can cover vertical structures, contributing both food and utility to the food forest.

Bitter Melon (*Momordica charantia*): Bitter melon is a tropical vine that produces edible fruits with a distinct bitter taste, often used in Asian cuisine. The plant thrives in USDA Zone 8 and grows quickly, providing vertical coverage while also contributing to food diversity. Bitter melon is rich in vitamins and antioxidants.

8. FUNGAL LAYER:

Pine Mushroom (*Tricholoma matsutake*): The pine mushroom, or matsutake, is a rare and highly valued edible fungus that grows in USDA Zone 8, typically near pine trees. It has a spicy, cinnamon-like aroma and is considered a delicacy in Japanese cuisine. This mushroom helps decompose dead wood and contributes to the nutrient cycling process in the forest ecosystem.

Puffball Mushroom (*Calvatia gigantea*): The puffball is an edible mushroom when harvested young, before the spores develop. It thrives in USDA Zone 8 and is often found in open areas or on the edges of forests. Puffballs contribute to soil health by breaking down organic matter, and their large size makes them an interesting addition to the fungal layer.

ZONE 8 PLAN EIGHT: FRUITY FUSION

This framework blends tropical fruits and complementary plants into a harmonious layered system. It's designed to fuse immediate edible rewards with long-term sustainability in a warm, humid setting.

1. CANOPY LAYER:

Shagbark Hickory (*Carya ovata*): Shagbark Hickory offers sweet, edible nuts that can be harvested in late summer or fall. This tree grows in a variety of soil types and provides habitat for a range of animals. Its hard wood can also be used for tools and building materials, adding value to the food forest.

Kentucky Coffee Tree (*Gymnocladus dioicus*): This tree's large canopy provides ample shade, while its seeds, when properly processed, offer a potential source of food. The seeds are roasted and can be used as a coffee substitute. Kentucky coffee tree is drought-tolerant and thrives in poor soils, making it a hardy choice for food forests.

2. UNDERSTORY LAYER:

Salal (*Gaultheria shallon*): Salal is an evergreen shrub producing edible berries often used in jams or dried for winter storage. It thrives in the shade and helps to retain moisture in the soil with its dense growth. The plant also provides habitat for pollinators and small animals.

Hazelnut (*Corylus americana*): Hazelnut is a multi-stemmed shrub that produces edible nuts, a valuable food source for humans and wildlife alike. It thrives in partial shade and helps create windbreaks in the understory, while also adding nitrogen to the soil.

3. SHRUB LAYER:

Siberian Pea Shrub (*Caragana arborescens*): This nitrogen-fixing shrub produces small, edible peas that can be used in soups or as animal feed. The plant is highly drought-tolerant and thrives in poor soils, making it an excellent choice for improving soil fertility in the food forest.

Black Hawthorn (*Crataegus douglasii*): This thorny shrub produces small, dark berries that are edible and often used in jams or teas. It tolerates a wide range of soil types and can handle partial shade. The shrub provides food for birds and other wildlife and adds an element of protection to the food forest.

4. HERBACEOUS LAYER:

Purple Tree Collard (*Brassica oleracea var. acephala*): A rare perennial brassica, Purple Tree Collard produces edible leaves that can be harvested year-round. It grows tall, often reaching heights of up to 6 feet, and prefers full sun to partial shade. The plant is highly nutritious and provides a reliable source of greens for the food forest.

Meadow Sweet (*Filipendula ulmaria*): The flowers and leaves of Meadow Sweet are edible and have a sweet, almond-like flavor, often used in teas or desserts. It prefers moist soils and partial shade, making it ideal for the herbaceous layer. Meadow Sweet also has medicinal uses and attracts pollinators with its fragrant blooms.

5. ROOT LAYER:

Parsnip (*Pastinaca sativa*): Parsnips are biennial root vegetables with long, tapered roots that develop a sweet flavor when exposed to frost. The deep roots help break up compacted soil, improving its overall structure. Parsnips are easy to grow in USDA Zone 8 and are typically harvested in the fall.

Turnip-rooted Chervil (*Chaerophyllum bulbosum*): This rare root crop offers sweet, nutty-flavored tubers that are traditionally used in soups or roasted. The plant is a biennial that thrives in moist, well-drained soils and contributes to soil health with its deep taproots. It also attracts beneficial insects.

6. GROUNDCOVER LAYER:

Trailing Arbutus (*Epigaea repens*): Trailing arbutus is a creeping evergreen groundcover with fragrant white or pink flowers. While not commonly grown for food, the leaves and flowers have been used medicinally. This plant prefers acidic, well-drained soils and provides year-round coverage in shady areas of the forest floor.

Bellflower (*Campanula rotundifolia*): Bellflower is a low-growing perennial with delicate, bell-shaped flowers. While not primarily grown for food, its leaves and flowers are edible and can be added to salads. Bellflower thrives in full sun to partial shade and helps prevent soil erosion with its spreading roots.

7. VERTICAL LAYER:

Hardy Passionfruit (*Passiflora incarnata*): Also known as maypop, hardy passionfruit produces edible fruits with a sweet, tangy flavor. This vigorous vine can cover trellises and fences and is well-suited to USDA Zone 8. The flowers attract pollinators, and the fruit is often eaten fresh or used in jams and desserts.

Wild Yam (*Dioscorea villosa*): Wild yam is a climbing vine that produces edible tubers, often used medicinally. This plant thrives in USDA Zone 8 and can cover vertical structures, providing both food and medicinal value to the food forest. Its roots are known for their anti-inflammatory properties.

8. FUNGAL LAYER:

Slippery Jack Mushroom (*Suillus luteus*): This edible mushroom grows in USDA Zone 8, often near pine trees. Slippery Jack has a slimy cap when wet and is commonly used in soups or stews. It forms symbiotic relationships with tree roots, aiding in nutrient exchange and enhancing the overall health of the food forest ecosystem.

Burnt Sugar Cap (*Lactarius quietus*): This edible mushroom is found in USDA Zone 8, typically under oak trees. The burnt sugar cap has a sweet, earthy flavor and contributes to soil health by breaking down leaf litter. It is also known for its symbiotic relationships with tree roots, helping to support nutrient cycling in the food forest.

Zone 8 Plan Nine: Rainforest Resilience Route

Inspired by the tenacity of rainforest ecosystems, this plan maps out a route to both abundant edibles and system strength. Its layered design is intended to mirror natural resilience, producing food even under intense humidity.

1. CANOPY LAYER:

Kousa Dogwood (*Cornus kousa*): The Kousa dogwood offers small, edible fruits that are sweet and packed with vitamin C. Its ornamental flowers, followed by red berries in late summer, create an attractive canopy that supports biodiversity. This species thrives in partial shade and can adapt to various soil conditions.

Mayhaw (*Crataegus opaca*): The Mayhaw produces small, tart berries that are used in jelly and syrup making. It thrives in wetland areas, adding diversity to the food forest's canopy. Its dense branching provides excellent nesting sites for birds, enhancing the biodiversity of the forest.

2. UNDERSTORY LAYER:

False Solomon's Seal (*Maianthemum racemosum*): This perennial plant grows well in shaded areas and produces small red berries that are edible when cooked. It is also known for its medicinal properties, and its spreading root system helps stabilize soil in the understory.

Blue Cohosh (*Caulophyllum thalictroides*): Blue Cohosh is a perennial herb known for its medicinal uses, though the plant itself is not considered edible. It thrives in the shaded understory, where it adds diversity to the forest floor and provides habitat for pollinators

3. SHRUB LAYER:

Western Serviceberry (*Amelanchier alnifolia*): Also

known as Saskatoon Berry, this shrub produces sweet, edible fruits that are rich in vitamins and can be eaten fresh or made into jams and pies. It is drought-tolerant and thrives in a variety of soil types, making it a resilient food forest plant.

Honeyberry (*Lonicera caerulea*): Honeyberry is a cold-hardy shrub that produces sweet, blueberry-like fruits. The berries are high in antioxidants and can be eaten fresh or used in desserts. Honeyberry prefers partial sun and well-drained soils, making it an excellent shrub for diverse food forests.

4. HERBACEOUS LAYER:

Stinging Nettle (*Urtica dioica*): Stinging Nettle has edible leaves that are highly nutritious and can be cooked in soups or used to make tea. The plant thrives in moist, shaded areas and is rich in vitamins and minerals. It also improves soil health by accumulating nutrients like nitrogen.

Prairie Verbena (*Glandularia bipinnatifida*): A low-growing, drought-tolerant wildflower that produces clusters of bright purple blooms, offering nectar to butterflies and bees. It's a tough plant that requires little water.

5. ROOT LAYER:

Rampion (*Phyteuma spicatum*): The edible roots of rampion are mild and sweet, often used in salads or cooked as a vegetable. This perennial plant is known for its blue, spiked flowers that attract pollinators. Rampion thrives in partial shade and helps improve biodiversity in the food forest.

Ulluco (*Ullucus tuberosus*): Ulluco is a perennial root vegetable with brightly colored tubers that have a slightly waxy texture. The tubers are used in various culinary dishes and offer a rare edible addition to the root layer. Ulluco prefers moist, well-drained soils and grows well in USDA Zone 8.

6. GROUNDCOVER LAYER:

Self-Heal (*Prunella vulgaris*): Self-heal is a creeping perennial herb with small purple flowers and edible leaves often used in salads or teas. This plant is highly medicinal, known for its anti-inflammatory properties, and thrives in both sun and partial shade. It forms a dense mat, making it an effective groundcover.

Creeping Thyme (*Thymus serpyllum*): Creeping thyme is a fragrant, low-growing herb that produces small purple flowers and is often used in cooking. It thrives in full sun and well-drained soils, making it an excellent groundcover for hot, dry areas. Creeping thyme attracts pollinators and provides culinary and medicinal benefits.

7. VERTICAL LAYER:

Bamboo Vine (*Smilax laurifolia*): Bamboo vine, also known as laurel greenbrier, produces edible shoots and berries. This fast-growing vine thrives in USDA Zone 8 and can cover fences or trellises, making it an excellent choice for vertical growth. Its berries are rich in antioxidants, and the young shoots can be cooked and eaten like asparagus.

Cape Honeysuckle (*Tecoma capensis*): Cape honeysuckle is a fast-growing vine that produces clusters of orange, tubular flowers that attract pollinators. While not traditionally grown for food, the flowers can be used to make a sweet syrup. This vine thrives in USDA Zone 8, where it can cover vertical structures and provide ornamental value.

8. FUNGAL LAYER:

Shaggy Mane (*Coprinus comatus*): A beneficial mushroom that breaks down organic matter and adds nutrients to the soil. It thrives in well-drained soils, supporting soil health and plant productivity.

Golden Chanterelle (*Cantharellus cibarius*): The golden chanterelle is a prized edible mushroom with a fruity aroma and peppery flavor. It thrives in USDA Zone 8, often growing in association with hardwood trees. Chanterelles help decompose organic material and form mycorrhizal relationships with tree roots, enhancing the food forest's nutrient dynamics.

Zone 8 Plan Ten: Verdant Vibes

Zone 8 Plan Ten: Verdant Vibes

This plan immerses you in a verdant world of tropical layers and lush edibles. It is crafted to harness the energy of a humid climate for a robust food forest that delights the senses and sustains itself.

1. CANOPY LAYER:

White Oak (*Quercus alba*): White Oaks produce acorns that are edible once leached, and they are a valuable food source for wildlife. This slow-growing tree thrives in a variety of soil types, including both dry and moist conditions. It provides long-lasting shade and is an important part of the ecosystem.

Yellow Poplar (*Liriodendron tulipifera*): The Yellow Poplar, or Tulip Poplar, offers edible seeds and large, fragrant flowers that attract pollinators. This tree is fast-growing and thrives in rich, moist soils. Its tall stature and large leaves provide valuable canopy cover in the food forest.

2. UNDERSTORY LAYER:

Woodland Strawberry (*Fragaria vesca*): Woodland Strawberries produce small, sweet, edible fruits and are an excellent ground cover for the understory. They spread quickly, creating a dense mat that prevents soil erosion and provides food for both humans and wildlife.

Heartleaf Ginger (*Asarum canadense*): Also known as Wild Ginger, this shade-loving plant produces aromatic rhizomes that can be used as a ginger substitute. It forms a dense ground cover in the understory and helps improve soil structure while offering habitat for small wildlife.

3. SHRUB LAYER:

Mountain Pepper (*Tasmannia lanceolata*): This rare, evergreen shrub produces small, edible berries and aromatic leaves used as a spice. The plant thrives in shady, moist conditions and adds a unique culinary element to the food forest. It also supports biodiversity by providing habitat for beneficial insects.

Cloudberry (*Rubus chamaemorus*): Cloudberry is a rare shrub that produces golden, edible berries known for their tart-sweet flavor. The berries are used in jams and desserts and are high in vitamins. Cloudberry thrives in moist, acidic soils and prefers partial shade, making it a unique addition to the shrub layer.

4. HERBACEOUS LAYER:

Sweet Cicely (*Myrrhis odorata*): Sweet Cicely produces edible leaves, seeds, and roots with a sweet, anise-like flavor. The plant thrives in partial shade and moist soils, making it an excellent herbaceous layer addition. It attracts pollinators and can also be used as a natural sweetener in cooking.

Rattlesnake Master (*Eryngium yuccifolium*): A hardy perennial that produces spiky flowers rich in nectar, attracting pollinators like bees and butterflies. It is drought-tolerant and provides habitat for pollinators and other wildlife.

5. ROOT LAYER:

Salsify (Goat's Beard) (*Tragopogon pratensis*): The roots of this perennial plant are edible, with a taste similar to oysters. Salsify's roots improve soil structure, and the plant's vibrant yellow flowers attract pollinators. This plant is drought-tolerant and thrives in a variety of soils.

Maca (*Lepidium meyenii*): Maca is a rare root crop with tubers that are often dried and ground into powder, providing a nutritious, energy-boosting food source. It grows well in USDA Zone 8 and offers benefits like improving soil structure and attracting pollinators with its small, white flowers.

6. GROUNDCOVER LAYER:

Wood Sorrel (*Oxalis acetosella*): Wood sorrel is a low-growing perennial with clover-like leaves and small, edible flowers. The leaves have a tart, lemony flavor and are often used in salads or as a garnish. Wood sorrel thrives in shady, moist areas, adding a unique edible groundcover option to the food forest.

Yerba Buena (*Clinopodium douglasii*): Yerba buena is a creeping perennial herb with small, mint-scented leaves that are often used in teas and medicinal remedies. It grows well in partial shade and moist soils, forming a dense groundcover that helps suppress weeds. Its fragrant leaves make it a valuable addition to the forest.

7. VERTICAL LAYER:

Chinese Yam (*Dioscorea polystachya*): Chinese yam is a vining plant that produces edible tubers often used in soups, stews, or stir-fries. The plant thrives in USDA Zone 8 and can cover vertical supports, contributing to the food forest's productivity. Its tubers are also known for their medicinal properties, especially in traditional Chinese medicine.

New Zealand Spinach (*Tetragonia tetragonioides*): New Zealand spinach is a heat-tolerant, vining plant that produces edible leaves rich in vitamins A and C. This plant thrives in USDA Zone 8 and can cover vertical structures while also providing a reliable source of greens during the hot summer months. It is often used as a substitute for traditional spinach.

8. FUNGAL LAYER:

Yellowfoot Mushroom (*Craterellus tubaeformis*): The yellowfoot mushroom is an edible fungus that grows well in USDA Zone 8, often found in mossy forests. It has a mild, earthy flavor and contributes to soil health by breaking down decaying organic matter. This mushroom also forms beneficial mycorrhizal associations with tree roots, supporting the food forest ecosystem.

Cinnabar Red Chanterelle (*Cantharellus cinnabarinus*): This vibrant red, edible mushroom thrives in USDA Zone 8 and grows in symbiosis with hardwood trees. Cinnabar red chanterelles have a delicate flavor and are often used in gourmet dishes. They contribute to soil enrichment by breaking down organic material and promoting mycorrhizal networks.

ZONE 9

Plant Hardiness Zone 9

Zone 9 is known for its warm, temperate conditions, with mild winters and long growing seasons, making it ideal for a wide variety of plant types, including subtropical and Mediterranean species. Winters are brief and generally mild, with occasional frost. The zone's minimum temperatures range between 20°F and 30°F (-6°C to -1°C). This makes it well-suited to plants that prefer warmth and sun exposure. Zone 9 stretches across parts of the U.S., including areas on the West Coast, Southwest, and Southeast.

Locations within Zone 9

Zone 9 includes regions with mild winters and warm to hot summers, with minimum temperatures ranging from 20°F to 30°F (-6°C to -1°C). The zone covers parts of the southern United States that are known for their moderate winter climates and warm growing conditions.

Below are the states and their specific areas that fall within USDA Hardiness Zone 9:

Alabama: Southern regions, such as Mobile, along the Gulf Coast.

Arizona: Entire Phoenix area and surrounding areas; parts of Tucson in the lower desert.

Arkansas: Southern areas, near Little Rock and along the Mississippi River.

California: Coastal regions, including Los Angeles, San Diego, Santa Barbara; Inland Empire areas like Riverside and San Bernardino; Southern coastal zones such as Santa Monica, Malibu, and Huntington Beach.

Florida: South Florida, including Miami, Fort

Lauderdale, and the Keys; Central Florida, including Orlando.

Georgia: Southern areas like Savannah and surrounding coastal regions.

Hawaii: Lowland, coastal, and lower elevation regions of all major islands.

Louisiana: Southern areas, including New Orleans and Baton Rouge.

Mississippi: Gulf Coast regions, such as Biloxi, Gulfport, and Pascagoula.

Nevada: Southern regions like Las Vegas and nearby areas.

New Mexico: Southern regions like Las Cruces and the Rio Grande Valley.

Oklahoma: Southern parts, particularly areas near the Texas border.

Oregon: Southern parts, including Medford, Ashland, and the Rogue Valley.

South Carolina: Southern areas, including Charleston and the coastal regions.

Texas: San Antonio and surrounding areas; central Austin and Houston's coastal and inland regions.

Washington: Southwestern areas near Portland, Oregon, and the Columbia River Gorge.

Zone 9 Climate

Zone 9 is characterized by hot, dry summers and mild winters with minimal snowfall. Coastal areas tend to have milder winters due to the ocean's influence, while inland areas may experience higher summer temperatures, often exceeding 90°F (32°C). The majority of rainfall occurs in the fall and winter months, while summers are typically dry. In some inland regions, summer droughts can be prolonged, so incorporating drought-tolerant plants and water conservation strategies is essential.

Soil Conditions in Zone 9

Soils in Zone 9 vary widely, ranging from sandy and well-drained to clay-heavy and compacted. Due to the long growing season, soil fertility can deplete quickly without careful management. Organic matter, such as compost, mulch, and cover crops, is necessary to improve soil health and structure. In dry summer months, irrigation is often necessary, especially for crops that require regular moisture.

Zone 9 Growing Season

Zone 9 enjoys a long growing season, typically running from early spring through late fall. Frost-sensitive plants can thrive throughout much of the year, as winter temperatures rarely fall below 20°F (-6°C). However, the summer heat can be challenging for plants not adapted to high temperatures, making water management and heat-tolerant plant selection critical. Light frost is uncommon but may occur in the colder parts of the zone during the winter, particularly inland.

Challenges of Zone 9

Despite the many growing opportunities, Zone 9 presents challenges. Heat stress in the summer can impact plants not suited to high temperatures. Proper planning for shade, water conservation, and the use of heat-tolerant plants is critical. Succulents and other heat-resistant plants can be included in landscaping to support the resilience of a food forest. Water conservation is also important, as summer drought conditions often require efficient irrigation systems, such as drip irrigation or soaker hoses, to maintain soil moisture. Mulching and using cover crops can help retain moisture and keep the soil cooler during the summer. Zone 9 is also prone to pests like aphids, whiteflies, caterpillars, and snails. Companion planting and promoting beneficial insects can help manage pests and support the health of a food forest.

Benefits of Zone 9 for Food Forests

The long growing season in Zone 9 allows for multiple harvests each year. The mild winters support fruit tree growth, and the extended summers provide ideal conditions for vegetables and herbs to thrive. The variety of subtropical and Mediterranean plants that can grow in Zone 9 enhances the diversity of food forests. Native plants well-adapted to the climate reduce the need for additional irrigation and fertilizers, making it easier to create a sustainable and productive growing environment.

Zone 9 provides an excellent opportunity for cultivating a wide variety of plants that thrive in warmer climates and longer growing seasons. With proper planning and attention to water management and heat, a Zone 9 food forest can be a diverse, productive, and resilient ecosystem.

Zone 9 Plan One: Subtropical Synergy

Tailored for the gentle warmth of subtropical climates, this scheme layers compatible species for maximum benefit. It balances a steady food harvest with an enduring commitment to ecological health.

1. CANOPY LAYER:

Mamoncillo (*Melicoccus bijugatus*): Mamoncillo, also known as Spanish lime, is a tropical fruit tree that thrives in USDA Zone 9. The small, green fruits have a sweet, tangy flesh and are often eaten fresh or made into beverages. This canopy tree can reach heights of up to 85 feet, providing shade and habitat for wildlife. It's drought-tolerant and can improve soil health through its deep root system.

Pili Nut (*Canarium ovatum*): The pili nut tree is native to Southeast Asia and can grow well in USDA Zone 9. The edible nuts are highly nutritious and used in various culinary applications. The tree produces large canopies, providing shade and acting as a windbreak in the food forest. It also enriches the soil with organic matter from its fallen leaves.

2. UNDERSTORY LAYER:

White Sapote (*Casimiroa edulis*): This small tree produces creamy, custard-like fruits with a sweet flavor that resembles a mix of banana, peach, and pear. The fruits are highly nutritious, rich in vitamins A and C. The tree thrives in warm climates and can tolerate light frosts, making it perfect for Zone 9. It also provides excellent shade and soil retention in the food forest.

Sea Grape (*Coccoloba uvifera*): This evergreen shrub produces clusters of edible grape-like fruits, which can be eaten fresh or made into jellies and wine. Sea Grape is salt-tolerant, making it ideal for coastal areas. Its large, round leaves are also attractive and offer habitat benefits for wildlife.

3. SHRUB LAYER:

Chilean Strawberry (*Fragaria chiloensis*): This rare strawberry species produces small, sweet fruits that are edible and highly prized for their flavor. It thrives in mild, coastal climates and can help stabilize soil, making it valuable for erosion control in a food forest.

Goji Berry (*Lycium barbarum*): This drought-tolerant shrub produces small, bright red berries that are packed with antioxidants, vitamins, and minerals. Goji berries are edible fresh or dried and are known for their health benefits. The shrub attracts pollinators and helps improve soil quality by preventing erosion.

4. HERBACEOUS LAYER:

Sweet Flag (*Acorus calamus*): The rhizomes of this aromatic herb are edible and have been used as a flavoring agent and in traditional medicine. Sweet Flag thrives in wetlands or moist environments, making it useful for controlling water flow and adding biodiversity to wet areas of a food forest.

Wild Tarragon (*Artemisia dracunculus*): Also known as "Russian Tarragon," the leaves of this perennial herb are edible and used to flavor dishes. Wild Tarragon thrives in warm, dry climates and is a drought-tolerant addition to the herbaceous layer, adding aromatic qualities and attracting pollinators.

5. ROOT LAYER:

Skirret (*Sium sisarum*): The roots of this perennial plant are edible, sweet, and starchy, often compared to parsnip. Skirret thrives in moist, fertile soils and brings high-yielding, delicious roots to the food forest, along with the ability to tolerate colder temperatures during the winter months.

Chinese Artichoke (*Stachys affinis*): Also known as "Crosne," the tubers of this plant are edible, small, and crunchy, with a nutty flavor. Chinese Artichoke grows well in warm climates, spreading easily through underground rhizomes, and improves soil health through its nitrogen-fixing capabilities.

6. GROUND COVER LAYER:

Salad Burnet (*Sanguisorba minor*): The leaves of Salad Burnet are edible and have a mild cucumber flavor, perfect for adding to salads or herbal drinks. It is a hardy perennial groundcover that thrives in dry, well-drained soils and tolerates both sun and shade. Its fern-like foliage helps suppress weeds and adds a delicate texture to the garden. Salad Burnet attracts pollinators and can be harvested almost year-round.

Wild Strawberry (*Fragaria vesca*): This species produces small, sweet strawberries that are edible and delicious when fresh. It forms a low-growing mat that provides excellent groundcover, helping to prevent soil erosion and improve soil health. Wild strawberries are highly resilient, adaptable to a variety of soil types, and a great companion for taller plants.

7. VERTICAL LAYER:

Hops (*Humulus lupulus*): Hops are a fast-growing, climbing vine suitable for USDA Zone 9, commonly used in brewing, with its flowers contributing to beer flavor and bitterness. In a food forest, hops add vertical structure and dense greenery. The shoots are also edible when young, similar to asparagus. Hops thrive in full sun and well-drained soil, providing shade and habitat for beneficial insects.

Akebia (*Akebia quinata*): Also known as chocolate vine, Akebia produces elongated, edible purple fruit with a sweet, custard-like pulp. It's a vigorous, twining vine that can cover trellises and fences, providing both ornamental value and edible yields. Akebia is adaptable to different soil types and can tolerate partial shade, making it a versatile addition to the vertical layer.

8. FUNGAL LAYER:

Black Morel (*Morchella elata*): Black morels are a prized edible mushroom known for their distinctive honeycomb-like caps and rich, earthy flavor. They grow well in temperate regions like USDA Zone 9, particularly in forested areas or near hardwood trees. Morels are excellent for adding fungal diversity to a food forest, as they help break down organic matter and improve soil health. Their fruiting bodies are harvested in spring and can be used fresh or dried in a variety of dishes.

Cauliflower Mushroom (*Sparassis crispa*): This mushroom, resembling a large, frilly cauliflower, grows at the base of coniferous trees in USDA Zone 9. It is edible and highly sought after for its delicate, nutty flavor and meaty texture. The Cauliflower Mushroom is a natural decomposer, playing a crucial role in breaking down dead wood and returning nutrients to the soil.

Zone 9 Plan Two: Balmy Bounty Blueprint

In a region defined by mild, balmy days, this blueprint organizes plant layers for an ongoing bounty. Its design focuses on combining immediate nutrition with a support system that flourishes in relaxed warmth.

1. CANOPY LAYER:

Indian Almond (*Terminalia catappa*): Also known as tropical almond, this tree produces edible seeds that are similar in taste to traditional almonds. It thrives in USDA Zone 9 and is a fast-growing species, providing ample canopy cover for other plants. Its deep roots help with soil stabilization, and its leaves can be used as a mulch to improve soil fertility.

Bunya Pine (*Araucaria bidwillii*): The bunya pine is an ancient, rare tree that produces large, edible nuts encased in cones. It grows well in USDA Zone 9 and can reach heights of over 100 feet, making it a dominant canopy species. The nuts are rich in starch and can be eaten raw or cooked, while the tree provides shelter and food for wildlife in the food forest.

2. UNDERSTORY LAYER:

Chilean Wine Palm (*Jubaea chilensis*): This rare, slow-growing palm produces edible seeds known as "coquitos," which resemble miniature coconuts. The palm's sap can also be harvested to make syrup, similar to maple syrup. It's drought-tolerant and can add vertical structure to the understory layer, while also providing habitat for wildlife.

Ice Cream Bean (*Inga edulis*): Known for its long pods filled with sweet, cottony pulp, this nitrogen-fixing tree improves soil fertility. The pulp is edible and has a mild vanilla flavor, making it a unique addition to a food forest. It thrives in warm climates, supporting soil health and producing shade for other understory plants.

3. SHRUB LAYER:

Honeyberry (*Lonicera caerulea*): This deciduous shrub produces elongated, blue berries that are sweet and tart, with a flavor similar to blueberries. The berries are edible and can be used in jams, desserts, and smoothies. Honeyberry is a cold-hardy plant, thriving in both temperate and subtropical zones, and adds early-season fruit to the food forest.

Buffaloberry (*Shepherdia argentea*): This nitrogen-fixing shrub produces small, edible red berries that are rich in vitamins and antioxidants. Buffaloberry is highly drought-tolerant and thrives in a variety of soils, making it an excellent plant for restoring degraded landscapes and improving soil fertility.

4. HERBACEOUS LAYER:

Miner's Lettuce (*Claytonia perfoliata*): The leaves of this low-growing herb are edible and highly nutritious, rich in vitamin C. Miner's Lettuce thrives in shaded, moist areas and provides an early-season salad crop, offering edible greens while enhancing soil moisture retention.

Orach (*Atriplex hortensis*): This rare leafy green has edible leaves that can be used in salads or cooked like spinach. Orach thrives in warm climates and can tolerate poor soils, making it a valuable addition to the herbaceous layer for its drought resistance and ability to improve soil health.

5. ROOT LAYER:

Earthnut Pea (*Lathyrus tuberosus*): The tubers of this leguminous plant are edible and were historically a staple crop in Europe. Earthnut Pea thrives in well-drained soils, fixing nitrogen into the soil, while providing small, nutritious tubers that can be roasted or eaten raw.

Hopniss (*Apios priceana*): This rare native legume produces edible tubers that are high in protein and can be cooked similarly to potatoes. Hopniss grows vigorously in moist soils and helps improve soil fertility by fixing nitrogen, making it a beneficial addition to the root layer.

6. GROUND COVER LAYER:

Yerba Buena (*Clinopodium douglasii*): Yerba Buena, also known as "Good Herb," is a trailing perennial groundcover known for its fragrant, mint-like leaves that are edible and used in teas or as flavoring. It thrives in moist, shaded areas, making it ideal for the forest floor in a food forest. The plant attracts beneficial insects and contributes to biodiversity.

Musk Strawberry (*Fragaria moschata*): Musk strawberries produce flavorful, aromatic berries that are smaller than commercial varieties but highly prized for their unique taste. This perennial forms a dense groundcover that helps retain soil moisture and offers fruit throughout the growing season. It prefers partial shade and well-drained soils, making it suitable for understory planting.

7. VERTICAL LAYER:

Chinese Yam (*Dioscorea polystachya*): This climbing vine produces underground tubers that are edible and nutritious, often used in stir-fries or stews. Chinese yam brings diversity to the vertical layer with its long, trailing stems and delicate leaves. It is drought-tolerant and grows well in a variety of soil conditions, contributing to food security in the forest.

Himalayan Blackberries (*Rubus armeniacus*): These brambles produce highly edible blackberries that are rich in vitamins and antioxidants. Though considered invasive in some areas, in a managed food forest, they can be trained to grow vertically along trellises. The dense foliage also attracts pollinators and provides shelter for wildlife, enriching the biodiversity.

8. FUNGAL LAYER:

Hawk's Wing Mushroom (*Sarcodon imbricatus*):

Known for its scaly, brown cap, this rare mushroom grows in symbiosis with conifer trees in temperate forests. It is edible, although slightly bitter, and is typically used in soups or dried for later use. Hawk's Wing helps improve soil conditions by forming mycorrhizal relationships with trees, enhancing nutrient exchange.

Maitake (*Grifola frondosa*): Also known as "Hen of the Woods," this mushroom grows in clusters at the base of oak trees. It is highly valued for its medicinal properties and savory taste, with both the cap and stem being edible. Maitake is a potent decomposer, aiding in breaking down tree debris and contributing to soil health in Zone 9 food forests.

Zone 9 Plan Three: Warm Wave Workshop

Riding the wave of warm, gentle breezes, this design plan uses thoughtful pairings across layers. It's all about generating a continuous flow of edibles while building a naturally sustainable environment.

1. CANOPY LAYER:

Brazil Nut (*Bertholletia excelsa*): The Brazil nut tree, known for its large, edible seeds, thrives in USDA Zone 9 and can grow up to 160 feet tall. The nuts are highly nutritious and contain healthy fats and protein. This tree plays a crucial role in the ecosystem by supporting pollinators like bees and maintaining biodiversity through its symbiotic relationships with other plants.

Kumquat (*Citrus japonica*): The kumquat tree produces small, tangy fruits that are edible with their peels, making them a unique citrus option for USDA Zone 9. This compact canopy tree can reach up to 15 feet in height, providing shade and a food source for pollinators. Its evergreen foliage adds year-round interest, and its fruit is rich in vitamin C and antioxidants.

2. UNDERSTORY LAYER:

Moringa (*Moringa oleifera*): This fast-growing tree produces highly nutritious leaves, flowers, and seed pods. Moringa is drought-tolerant and thrives in tropical and subtropical climates. The leaves are a rich source of vitamins and minerals, offering a sustainable food source, while the tree itself enhances soil health and structure.

Pitomba (*Eugenia luschnathiana*): This small, rare fruit tree produces bright orange fruits with a sweet and tart flavor, often compared to apricots. The fruits are edible and can be eaten fresh or used in desserts. Pitomba is well-suited to tropical and subtropical climates, providing a rare fruit option for Zone 9 food forests.

3. SHRUB LAYER:

Evergreen Huckleberry (*Vaccinium ovatum*): This hardy, evergreen shrub produces small, sweet, black berries that are edible and often used in jams and pies. Evergreen huckleberry thrives in shady areas, providing year-round structure and habitat for wildlife, while adding edible berries to the food forest.

Dwarf Pawpaw (*Asimina parviflora*): This rare shrub produces small, custard-like fruits with a sweet, tropical flavor. The fruits are rich in vitamins and can be eaten fresh or used in desserts. The plant is highly adaptable to warm, humid climates and provides shade and shelter in the food forest.

4. HERBACEOUS LAYER:

Papalo (*Porophyllum ruderale*): This heat-tolerant herb is native to Central America and has edible leaves used as a cilantro substitute in cooking. Papalo thrives in hot, dry conditions, adding unique flavors to culinary dishes and providing habitat for beneficial insects in the food forest.

Purple Deadnettle (*Lamium purpureum*): The young leaves and flowers of this herb are edible and can be used in salads or as a garnish. Purple Deadnettle is highly adaptable, growing in a variety of soil types, and it attracts pollinators while providing early-season edible greens.

5. ROOT LAYER:

Lesser Celandine (*Ficaria verna*): The small tubers of this plant are edible when cooked and were traditionally foraged in Europe. Lesser Celandine prefers moist, shaded environments and acts as a good groundcover plant, stabilizing soil and providing early spring flowers for pollinators.

Maca (*Lepidium meyenii*): This ancient root

vegetable, originally from the Andes, has edible roots often used as a superfood for their high nutritional content. Maca thrives in well-drained, sunny areas and is known for boosting energy, making it a hardy, drought-tolerant plant for Zone 9.

6. GROUND COVER LAYER:

Gotu Kola (*Centella asiatica*): Gotu Kola is a low-growing herb with round, edible leaves that are used in traditional medicine and as a culinary herb in salads or teas. It thrives in moist, shady environments and is perfect for covering bare soil in humid conditions. Gotu Kola improves soil health and attracts beneficial insects to the food forest.

Wintergreen (*Gaultheria procumbens*): Wintergreen is a low-growing evergreen groundcover that produces edible red berries and aromatic leaves, which are often used for tea or flavoring. It thrives in acidic, well-drained soils and provides year-round greenery. Wintergreen helps with soil retention and creates a cooling microclimate in shady forest areas.

7. VERTICAL LAYER:

Kiwi Vine (*Actinidia deliciosa*): Kiwi vines are hardy in USDA Zone 9 and provide delicious, nutrient-rich fruits. These vigorous climbers need strong support structures and benefit from pruning to control their growth. They thrive in full sun and well-drained soil, making them an excellent choice for vertical layers that offer both food and aesthetic appeal.

Japanese Wineberry (*Rubus phoenicolasius*): This relative of raspberries produces small, sweet red fruits encased in a bristly calyx. The plant grows well in Zone 9, forming dense, arching canes that can be trained on trellises. The fruit is rich in antioxidants and vitamins, while the canes can create a natural barrier in the food forest.

8. FUNGAL LAYER:

Shimeji Mushroom (*Hypsizygus tessellatus*): Shimeji mushrooms grow well in temperate regions like USDA Zone 9 and are known for their small, clustered fruiting bodies with a firm texture. Both brown and white varieties are edible and are commonly used in stir-fries and soups. These mushrooms are important for their role in decomposing dead wood, adding organic matter to the soil.

Blewit (*Clitocybe nuda*): Blewits are purple-tinged mushrooms that thrive in mulched areas and decaying leaf litter. They are edible when cooked, with a mild, nutty flavor. In a food forest, they act as a decomposer, breaking down organic materials to improve soil fertility and structure.

Zone 9 Plan Four: Gentle Grove Guide

This guide creates a softly lit grove of layered edibles perfect for a subtropical setting. Its focus is on gently integrating food production with the subtle rhythms of nature to ensure a balanced system.

1. CANOPY LAYER:

White Sapote (*Casimiroa edulis*): The white sapote is a tropical fruit tree that produces creamy, custard-like fruits. It grows well in USDA Zone 9 and can reach up to 60 feet in height. The tree is drought-tolerant and provides ample shade, while its fruits are rich in vitamin C and fiber, making it a valuable addition to the food forest for both humans and wildlife.

Indian Bael (*Aegle marmelos*): The bael tree produces hard-shelled fruits with sweet, aromatic pulp that is edible and often used in drinks and desserts. It thrives in USDA Zone 9 and can grow up to 50 feet tall, providing a dense canopy. The tree is drought-tolerant, and its deep roots help improve soil structure, while the fruits are valued for their medicinal properties.

2. UNDERSTORY LAYER:

Wampi (*Clausena lansium*): This tropical fruit tree produces small, grape-like fruits with a tangy, sweet flavor. Wampi fruits are eaten fresh or made into preserves, and the tree is highly adaptable to warm climates. Its dense foliage provides shade and helps support the understory ecosystem.

Mamoncillo (*Melicoccus bijugatus*): Also known as Spanish lime, this small tree produces clusters of green fruits with a sweet and tart pulp. The fruits are popular in tropical regions and can be eaten fresh or used in drinks. Mamoncillo is a good nitrogen-fixer, helping to improve soil fertility in a food forest.

3. SHRUB LAYER:

Roughleaf Dogwood (*Cornus drummondii*): This shrub produces small, white berries that are edible when cooked and often used in preserves. Roughleaf Dogwood is drought-tolerant and can thrive in poor soils, making it ideal for stabilizing slopes and improving soil quality in a food forest.

Brazilian Pepper Tree (*Schinus terebinthifolia*): This evergreen shrub produces small, red fruits often called "pink peppercorns," which are edible and used as a spice. Brazilian Pepper Tree is highly adaptable to warm climates, and its dense foliage provides shelter and habitat for wildlife in a food forest.

4. HERBACEOUS LAYER:

Samphire (*Crithmum maritimum*): Also known as "Rock Samphire," this coastal herb has edible leaves with a salty flavor and is often used in salads or pickled. Samphire thrives in sandy soils and salt-tolerant environments, making it an excellent plant for adding diversity and edibility to coastal or arid food forests.

Goldenrod (*Solidago odora*): The leaves and flowers of this herb are edible and used to make tea with a sweet, anise-like flavor. Goldenrod is drought-tolerant and thrives in poor soils, attracting pollinators and adding medicinal benefits to the food forest with its anti-inflammatory properties.

5. ROOT LAYER:

Scorzonera (*Scorzonera hispanica*): The edible roots of this plant, also known as "Black Salsify," are prized for their sweet, delicate flavor. Scorzonera grows well in full sun and well-drained soils, offering a nutritious food source rich in dietary fiber and supporting pollinators with its flowers.

Chufa (Cyperus esculentus var. sativus): Also known as "Tiger Nuts," the small tubers of this sedge plant are edible and often used to make beverages or eaten raw. Chufa thrives in warm, sandy soils, providing edible roots while also serving as a natural groundcover to prevent soil erosion.

6. GROUND COVER LAYER:

Creeping Jenny (*Lysimachia nummularia*): While not commonly grown for food, Creeping Jenny's leaves are occasionally used as a garnish or in salads. It's a fast-spreading groundcover that thrives in moist, shaded areas, and its bright green foliage adds vibrancy to the forest floor. This plant is useful for soil stabilization and can grow in a wide range of soil types.

Pennyroyal (*Mentha pulegium*): Pennyroyal is a low-growing groundcover with small, aromatic leaves used in herbal teas and traditional medicine. It prefers moist soils and partial shade, making it an ideal candidate for damp areas in the food forest. Pennyroyal can repel certain pests, which is beneficial in companion planting.

7. VERTICAL LAYER:

Ceylon Spinach (*Basella rubra*): Also known as Malabar spinach, this fast-growing, climbing vine produces edible, succulent leaves that are used in salads and cooking. It thrives in hot climates and can quickly cover a trellis or other vertical structure. The leaves are rich in vitamins A and C, and the plant is ornamental with its attractive, reddish-purple stems.

Balloon Vine (*Cardiospermum halicacabum*): Known for its balloon-like seed pods, this vine is a fast grower in Zone 9. Its young shoots, leaves, and seeds are edible and used in various traditional cuisines. Balloon vine can provide quick green coverage and is useful for adding vertical interest, along with supplying edible parts that can contribute to a food forest's diversity.

8. FUNGAL LAYER:

Chanterelle (*Cantharellus cibarius*): Chanterelles are highly prized for their bright yellow color, rich flavor, and meaty texture. They grow in symbiosis with hardwoods in USDA Zone 9, forming mycorrhizal relationships that enhance nutrient cycling. These mushrooms are excellent for both culinary use and soil improvement, making them a valuable addition to the fungal layer.

Elm Oyster Mushroom (*Hypsizygus ulmarius*): Despite its name, the elm oyster is not related to true oysters but grows in clusters on dead or decaying wood, particularly elms and other hardwoods. It is edible, with a firm texture and mild flavor, and contributes to the decomposition process in the food forest, breaking down lignin in wood to release nutrients.

Zone 9 Plan Five: Mellow Mosaic Method

Celebrating the mellow vibe of a warm climate, this method arranges a mosaic of layered, complementary plants. It aims to produce an inviting array of edibles that work together for long-term sustainability.

1. CANOPY LAYER:

Macadamia Nut (*Macadamia integrifolia*): The macadamia nut tree is known for its edible nuts, which are rich in healthy fats and have a creamy texture. It grows well in USDA Zone 9, reaching heights of up to 40 feet. The tree provides significant canopy cover, helps prevent soil erosion, and attracts pollinators, making it an important addition to the food forest.

Rambutan (*Nephelium lappaceum*): Rambutan is a tropical fruit tree that thrives in USDA Zone 9, producing spiky, red fruits with juicy, sweet flesh. The tree can grow up to 80 feet tall, providing a thick canopy that offers shade and habitat for wildlife. Its fruits are rich in vitamin C and antioxidants, and the tree helps improve soil quality through its organic matter contributions.

2. UNDERSTORY LAYER:

Chilean Myrtle (*Luma apiculata*): This small evergreen tree produces edible black berries with a sweet flavor, often used in jams and jellies. Its dense, glossy leaves create a thick understory layer, while the tree also offers year-round wildlife habitat. It thrives in mild, frost-free climates and is highly ornamental.

Grumichama (*Eugenia brasiliensis*): This rare, tropical fruit tree produces cherry-like fruits that are sweet and juicy. The fruits are excellent for fresh eating, as well as making jams and desserts. The tree is highly ornamental with dark, glossy leaves and thrives in warm climates, providing shade and shelter for other understory plants.

3. SHRUB LAYER:

Texas Sage (*Leucophyllum frutescens*): Also known as "Cenizo," this drought-tolerant shrub produces edible flowers that can be used in salads. It is an ornamental shrub with silver-gray foliage and purple blooms, thriving in dry, arid climates. Texas Sage helps improve soil structure and attracts pollinators to the food forest.

Strawberry Tree (*Arbutus unedo*): This evergreen shrub produces small, red, strawberry-like fruits that are sweet and can be eaten fresh or made into jams and liqueurs. The tree is drought-tolerant and thrives in Mediterranean climates, providing year-round structure and adding biodiversity to the food forest.

4. HERBACEOUS LAYER:

Winter Savory (*Satureja montana*): This perennial herb has edible leaves that are used as a seasoning for meats and stews. Winter Savory thrives in warm, sunny climates and is highly drought-tolerant, making it a valuable addition for its culinary use and its ability to repel insect pests in the food forest.

Purslane (*Portulaca oleracea*): The succulent leaves and stems of this herb are edible and rich in omega-3 fatty acids, making it a nutritious addition to salads. Purslane thrives in warm, dry climates and is drought-tolerant, providing ground cover and improving soil moisture retention in the food forest.

5. ROOT LAYER:

Oca (*Oxalis tuberosa*): The tubers of this perennial plant are edible, with a tangy, lemony flavor. Oca grows best in warm climates with well-drained soil, and it contributes to soil health by loosening compacted soil as the tubers grow, offering a unique and colorful root crop.

Parsnip (*Pastinaca sativa*): This biennial plant has edible roots with a sweet, earthy flavor that improves after frost exposure. Parsnip thrives in deep, well-drained soils and provides a nutrient-dense root vegetable while enhancing soil aeration and attracting pollinators in its second year of growth.

6. GROUND COVER LAYER:

Sweet Woodruff (*Galium odoratum*): Sweet Woodruff is a fragrant, edible groundcover whose leaves and flowers are used in teas, wines, and desserts. It thrives in moist, shady environments and forms a dense carpet of greenery, perfect for controlling weeds and preventing soil erosion. Sweet Woodruff is also known for its ability to attract beneficial insects.

Houttuynia (*Houttuynia cordata*): This low-growing herb produces heart-shaped leaves that are edible and used in Asian cuisine for their peppery, citrusy flavor. Houttuynia thrives in wet, shady environments and forms a dense mat, helping to suppress weeds and improve soil moisture retention. Its unique foliage adds visual interest to the groundcover layer.

7. VERTICAL LAYER:

Chilean Nettle Tree (*Solanum crispum*): Despite its common name, this vine is not related to nettles. It produces small, edible berries that can be eaten when fully ripe. The vine is fast-growing and can provide vertical coverage, offering food and habitat for beneficial insects.

Magnolia Vine (*Schisandra chinensis*): This climbing vine is valued for its edible red berries, known as "five-flavor fruit," which are used in traditional Chinese medicine. It thrives in partial shade, making it a good choice for layering in a food forest. The berries can be eaten fresh or dried, and the vine adds medicinal value to the forest ecosystem.

8. FUNGAL LAYER:

Porcini (*Boletus edulis*): Also known as "King

Bolete," this mushroom is revered for its robust flavor and meaty texture. It forms mycorrhizal relationships with trees, particularly oaks and pines, making it a valuable fungal addition to a food forest in Zone 9. Porcini mushrooms help enhance soil health while offering a delicious, edible fruiting body.

Hedgehog Mushroom (*Hydnum repandum*): Named for the small, tooth-like structures under its cap, the hedgehog mushroom is edible and has a sweet, nutty flavor. It grows in symbiosis with trees, often in shady, wooded areas, and contributes to nutrient cycling by decomposing organic matter. It's a rare but beneficial species for the fungal layer in a food forest.

Zone 9 Plan Six: Sunshine Sprouts

Bathed in abundant, gentle sunshine, this strategy pairs sprouting edibles in every layer. It is designed to boost early growth and maintain a reliable food source while nurturing the ecosystem's foundation.

1. CANOPY LAYER:

Ice Cream Bean (*Inga edulis*): This fast-growing canopy tree produces long pods filled with sweet, cotton-like pulp, known as ice cream beans. It thrives in USDA Zone 9 and can grow up to 100 feet tall. The tree is nitrogen-fixing, improving soil fertility, and its large canopy provides shade and wind protection, benefiting the overall food forest ecosystem.

Monkey Pod Tree (*Samanea saman*): Also known as rain tree, this fast-growing tree produces edible seeds that can be roasted and eaten. It thrives in USDA Zone 9 and can reach heights of up to 80 feet, creating a broad canopy. The tree is nitrogen-fixing, enriching the soil, and provides shelter for wildlife, making it an important component of the food forest.

2. UNDERSTORY LAYER:

Natal Plum (*Carissa macrocarpa*): This small, spiny shrub produces bright red fruits that are sweet and tangy, often used in jams and desserts. The plant is drought-tolerant and salt-tolerant, making it ideal for coastal areas. It also provides dense cover for wildlife, making it valuable for a diverse ecosystem.

Cocoplum (*Chrysobalanus icaco*): This native shrub produces small, edible fruits that can be eaten fresh or made into jellies. Cocoplum is tolerant of a wide range of soils, including sandy coastal conditions, and provides habitat for wildlife. It also acts as a low-maintenance understory shrub, improving soil structure.

3. SHRUB LAYER:

Chinese Quince (*Pseudocydonia sinensis*): This deciduous shrub produces large, fragrant fruits that are edible when cooked and often used in jams and desserts. Chinese Quince is highly ornamental, with attractive bark and fragrant flowers, making it a valuable addition to the shrub layer of the food forest.

Saffron Plum (*Sideroxylon celastrinum*): This rare, drought-tolerant shrub produces small, edible purple fruits with a sweet, plum-like flavor. Saffron Plum is well-suited for warm climates and provides habitat for wildlife, while its fruits add a unique edible component to the food forest.

4. HERBACEOUS LAYER:

Wild Bergamot (*Monarda fistulosa*): The leaves and flowers of this herb are edible and can be used to make tea or as a flavoring in dishes. Wild Bergamot thrives in dry, sunny areas, attracting pollinators and beneficial insects, while providing medicinal benefits and edible flowers for the herbaceous layer.

Sea Beet (*Beta vulgaris subsp. maritima*): This coastal herb has edible leaves and roots that can be used in salads, cooked dishes, or as a beet substitute. Sea Beet thrives in saline soils and sandy environments, making it ideal for coastal food forests while providing nutrient-rich food sources.

5. ROOT LAYER:

Mashua (*Tropaeolum tuberosum*): The tubers of this plant are edible and known for their spicy, peppery flavor. Mashua thrives in warm climates, climbing vertically or sprawling across the ground, and is highly productive, adding nitrogen-fixing benefits and pest control to the food forest.

Canna (*Canna indica*): The rhizomes of this plant are edible and starchy, often boiled or roasted.

Canna thrives in hot, sunny environments and wet soils, making it ideal for low-lying areas of a food forest. It also offers large, striking flowers that attract pollinators and beneficial insects.

6. GROUND COVER LAYER:

Mallow (*Malva neglecta*): Mallow is a low-growing herb with edible leaves and flowers, commonly used in salads or as a medicinal plant. It thrives in poor, well-drained soils and can tolerate full sun or partial shade. Mallow helps improve soil structure and brings diversity to the food forest's herbaceous layer.

Ground Ivy (*Glechoma hederacea*): Also known as Creeping Charlie, this plant produces edible leaves that can be used in salads or herbal teas. It forms a thick mat of groundcover, thriving in both sun and shade, and helps control soil erosion. Ground Ivy is resilient and can survive in a variety of soil conditions, making it a great addition to the forest floor.

7. VERTICAL LAYER:

Snake Gourd (*Trichosanthes cucumerina*): Snake gourd is a tropical vine that produces long, edible gourds used in soups and stir-fries. It grows well in Zone 9, requiring a trellis or support to manage its fast growth. This vine brings a unique, exotic element to the food forest, along with providing a food source.

Tropical Pumpkin (*Cucurbita moschata*): A sprawling vine that can be trained to grow vertically, tropical pumpkin produces large, edible fruit commonly used in pies and stews. The plant thrives in hot climates and well-drained soils, adding both nutrition and visual interest to the food forest. It also serves as a ground cover if allowed to sprawl.

8. FUNGAL LAYER:

Parasol Mushroom (*Macrolepiota procera*): This tall, distinctive mushroom has a large, umbrella-like cap and is edible when cooked, with a nutty flavor. It prefers grassy areas and edges of forests, making it suitable for the outer edges of a food forest. Parasol mushrooms contribute to the fungal diversity and help break down organic materials, enriching the soil.

Coral Tooth Fungus (*Hericium coralloides*): This rare fungus grows in a cascading, coral-like form on decaying hardwoods. It is edible, with a mild flavor and meaty texture, and is known for its medicinal properties. In a food forest, it acts as a decomposer, breaking down dead wood and returning nutrients to the soil.

Zone 9 Plan Seven: Radiant Regimen

Drawing inspiration from lush, subtropical canopies, this regimen layers species to reflect nature's inherent radiance. It delivers a blend of immediate harvest and enduring support for a resilient food forest.

1. CANOPY LAYER:

Tamarind (*Tamarindus indica*): Tamarind is a tropical tree that produces pods filled with tangy, edible pulp often used in cooking. It grows well in USDA Zone 9 and can reach up to 80 feet in height, offering a broad canopy for shade. The tree is drought-tolerant and provides benefits to the food forest by improving soil fertility through nitrogen fixation.

Cempedak (*Artocarpus integer*): Cempedak is a rare tropical tree that produces large, elongated fruits with sweet, edible flesh similar to jackfruit. It thrives in USDA Zone 9 and can grow up to 65 feet tall, providing dense canopy cover. The tree is known for its rapid growth and ability to thrive in nutrient-poor soils, making it a valuable addition to food forests in tropical climates.

2. UNDERSTORY LAYER:

Mabolo (*Diospyros blancoi*): Also known as Velvet Apple, this tropical tree produces large, velvety-skinned fruits with a creamy, custard-like flesh. The fruits are rich in nutrients and can be eaten fresh or used in desserts. Mabolo grows well in warm, humid climates and is a rare addition to the food forest.

Cedar Bay Cherry (*Eugenia reinwardtiana*): This small shrub produces bright red fruits with a sweet, tangy flavor, often used in jams, jellies, and desserts. The plant thrives in tropical and subtropical regions and is highly adaptable to coastal conditions. Its dense growth provides cover and habitat for wildlife.

3. SHRUB LAYER:

Lemonade Berry (*Rhus integrifolia*): This evergreen shrub produces small, red berries that can be soaked in water to make a tangy lemonade-flavored drink. Lemonade Berry is highly drought-tolerant and thrives in arid, coastal regions, providing food and habitat for birds and insects.

Miracle Fruit (*Synsepalum dulcificum*): This tropical shrub produces small, red berries that alter the taste perception, making sour foods taste sweet. The berries are edible and often used as a novelty fruit. Miracle Fruit thrives in warm, humid climates and provides a unique edible addition to the food forest.

4. HERBACEOUS LAYER:

Giant Fennel (*Ferula communis*): Not to be confused with common fennel, the stems and shoots of Giant Fennel are edible when cooked. This tall, striking herb thrives in dry, rocky soils and adds vertical structure to the herbaceous layer while attracting pollinators and beneficial insects.

Vervain (*Verbena hastata*): The leaves and flowers of this herb are edible and used in teas with calming properties. Vervain thrives in moist soils and attracts pollinators, providing medicinal benefits and a calming tea source for the food forest.

5. ROOT LAYER:

Arrowleaf Elephant Ear (*Xanthosoma sagittifolium*): The tubers of this tropical plant are edible and similar in taste to taro. Arrowleaf Elephant Ear thrives in warm, humid environments and is valued for its large, attractive leaves, making it an ornamental as well as a food-producing plant.

Tuberous Pea (*Amphicarpaea bracteata*): The tubers of this nitrogen-fixing plant are edible, with a taste similar to peanuts. Tuberous Pea thrives in moist soils and shaded environments, offering a valuable food source while improving soil fertility and providing cover in the food forest understory.

6. GROUND COVER LAYER:

Self-Heal (*Prunella vulgaris*): Self-Heal is a low-growing herb with edible leaves and flowers that are often used in teas or as medicinal remedies. It thrives in moist, shaded areas and forms a dense groundcover that protects soil from erosion. Self-Heal attracts pollinators, enhancing biodiversity in the food forest.

Creeping Raspberry (*Rubus calycinoides*): Creeping Raspberry forms a durable, mat-like groundcover and produces small, edible raspberry-like fruits. It thrives in well-drained soils and prefers full sun but can tolerate partial shade. This plant helps with soil stabilization and adds ornamental value with its textured foliage.

7. VERTICAL LAYER:

Vanilla Orchid (*Vanilla planifolia*): A tropical vine that produces vanilla beans and thrives in warm, humid conditions. It provides shade for ground-level plants and attracts bees with its beautiful, fragrant flowers.

Bitter Melon (*Momordica charantia*): This fast-growing vine is commonly grown in tropical and subtropical regions, producing edible, bitter fruits that are used in Asian and Indian cuisines. Bitter melon thrives in hot, humid climates and can quickly cover trellises or other vertical supports. Its fruits are known for their medicinal properties, particularly in regulating blood sugar levels.

8. FUNGAL LAYER:

Puffball Mushroom (*Calvatia gigantea*): Giant puffballs are round, white mushrooms that can grow to an impressive size. When young and firm, they are edible and have a mild, earthy flavor. Puffballs play an essential role in decomposing dead plant material and improving soil structure in a food forest.

Black Trumpet Mushroom (*Craterellus cornucopioides*): Also known as "Horn of Plenty," this funnel-shaped mushroom is prized for its smoky, rich flavor. It grows in symbiotic relationships with hardwoods in USDA Zone 9. Black trumpets are valuable for their role in mycorrhizal networks, enhancing tree health while providing an edible yield.

Zone 9 Plan Eight: Eco Elysium Edifice

Zone 9 Plan Eight: Eco Elysium Edifice

I designed this edifice to represent a mini paradise where every layer contributes to both food production and ecosystem harmony. It's intended to provide a sanctuary of edibles and natural balance in a subtropical climate.

1. CANOPY LAYER:

Baobab (*Adansonia digitata*): The baobab tree, often called the "tree of life," produces edible fruits high in vitamin C and antioxidants. It grows well in USDA Zone 9 and can reach impressive heights of up to 100 feet. This tree is highly drought-tolerant and its large trunk stores water, helping to regulate the microclimate in the food forest.

Pindo Palm (*Butia capitata*): The pindo palm, also known as jelly palm, produces small, yellow-orange fruits that are edible and often made into jellies or wine. It thrives in USDA Zone 9 and can grow up to 25 feet tall, providing a smaller canopy option. The tree is drought-tolerant and its fruits attract pollinators, contributing to biodiversity in the food forest.

2. UNDERSTORY LAYER:

Monstera (*Monstera deliciosa*): Known for its dramatic, split leaves, this tropical vine produces unique, pineapple-flavored fruits. The fruits are edible once fully ripened, adding an exotic touch to the food forest. Monstera thrives in warm, humid climates and can provide a lush, ornamental understory while climbing on trees.

Jaboticaba (*Plinia cauliflora*): This unique tree produces grape-like fruits directly on its trunk, offering a striking visual and edible feature. The fruits are sweet and often used in jellies, wines, and liqueurs. Jaboticaba thrives in warm, humid climates and provides excellent shade in the understory layer.

3. SHRUB LAYER:

Texas Persimmon (*Diospyros texana*): This small shrub produces small, black fruits that are sweet and edible when fully ripe. Texas Persimmon is drought-tolerant and well-suited to arid regions, providing fruit for wildlife and adding to the diversity of the shrub layer in the food forest.

Yellowhorn (*Xanthoceras sorbifolium*): This rare shrub produces small, edible nuts that can be eaten roasted or ground into flour. Yellowhorn is highly drought-tolerant and thrives in poor soils, making it an excellent plant for adding edible nuts and enhancing soil fertility in the food forest.

4. HERBACEOUS LAYER:

Horehound (*Marrubium vulgare*): The leaves of this herb are edible and commonly used to make herbal teas or candies. Horehound thrives in dry, poor soils and is drought-tolerant, providing medicinal value to the food forest and attracting pollinators with its small white flowers.

Salad Burnet (*Sanguisorba minor*): The young leaves of this herb are edible and add a cucumber-like flavor to salads. Salad Burnet thrives in dry, well-drained soils and is drought-tolerant, making it a valuable addition to the herbaceous layer for its edible leaves and its ability to attract pollinators.

5. ROOT LAYER:

Daylily (*Hemerocallis fulva*): The tubers of this plant are edible and can be roasted or boiled, offering a potato-like texture. Daylily grows well in a wide range of soils and provides seasonal flowers that attract pollinators while offering additional edible buds and shoots.

Yellow Nutsedge (*Cyperus esculentus*): The tubers of this sedge plant are edible and commonly used to make a sweet, milky beverage. Yellow Nutsedge grows vigorously in warm, sandy soils and improves soil structure, acting as a beneficial groundcover while providing a nutritious root crop.

6. GROUNDCOVER LAYER:

Chocolate Mint (*Mentha x piperita f. citrata*): Chocolate Mint is a low-growing herb with a distinct chocolatey aroma and flavor, commonly used in teas and desserts. It spreads quickly, providing excellent groundcover, especially in moist, partially shaded areas. Chocolate Mint attracts pollinators and can act as a pest repellent.

Partridgeberry (*Mitchella repens*): Partridgeberry is a creeping evergreen groundcover that produces edible red berries, which are mildly sweet. It thrives in acidic, shady environments, making it perfect for the forest floor. This plant is also beneficial for soil stabilization and is known for its decorative value with bright berries and delicate leaves.

7. VERTICAL LAYER:

Tindora (*Coccinia grandis*): Also called ivy gourd, tindora is a climbing vine that produces small, cucumber-like fruits that are edible when young. It grows quickly and can provide dense green coverage on trellises or fences. The fruits are used in a variety of Asian dishes and are high in vitamins and minerals.

Purple Hyacinth Bean (*Lablab purpureus*): This vigorous climbing vine produces edible pods and beans when cooked properly. It's a beautiful plant with deep purple stems and flowers, adding both aesthetic and edible value to the vertical layer. Purple hyacinth bean fixes nitrogen in the soil, benefiting the surrounding plants.

8. FUNGAL LAYER:

Deer Mushroom (*Pluteus cervinus*): Deer mushrooms grow on decaying hardwood and are edible when young, with a mild, earthy taste. They are important in a food forest for breaking down wood and contributing to the nutrient cycle. Their presence adds to the fungal diversity and supports overall ecosystem health.

Velvet Foot Mushroom (*Flammulina velutipes*): Also known as "Enoki," this mushroom is prized for its delicate texture and mild flavor, often used in Asian cuisine. It grows in clusters on decaying wood, particularly in cooler months, making it a good candidate for fall and winter harvests in USDA Zone 9. Velvet foot helps decompose wood, enriching the soil.

Zone 9 Plan Nine: Layered Lush Lab

Think of this plan as a living laboratory where layered experiments yield robust, edible results. It encourages innovation in pairing species to produce a system that is as nourishing as it is sustainable.

1. CANOPY LAYER:

African Breadfruit (*Treculia africana*): African breadfruit is a rare tree that produces large, starchy fruits, which are edible and used in cooking like traditional breadfruit. It grows well in USDA Zone 9, reaching heights of up to 100 feet. The tree is highly productive and provides food, shelter, and soil improvement through its nutrient-rich leaf litter.

Soursop (*Annona muricata*): Soursop is a tropical fruit tree that produces large, spiky fruits with soft, sweet-sour flesh. It thrives in USDA Zone 9 and can grow up to 30 feet tall. The tree is valued for its medicinal properties, and its fruit is rich in vitamin C and fiber, making it a valuable addition to the food forest for both humans and wildlife.

2. UNDERSTORY LAYER:

Star Apple (*Chrysophyllum cainito*): This tropical tree produces star-shaped, sweet fruits that are creamy and refreshing. The fruits are eaten fresh or used in desserts, and the tree itself is highly ornamental with glossy leaves. Star Apple is well-suited for tropical climates and provides shade and shelter for other plants in the understory.

Lucuma (*Pouteria lucuma*): This small tree produces golden, sweet fruits with a unique, custard-like flavor often used in desserts and smoothies. The fruits are rich in vitamins and antioxidants, making them highly nutritious. Lucuma thrives in warm climates and adds an exotic touch to the understory of a food forest.

3. SHRUB LAYER:

Pigeon Pea (*Cajanus cajan*): This nitrogen-fixing shrub produces edible seeds that are commonly used in soups, stews, and curries. Pigeon Pea is drought-tolerant and thrives in warm, tropical climates, improving soil health and providing high-protein food for humans and wildlife.

Tea Plant (*Camellia sinensis*): This small shrub produces leaves that are harvested to make tea. The leaves are edible and can be used to make green, black, or white tea. Tea Plant thrives in warm, humid climates and provides an additional product for a diverse, multifunctional food forest.

4. HERBACEOUS LAYER:

Society Garlic (*Tulbaghia violacea*): The leaves and flowers of this herb are edible and have a mild garlic flavor, often used in salads and soups. Society Garlic thrives in hot, dry climates and adds a decorative and edible component to the food forest while repelling pests with its garlic-like scent.

Costmary (*Tanacetum balsamita*): The leaves of this herb are edible and have a minty flavor, often used in teas and flavoring for desserts. Costmary thrives in well-drained soils and sunny locations, providing medicinal benefits and attracting pollinators to the herbaceous layer of the food forest.

5. ROOT LAYER:

Ground Ivy (*Glechoma hederacea*): The creeping roots of this plant are edible and can be used to make teas with medicinal benefits. Ground Ivy thrives in moist, shaded environments, offering groundcover and erosion control, while also attracting pollinators with its small, purple flowers.

Tigernut Sedge (*Cyperus rotundus*): The tubers of this plant are edible and historically used for their high oil content. Tigernut Sedge thrives in warm, dry conditions and poor soils, improving soil structure and providing a resilient, drought-tolerant source of edible tubers.

6. GROUNDCOVER LAYER:

Wild Ginger (*Zingiber mioga*): Not to be confused with the North American wild ginger, this Asian variety produces edible flower buds and shoots. It thrives in shady, moist areas, making it ideal for the groundcover layer in a food forest. Wild Ginger adds culinary diversity and improves soil moisture retention.

Roman Chamomile (*Chamaemelum nobile*): A perennial ground cover with daisy-like flowers that attract pollinators. Chamomile helps deter pests and enriches the soil with its organic matter.

7. VERTICAL LAYER:

Butterfly Pea (*Clitoria ternatea*): Known for its vibrant blue flowers, butterfly pea is a climbing vine that is both ornamental and functional. The flowers are used in teas and food coloring, and the plant thrives in warm climates. It can easily be trained on a trellis, offering a splash of color and an edible product to the food forest.

Scarlet Gourd (*Coccinia grandis*): This vining plant produces edible gourd-like fruits and has striking red flowers that attract pollinators. The young shoots, leaves, and fruits are all edible and used in various cuisines. Scarlet gourd grows quickly and provides vertical coverage, making it useful in food forests for both aesthetic and practical purposes.

8. FUNGAL LAYER:

Bay Bolete (*Imleria badia*): This edible mushroom grows under conifers and hardwoods, forming mycorrhizal relationships that improve tree health. Its fruiting body is brown with a soft texture and nutty flavor, making it a valuable culinary addition. Bay boletes contribute to the fungal biodiversity of the forest floor, supporting nutrient exchange between plants and soil.

Candy Cap Mushroom (*Lactarius rubidus*): Known for its sweet, maple-like aroma, the candy cap mushroom is edible and often used in desserts or to flavor syrups. It grows in symbiosis with trees, helping to enhance soil fertility through nutrient cycling. This mushroom adds a unique flavor element to the fungal layer of the food forest.

Zone 9 Plan Ten: Harmony Harvest Hub

A hub of balanced growth, this plan integrates food production with ecological harmony through careful layering. It is designed to harvest nutritional benefits while building a system that naturally supports itself.

1. CANOPY LAYER:

Soursop (*Annona muricata*): Soursop is a tropical fruit tree that produces large, spiky fruits with soft, sweet-sour flesh. It thrives in USDA Zone 9 and can grow up to 30 feet tall. The tree is valued for its medicinal properties, and its fruit is rich in vitamin C and fiber, making it a valuable addition to the food forest for both humans and wildlife.

Jackfruit (*Artocarpus heterophyllus*): The jackfruit tree produces massive, edible fruits with sweet, fibrous flesh. It thrives in USDA Zone 9 and can reach up to 80 feet tall, providing a significant canopy for shade. The tree is highly productive, and its fruits are used in both savory and sweet dishes, while its large leaves provide organic matter for improving soil fertility.

2. UNDERSTORY LAYER:

Mulberry (*Morus nigra*): This small tree produces dark, sweet, and juicy berries that are highly nutritious and packed with antioxidants. The fruits can be eaten fresh, dried, or used in jams and desserts. Mulberry trees grow quickly and provide excellent shade in the understory, while also attracting pollinators and wildlife.

Curry Leaf Tree (*Murraya koenigii*): This small, aromatic tree produces edible leaves that are commonly used in cooking for their distinctive flavor. The tree is highly drought-tolerant and thrives in warm, subtropical climates. Its fragrant leaves add culinary value to the food forest, while the tree itself helps improve soil health.

3. SHRUB LAYER:

Aronia Berry (*Aronia arbutifolia*): This deciduous shrub produces small, edible berries that are rich in antioxidants and commonly used in juices, jams, and wines. Aronia Berry is highly adaptable to various soil types and conditions, providing both food and wildlife habitat in a food forest.

Caneberry (*Rubus ursinus*): This trailing shrub produces small, sweet, blackberry-like fruits that are edible and often used in jams and desserts. Caneberry thrives in well-drained soils and can be grown along edges of the food forest, providing fruit while supporting pollinators and wildlife.

4. HERBACEOUS LAYER:

Marsh Marigold (*Caltha palustris*): The young leaves and flower buds of this herb are edible when cooked and often used in spring dishes. Marsh Marigold thrives in wet, marshy environments, providing edible greens while improving water retention and attracting pollinators in moist areas of the food forest.

Good King Henry (*Blitum bonus-henricus*): This rare perennial herb produces edible leaves that can be used as a spinach substitute and edible flower buds similar to broccoli. Good King Henry thrives in nutrient-poor soils and tolerates drought, making it a valuable food source for both people and pollinators.

5. ROOT LAYER:

Siberian Onion (*Allium ochotense*): The bulbous roots of this plant are edible and have a strong onion flavor. Siberian Onion thrives in well-drained soils and full sun, offering an alternative to more common onions while attracting pollinators and providing nutrient-dense bulbs.

Lovage (*Levisticum officinale*): The roots of this herbaceous plant are edible and have a flavor similar to celery. Lovage thrives in sunny, well-drained areas, providing a perennial food source for both the roots and leaves while adding height and aromatic qualities to the food forest.

6. GROUND COVER LAYER:

Creeping Sage (*Salvia sonomensis*): Creeping Sage is a drought-tolerant groundcover that produces edible leaves, which are often used as a seasoning in culinary dishes. It thrives in well-drained soils and sunny locations but can tolerate partial shade. Creeping Sage is beneficial for erosion control and attracting pollinators.

Dwarf Mondo Grass (*Ophiopogon japonicus*): Although not commonly eaten, Dwarf Mondo Grass can be used in herbal medicine. This evergreen groundcover thrives in shady, moist environments and provides excellent soil protection from erosion. Its dense, tufted growth is ideal for filling gaps in the forest floor.

7. VERTICAL LAYER:

Perennial Cucumber (*Cyclanthera pedata*): Also known as Achocha, this vine produces cucumber-like fruits that are edible and often used in salads and stir-fries. It is a fast-growing plant in USDA Zone 9 and can provide abundant yields of fresh, crunchy fruits. The vine is hardy and can cover fences or trellises with ease.

Red Sissoo Spinach (*Alternanthera sessilis*): A fast-growing vine with edible leaves, red sissoo spinach thrives in warm, tropical climates. The leaves are rich in vitamins and can be used raw in salads or cooked. Its growth habit makes it an excellent choice for covering trellises or vertical structures in a food forest.

8. FUNGAL LAYER:

Pheasant's Back Mushroom (*Polyporus squamosus*):

This mushroom, with its distinct scaly pattern, grows on dead or decaying hardwood. It is edible when young, with a mild, cucumber-like taste. Pheasant's Back mushrooms help decompose dead wood, enriching the soil and creating habitat for beneficial organisms in the food forest.

Winter Chanterelle (*Craterellus tubaeformis*): This small, trumpet-shaped mushroom grows in cool, damp conditions and is edible, with a slightly fruity flavor. It plays an important role in the fungal layer of the food forest by forming mycorrhizal networks with trees and contributing to nutrient cycling. Its late-season fruiting provides an extended mushroom harvest.

ZONE 10

Plant Hardiness Zone 10

Zone 10 is among the warmest regions on the USDA Plant Hardiness Zone map, with average annual minimum temperatures ranging between 30°F to 40°F (-1.1°C to 4.4°C). This zone features long, hot growing seasons with mild winters, making it perfect for tropical and subtropical plants that thrive in prolonged heat and are not frost-tolerant.

Locations within Zone 10

Arizona: Southern parts, including Tucson, Phoenix, and Yuma.

California: Coastal regions of Southern California, such as Los Angeles, San Diego, and Santa Barbara, as well as inland areas like Palm Springs and parts of the Mojave Desert.

Florida: Southern Florida, including Miami, Fort Lauderdale, and Key West.

Hawaii: Most of the state, particularly coastal and lowland regions across the islands.

Louisiana: Coastal zones, including New Orleans and the southern parts of the state.

Texas: Southern Texas, including Brownsville, McAllen, and the Lower Rio Grande Valley.

Climate Characteristics of Zone 10

Temperature: Zone 10 has consistently warm temperatures, with summer highs often surpassing 90°F (32°C). Winters are mild, with nighttime temperatures rarely dropping below 40°F (4°C). These warm conditions are ideal for heat-loving plants, including many tropical crops. Rainfall: Rainfall varies by region in Zone 10. Southern California, for instance, has dry summers and wet winters due to its Mediterranean climate. Florida and parts of Texas experience more consistent rainfall but still face occasional droughts.

Frost: Frost is infrequent in Zone 10 and only occurs on the coldest winter nights. Most frost-sensitive plants are safe from freezing, making this zone suitable for tropical and subtropical crops.

WIND AND HUMIDITY

Wind: Coastal areas are prone to strong winds, particularly in southern California, the Gulf Coast, and Hawaii. While these winds can cool the environment, they can also harm plants. Using trellises, windbreaks, or strategically placed trees can help shield plants from wind damage.

Humidity: Coastal regions tend to be humid, benefiting tropical plants but also encouraging fungal diseases. In contrast, inland desert areas experience dry conditions, requiring careful water management for successful plant growth.

Soil Types and Conditions in Zone 10

Zone 10 soils range from sandy and alkaline in coastal areas to slightly more fertile and acidic inland. Irrigation is crucial during dry periods, especially in the summer. Organic matter, such as compost, can improve soil structure and fertility. Good drainage is essential for growing healthy plants in this zone.

Growing Seasons in Zone 10

Zone 10 has a year-round growing season, but the hottest months, from late spring to early fall, can be challenging for plants that prefer cooler temperatures or shade. Shade structures or heat-tolerant plants are recommended for successful growth during this period.

Common Challenges in Zone 10

Water Management: Extended heat and sporadic rainfall make water conservation essential. Drip irrigation, rainwater harvesting, and drought-tolerant plants are necessary for sustainable food forest management. Mulching and soil amendments can help retain moisture.

Pests and Diseases: The warm, humid conditions create a breeding ground for pests like aphids, whiteflies, and mealybugs. Regular monitoring is needed to manage infestations. Fungal diseases also thrive in the heat and humidity, so ensuring proper air circulation and using organic fungicides may be required.

Soil Erosion: Sandy or dry soils in Zone 10 are prone to erosion, especially during heavy rain or strong winds. Groundcover plants, mulch, and proper water management can help minimize erosion.

Salt and Coastal Conditions: Coastal regions face salt exposure, which can harm plants. Salt-tolerant varieties and windbreaks are important for protecting gardens in these areas.

Zone 10's mild winters, long growing seasons, and abundant sunlight make it an excellent environment for tropical and subtropical plants. However, careful water management and pest control are essential for maintaining a healthy, thriving garden.

Zone 10 Plan One: Edible Expedition

Embark on an expedition into equatorial abundance with this daring plan. It layers tropical and warm-climate species to deliver a bold harvest while setting the stage for long-term ecological support.

1. CANOPY LAYER:

Pulasan (*Nephelium ramboutan-ake*): Pulasan is a rare tropical fruit tree closely related to rambutan, thriving in USDA Zone 10. Its sweet, juicy flesh surrounds a large seed, and the fruit is similar in taste to lychee. Pulasan trees grow up to 40 feet tall, forming a valuable part of the canopy with dense shade. They attract pollinators and are prized for both their delicious fruit and their role in protecting soil and smaller plants from excessive sun exposure.

Ice Cream Bean (*Inga edulis*): This leguminous tree grows up to 60 feet and produces long, bean-like pods filled with sweet, cottony pulp, which is edible. The seeds are also edible when cooked. In addition to providing delicious fruit, the Ice Cream Bean tree enriches the soil with nitrogen, improving fertility in the food forest. It is fast-growing and offers excellent shade for understory plants.

2. UNDERSTORY LAYER:

Black Pepper (*Piper nigrum*): Black pepper is a tropical vine that thrives in USDA Zone 10, making it an excellent understory plant. The edible part is the peppercorn, which is harvested and dried to produce black, white, or green pepper. It climbs by twining around supports and provides a steady source of spice while adding a tropical look to the food forest. It also attracts beneficial insects and helps in companion planting.

Allspice (*Pimenta dioica*): Native to the Caribbean, allspice is an evergreen shrub that produces aromatic berries used as a spice in cooking. The berries, leaves, and even the bark contain essential oils that are highly valued. It thrives in partial shade, making it a good choice for the understory. It adds diversity to the forest ecosystem and helps enhance soil health.

3. SHRUB LAYER:

Indian Jujube (*Ziziphus mauritiana*): Indian Jujube is a drought-tolerant shrub that thrives in USDA Zone 10, producing small, sweet, and tangy fruits. The fruits are eaten fresh, dried, or made into jellies and are rich in vitamin C. This hardy plant adds biodiversity to the food forest and can tolerate poor soil conditions while providing edible fruit.

Curry Leaf Tree (*Murraya koenigii*): This aromatic shrub is famous for its leaves, used extensively in South Asian cuisine. It grows well in tropical climates like Zone 10, reaching up to 6-15 feet. Besides its culinary uses, curry leaves are known for their medicinal properties, including digestive and anti-inflammatory benefits.

4. HERBACEOUS LAYER:

Society Garlic (*Tulbaghia violacea*): Society garlic produces edible flowers and leaves, both of which have a mild garlic flavor, making it a versatile addition to salads and savory dishes. This herbaceous perennial thrives in sunny locations, attracting beneficial pollinators to the food forest. It is drought-tolerant and can help repel pests with its garlic-like scent, contributing to the overall health of the forest ecosystem.

Mashua (*Tropaeolum tuberosum*): Mashua is a rare Andean tuber with edible roots that are high in antioxidants and vitamin C. The plant also has edible flowers with a spicy, peppery taste. This climbing herbaceous perennial thrives in partial shade, making it a good fit for the understory of a food forest. It helps deter pests due to its natural insect-repellent properties.

5. ROOT LAYER:

Skirret (*Sium sisarum*): Skirret is a lesser-known root vegetable with edible, sweet, and nutty-tasting roots. The roots are rich in vitamins and minerals, and they grow in clusters, making them easy to harvest. Skirret thrives in moist, well-drained soils and adds a unique, nutritious root crop to the food forest. It also helps improve soil structure with its fibrous roots.

Shallot *(Allium cepa var. aggregatum)*: Shallots are a close relative of garlic and onions, with edible bulbs that have a milder flavor. They thrive in the warmer, subtropical climate of Zone 10 and can be used both fresh and in cooked dishes. Shallots are excellent for natural pest deterrence, contributing to the health of the food forest ecosystem.

6. GROUNDCOVER LAYER:

Dymondia (*Dymondia margaretae*): Dymondia is an attractive, low-growing groundcover with silver-green leaves and tiny yellow flowers. Its edible leaves can be used in salads or as an herb in cooking, offering a mild, slightly bitter flavor. Dymondia is drought-tolerant and thrives in full sun to partial shade, providing excellent coverage for the ground layer of a food forest, while also helping prevent soil erosion and retaining moisture.

Sweet Potato Vine (*Ipomoea batatas*): The sweet potato vine is a fast-growing groundcover that produces edible tubers, rich in starch and vitamins. The vine itself can also be used as a nutritious green in salads and stir-fries. This versatile plant grows well in the warm Zone 10 climate, spreading quickly to provide ample ground coverage, while its deep roots can help prevent soil erosion and improve soil quality.

7. VERTICAL LAYER:

Dragon Fruit (*Hylocereus spp.*): Dragon fruit, also

known as pitaya, is a vibrant, tropical cactus that produces striking, edible fruit. The flesh of the fruit is sweet and packed with nutrients, and the plant is relatively low-maintenance, thriving in hot, dry conditions. As a climbing cactus, it provides vertical coverage for the food forest, and its flowers attract pollinators like bats and moths, making it beneficial to the ecosystem.

Rattan Palm (*Calamus rotang*): Rattan palm is a tropical vine that produces edible shoots, which are eaten in Southeast Asia and are known for their tender texture. This vine can grow tall and provide valuable vertical space in the food forest. It is a perfect candidate for climbing and producing edible stems while helping to support other plants by acting as a living trellis.

8. FUNGAL LAYER:

Turkey Tail Mushroom (*Trametes versicolor*): Turkey tail is a fascinating polypore fungus known for its colorful, fan-shaped caps that resemble the tail of a turkey. While not typically consumed for flavor, it has potent medicinal properties, especially for immune support and anti-inflammatory benefits. This fungus thrives on decaying wood, making it an excellent decomposer in a food forest ecosystem, helping to recycle nutrients into the soil and promoting soil health.

Chaga Mushroom (*Inonotus obliquus*): Chaga mushroom is a parasitic fungus that grows on birch trees and is known for its rich antioxidant properties. Although not typically eaten fresh, it is commonly used to make teas that are said to boost immune function and reduce inflammation. In the food forest, chaga contributes to nutrient cycling by breaking down woody material and promoting healthier forest dynamics.

Zone 10 Plan Two: Tropic Tonic

Infused with the energy of warmth, this plan employs tactical pairings for maximum flavor and resilience. It's dedicated to crafting a system that provides both immediate nutrition and lasting sustainability.

1. CANOPY LAYER:

Candle Nut Tree (*Aleurites moluccanus*): Also known as Kukui Nut, this tree grows up to 50-80 feet tall in USDA Zone 10 and produces nuts that can be eaten after cooking or used for oil extraction. The nuts were historically used as candles, giving the tree its common name. Its dense canopy provides shade, and its nutrient-rich leaves improve the soil as they decompose, while the tree also serves as habitat for wildlife.

Jaboticaba (*Plinia cauliflora*): Jaboticaba, a slow-growing tree native to Brazil, is known for its unique growth habit of producing fruit directly on its trunk and branches. The small, grape-like fruits are edible and prized for their sweet flavor. The tree reaches up to 40 feet in height, contributing to the food forest canopy while offering both fruit and aesthetic appeal. Its slow growth and dense foliage make it a durable, long-term canopy provider.

2. UNDERSTORY LAYER:

Pitanga (*Eugenia uniflora var. Pitanga*): Known as Surinam cherry's wild relative, Pitanga produces small, ribbed, red fruits with a sweet-tart flavor. This shrub is adaptable to USDA Zone 10 and prefers partial shade. The fruit is edible and can be eaten raw or made into jams and jellies. It's a rare and attractive understory plant that also attracts pollinators.

Vanilla Orchid (*Vanilla planifolia*): Vanilla is a vining orchid that produces edible vanilla beans, one of the most sought-after spices in the world. The plant thrives in the warm, humid environment of USDA Zone 10 and requires a tree or trellis for support. Its fragrant flowers attract pollinators, and the beans can be harvested for culinary use, adding significant value to a food forest.

3. SHRUB LAYER:

Barbados Gooseberry (*Pereskia aculeata*): A rare and edible cactus shrub that produces small, tart fruits similar to gooseberries. The leaves are also edible and used in salads or cooked as a vegetable. It grows vigorously in tropical and subtropical zones, offering unique foliage and fruit for the food forest.

Candelabra Tree (*Senna didymobotrya*): This nitrogen-fixing shrub is a great addition to a food forest, especially for enriching the soil. Though not edible itself, it supports other plants by improving soil fertility and is useful for wildlife. It thrives in warm climates like USDA Zone 10.

4. HERBACEOUS LAYER:

New Zealand Spinach (*Tetragonia tetragonioides*): New Zealand spinach is an edible ground-hugging herb that can be used as a substitute for spinach. Its leaves are rich in vitamins A and C, and the plant is heat-tolerant, thriving in warm climates like Zone 10. It is an excellent source of leafy greens year-round and helps suppress weeds due to its low-growing habit.

Purslane (*Portulaca oleracea*): Purslane is a succulent herb with edible leaves that are rich in omega-3 fatty acids, making it a nutritious addition to salads and stir-fries. It is drought-resistant and grows well in poor soils, helping to improve soil quality with its deep roots. Purslane also attracts pollinators, enhancing biodiversity in the food forest.

5. ROOT LAYER:

Crosne (*Stachys affinis*): Crosne, also known as Chinese artichoke, is a unique edible root vegetable with a nutty flavor and a crunchy texture. The tubers are small and knobby, but packed with fiber and nutrients. This plant grows well in well-drained soils, making it a perfect addition to the root layer of a food forest, adding diversity and a fun harvesting experience.

Daikon Radish (*Raphanus sativus var. longipinnatus*): Daikon radish produces large, edible roots that have a mild, peppery flavor and are often used in Asian cuisines. The roots are rich in fiber, antioxidants, and vitamin C, offering a nutritious food source. Daikon radish helps aerate the soil with its long, deep roots, improving soil quality in the food forest.

6. GROUNDCOVER LAYER:

Peanut (*Arachis hypogaea*): The peanut is an edible groundcover that produces underground pods, which are the edible part of the plant. It grows best in well-drained, sandy soils and thrives in warm climates like Zone 10. Peanuts help fix nitrogen in the soil, improving soil fertility, making them a beneficial addition to the food forest ecosystem.

Creeping Sorrel (*Rumex acetosella*): Creeping sorrel is a low-growing herb with tangy, edible leaves that can be used in salads or as a garnish. It thrives in well-drained soils and can handle a variety of growing conditions, making it ideal for a groundcover layer. This plant helps suppress weeds and contributes to the biodiversity of the food forest.

7. VERTICAL LAYER:

Chilean Guava (*Ugni molinae*): The Chilean guava is a small shrub or vine-like plant that produces small, edible berries. These sweet berries can be eaten fresh or used in jams and preserves. This plant thrives in the subtropical climate of Zone 10 and can climb or trail along structures, adding vertical beauty and productivity to the food forest.

Loofah (*Luffa aegyptiaca*): Loofah is a tropical vine that produces edible young fruit, which can be eaten like a squash, and mature fruit, which is harvested to make natural sponges. This plant grows quickly and can easily be trained to climb trellises or structures, offering edible fruits while also contributing to the vertical layer of the food forest.

8. FUNGAL LAYER:

Enoki Mushroom (*Flammulina velutipes*): Enoki mushrooms are delicate, edible fungi with a mild, slightly sweet flavor, often used in soups and salads. They grow in clusters on decaying wood, making them a great fit for the fungal layer of a food forest. Enoki mushrooms can help decompose wood, improving soil fertility and supporting the overall health of the forest ecosystem.

Chanterelle Mushroom (*Cantharellus cibarius*): The chanterelle is a prized edible mushroom with a distinctive trumpet shape and rich, fruity flavor. It is typically found in the wild growing under hardwood trees and prefers slightly acidic soil, making it suitable for the fungal layer of a food forest. Besides providing delicious, foraged food, chanterelles contribute to soil health by forming mycorrhizal relationships with trees, promoting nutrient exchange.

Zone 10 Plan Three: Fervent Flora Framework

A framework built on the fervor of equatorial growth, this plan layers exotic edibles in a strategic formation. It's designed to harness the region's energy for a continuously productive and self-sustaining food forest.

1. CANOPY LAYER:

Mammee Apple (*Mammea americana*): This tropical tree grows up to 60 feet tall and produces large, round, orange-brown fruits with sweet, apricot-like flesh. Both the flesh and seeds are edible, and the fruit is used in a variety of culinary dishes. The Mammee Apple tree provides dense shade and helps retain moisture in the forest, while its fallen leaves contribute to soil fertility.

Langsat (*Lansium parasiticum*): Langsat is a tropical fruit tree that grows up to 40-50 feet tall in USDA Zone 10. It produces clusters of small, sweet, translucent fruits with a slight tartness. The fruits are rich in vitamins and antioxidants, making them a valuable edible resource in a food forest. This tree provides essential canopy cover while improving the biodiversity of the ecosystem.

2. UNDERSTORY LAYER:

Clove Tree (*Syzygium aromaticum*): Clove trees grow well in USDA Zone 10, producing aromatic flower buds that are harvested as cloves, a valuable spice. This slow-growing tree prefers dappled light and works well in the understory layer. Its leaves and bark also contain essential oils, making it a multipurpose plant in the food forest.

Jaboticaba (*Plinia cauliflora*): Jaboticaba, also known as the Brazilian grape tree, produces edible grape-like fruits directly on its trunk. This slow-growing tree thrives in partial shade, making it ideal for the understory layer. The fruits are eaten fresh or used in making wines and jellies, and the tree is known for its ability to tolerate drought once established.

3. SHRUB LAYER:

Natal Plum (*Carissa macrocarpa*): Natal Plum is an evergreen shrub that produces edible, cranberry-like fruits rich in antioxidants. It grows in hot, coastal environments and tolerates salt and drought conditions, making it perfect for Zone 10. The plant's dense growth habit provides excellent windbreaks or hedges for the food forest.

False Guava (*Psidium cattleianum*): This rare fruiting shrub produces red or yellow fruits that are edible and have a tart, guava-like flavor. It thrives in tropical and subtropical regions and is valued for its tolerance to poor soils and resistance to pests, making it a resilient option for food forests.

4. HERBACEOUS LAYER:

Malabar Spinach (*Basella rubra*): This climbing herb produces edible leaves that are thick and slightly mucilaginous, making them perfect for soups, stews, and salads. It thrives in hot, humid climates and grows rapidly, providing a continuous harvest of greens throughout the year. Malabar spinach also adds vertical interest to the herbaceous layer and helps increase productivity in smaller spaces.

Okra (*Abelmoschus esculentus*): Okra is a tropical herbaceous plant with edible pods that are rich in vitamins A and C, and are commonly used in soups and stews. The plant thrives in hot climates, and its tall stature can provide shade to other understory plants. It also attracts pollinators and helps create a microclimate in the food forest.

5. ROOT LAYER:

Salsify (*Tragopogon porrifolius*): Salsify, also known as oyster plant, has edible roots that are prized for their oyster-like flavor. The plant thrives in Zone 10's sunny conditions and can be a great addition to the root layer. Salsify's deep taproot helps break up compacted soil, improving drainage and aeration.

Maca (*Lepidium meyenii*): Maca is a rare root crop that grows well in high altitudes and cooler regions but can also thrive in warmer climates like Zone 10 with appropriate care. The root is edible and has a mildly sweet, nutty flavor. It is highly valued for its health benefits, including boosting energy and balancing hormones, and is a great crop for medicinal and culinary use in a food forest.

6. GROUNDCOVER LAYER:

Minutina (*Veronica filiformis*): Minutina, also known as Italian field salad, is a low-growing edible groundcover with tender leaves that can be used in fresh salads. It thrives in the mild temperatures of Zone 10 and can handle partial shade. The plant adds valuable green matter to the food forest while also providing ground coverage and helping to maintain soil moisture.

Comfrey (*Symphytum officinale*): Comfrey is a deep-rooted perennial that can spread across the ground, producing edible leaves that are packed with nutrients and can be used in teas or as a fertilizer. Its roots help break up compacted soil, and it attracts pollinators with its beautiful flowers. Comfrey's large leaves decompose quickly, adding organic matter to the soil, benefiting the overall health of the food forest.

7. VERTICAL LAYER:

Chilean Bellflower (*Phygelius capensis*): Known for its striking tubular flowers, the Chilean bellflower can be grown as a climbing plant in Zone 10. While primarily ornamental, it can produce edible flowers and leaves that have been used in traditional herbal remedies. This fast-growing climber adds vibrant color and vertical coverage while contributing to the overall biodiversity of the food forest.

Grape (*Vitis vinifera*): Grape vines are well-known for their edible fruit, which can be eaten fresh, dried, or made into juices and jams. These vines thrive in the warm climate of Zone 10, offering a reliable source of fruit while also providing vertical coverage. Grapevines improve the soil by reducing erosion and attract beneficial insects, supporting the overall health of the food forest.

8. FUNGAL LAYER:

Cordyceps Mushroom (*Cordyceps sinensis*): Known for its unique and medicinal properties, cordyceps is a parasitic fungus that grows on insects, particularly caterpillars. While not often eaten fresh, it is highly valued for its ability to boost energy and endurance. In a food forest, cordyceps can enhance biodiversity by breaking down organic material and contributing to natural pest control through its lifecycle.

Coriolus Mushroom (*Trametes versicolor*): Also known as Turkey Tail, this fungus is renowned for its medicinal uses, particularly its support for the immune system. It grows on decaying logs and branches, recycling organic material into usable nutrients for other plants in the food forest. This mushroom is particularly useful for maintaining soil health and fostering a balanced ecosystem by decomposing wood and plant matter.

This initiative brings the vibrant spirit of island ecosystems into a layered design plan. It pairs complementary species to infuse the system with unique flavors and robust ecological support, ideal for equatorial settings.

1. CANOPY LAYER:

Monkeypod Tree (*Albizia saman*): This fast-growing tree can reach up to 80 feet tall and offers valuable shade in a food forest. Its pods are edible when young and provide a sweet pulp that can be consumed raw or cooked. Monkeypod trees are excellent nitrogen-fixers, enriching the soil, and their large, spreading canopy provides shade and protection for smaller plants in the forest.

Bael Tree (*Aegle marmelos*): Bael trees grow up to 50 feet tall in USDA Zone 10 and produce large, round fruits with a hard shell and sweet, aromatic pulp. The fruit is commonly used in drinks and traditional medicine. Bael trees thrive in hot, dry climates and bring resilience to the food forest, contributing to its biodiversity while providing valuable edible fruit and medicinal benefits.

2. UNDERSTORY LAYER:

Soursop (*Annona muricata*): Soursop is a tropical fruit tree that produces large, spiky fruits with a creamy, sweet-sour flesh. This plant thrives in partial shade, making it suitable for the understory. The fruit is edible and used to make beverages, desserts, and traditional medicines. Its broad leaves help create microclimates in the food forest.

Baobab (*Adansonia digitata*): Known as the "tree of life," the baobab can thrive in USDA Zone 10 with careful placement in the understory when young. Its fruit pulp is highly nutritious, rich in vitamin C, and edible, often used in smoothies and health supplements. Baobab trees also contribute to soil fertility by fixing nitrogen.

3. SHRUB LAYER:

Bitter Melon Vine Shrub (*Momordica charantia*): This unique climbing shrub produces bitter, edible fruits that are commonly used in Asian cuisine for their medicinal benefits, including lowering blood sugar levels. The plant thrives in warm, tropical climates like Zone 10 and offers both culinary and health benefits.

Grumichama (*Eugenia brasiliensis*): This slow-growing tropical shrub produces small, black, cherry-like fruits that are sweet and aromatic. It thrives in USDA Zone 10 and adds unique flavors to the food forest while offering a compact, manageable size for understory planting.

4. HERBACEOUS LAYER:

Perilla (*Perilla frutescens*): Perilla is an herb with edible leaves and seeds that are commonly used in Asian cuisine for their unique flavor. The plant is rich in omega-3 fatty acids and antioxidants, adding nutritional diversity to the food forest. It grows well in partial shade and attracts beneficial insects, supporting the overall ecosystem.

Fenugreek (*Trigonella foenum-graecum*): Fenugreek is a nitrogen-fixing herb that improves soil fertility while providing edible seeds and leaves. The seeds are commonly used as a spice, while the leaves are eaten as a green vegetable. This annual herb thrives in warm climates and plays an important role in enhancing the nutrient cycle within the food forest.

5. ROOT LAYER:

Celeriac (*Apium graveolens var. rapaceum*): Celeriac, or celery root, produces edible roots that have a flavor similar to celery and are commonly used in soups and stews. This plant thrives in rich, moist soils and helps enhance the flavor profile of your food forest. It also attracts beneficial insects and pollinators with its flowers.

Chufa (*Cyperus esculentus*): Chufa, also known as tiger nut or earth almond, is a hardy perennial that produces small, edible tubers that are sweet and nutty. Chufa thrives in Zone 10's hot, sunny conditions and has a deep root system that can help prevent soil erosion. The tubers are often used to make horchata, a traditional beverage, and can be harvested for both food and medicinal uses.

6. GROUNDCOVER LAYER:

Dandelion (*Taraxacum officinale*): Dandelion is a well-known edible plant, with all parts of the plant being edible – leaves, roots, and flowers. The leaves can be used in salads, and the roots can be roasted for coffee substitutes. Dandelion thrives in disturbed soils and helps break up compacted ground, making it a useful groundcover that adds both beauty and health benefits to the food forest.

Lemongrass (*Cymbopogon citratus*): Lemongrass is a fragrant herb that serves as both a culinary delight and an excellent groundcover. The stalks are edible and widely used for their citrusy flavor in soups and teas. Lemongrass helps repel pests like mosquitoes, and its dense growth provides good coverage for the ground layer while thriving in Zone 10's warm climate.

7. VERTICAL LAYER:

Tromboncino Squash (*Cucurbita moschata*): Tromboncino squash is a unique, long-fruited variety of squash that can be eaten at different stages—young and tender or fully mature. This vigorous climber provides an excellent vertical presence, allowing for productive fruiting while maximizing space. It is heat-tolerant and grows well in the subtropical conditions of Zone 10.

Chayote Vine (*Sechium edule*): Known for its edible, green fruits, which are often used in cooking as a vegetable, chayote vine is a climber that provides both vertical coverage and edible production. The vines produce tender, edible shoots and leaves, which can be used in salads or stir-fries. This tropical plant thrives in the heat of Zone 10 and adds both texture and flavor to the food forest.

8. FUNGAL LAYER:

Pine Mushrooms (*Tricholoma magnivelare*): Pine mushrooms, also called matsutake, are highly sought-after for their unique and aromatic flavor. They grow symbiotically with pine trees, forming mycorrhizal relationships that benefit both the fungi and the tree. In the food forest, they promote tree health, improve soil fertility, and offer a rare and valuable edible mushroom.

Shaggy Mane Mushroom (*Coprinus comatus*): Shaggy mane mushrooms are edible fungi with a delicate flavor, often used in cooking when young. These mushrooms grow in grassy areas and along decomposing organic matter, contributing to nutrient cycling. As they break down organic material, they improve soil structure and fertility, which helps the overall health of the food forest.

Zone 10 Plan Five: Solar Spice Spectrum

Focusing on a dazzling spectrum of sun-loving spices and edibles, this plan layers species to create a dynamic taste profile. Its purpose is to blend culinary delight with a resilient, sustainable structure.

1. CANOPY LAYER:

Malay Apple (*Syzygium malaccense*): Malay Apple, also known as Mountain Apple, grows up to 60 feet tall and produces crisp, sweet fruits with a refreshing, watery texture. The entire fruit is edible, and it is commonly eaten fresh or used in salads and jams. This tree provides a dense canopy in tropical food forests, offering shade and protection to understory plants.

Santol (*Sandoricum koetjape*): The Santol tree reaches heights of up to 50 feet and produces round, yellow-orange fruits with a sweet, sour flavor. Both the flesh and seeds are edible, with the fruit often used in sauces or jams. The tree's dense foliage creates a thick canopy, helping to regulate moisture levels and provide habitat for beneficial insects and wildlife.

2. UNDERSTORY LAYER:

Miracle Fruit (*Synsepalum dulcificum*): Miracle fruit produces small red berries that temporarily alter taste perception, making sour foods taste sweet. This rare plant grows well in the shade of larger trees, making it ideal for the understory layer in Zone 10. It also attracts pollinators and enhances biodiversity in the food forest.

Black Sapote (*Diospyros texana*): Also known as the chocolate pudding fruit, Black Sapote produces dark, sweet fruits with a texture similar to pudding. This tree thrives in the understory layer of tropical food forests and prefers partial shade. The fruits are highly nutritious and can be used in desserts or eaten fresh.

3. SHRUB LAYER:

Surinam Cherry (*Eugenia uniflora*): Known for its vibrant red, ribbed fruits, Surinam Cherry grows well in hot, tropical climates. The fruits are eaten fresh or used in jams and jellies, and the plant is also ornamental, providing colorful foliage and edible fruits.

Evergreen Huckleberry (*Vaccinium ovatum*): This shrub produces small, dark, edible berries that are rich in antioxidants. It is adapted to coastal and mountainous regions, making it suitable for USDA Zone 10. The plant also attracts pollinators and wildlife, adding value to the ecosystem.

4. HERBACEOUS LAYER:

Cuban Oregano (*Plectranthus amboinicus*): Cuban oregano is a succulent herb with thick, aromatic leaves that are used as a flavoring in soups and stews. It grows well in hot, sunny environments and is drought-tolerant, making it ideal for Zone 10. The plant's strong scent helps repel insects, contributing to natural pest control in the food forest.

Agretti (*Salsola soda*): Agretti is a rare herb with edible leaves that are rich in minerals and have a slightly salty taste, often used in Italian cuisine. It thrives in saline or poor soils, making it a good plant for areas with challenging conditions. Agretti also helps improve soil structure with its deep roots, contributing to soil health in the food forest.

5. ROOT LAYER:

Mitsuba (*Cryptotaenia japonica*): Mitsuba is an edible perennial herb with root tubers and fragrant leaves. The roots are tender and mild in flavor, similar to a cross between celery and parsley. Mitsuba thrives in shaded areas of the food forest, making it an ideal candidate for the root layer under taller plants.

Tumeric (*Curcuma longa*): Turmeric is an edible root known for its vibrant orange color and medicinal properties. The root is widely used as a spice and for its anti-inflammatory effects. In Zone 10, turmeric thrives in rich, moist soils and can be harvested for its rhizomes, making it both a culinary and medicinal addition to the food forest.

6. GROUNDCOVER LAYER:

Moss Phlox (*Phlox subulata*): Moss phlox is a low-growing perennial with vibrant flowers and edible leaves. The plant thrives in full sun, spreading rapidly to cover the ground. It is known to help control erosion and improve soil structure, adding beauty and nutrition to the food forest's ground layer.

Chili Pepper (*Capsicum annuum*): Although traditionally a vertical plant, chili pepper can also serve as a low-growing groundcover in warm climates like Zone 10, producing edible fruits. The peppers can range from mild to spicy and are rich in antioxidants and vitamins. The plant attracts pollinators and contributes to pest control in the food forest.

7. VERTICAL LAYER:

Black-eyed Pea (*Vigna unguiculata*): Black-eyed peas are a climbing legume that produces edible pods, and the seeds are highly nutritious. This vine provides nitrogen fixation to improve soil quality, supporting other plants in the food forest. It thrives in warm climates like Zone 10, making it an excellent option for both vertical growth and improving soil health.

Climbing Fig (*Ficus pumila*): The climbing fig is a small, fast-growing vine that can produce edible fruit when grown in ideal conditions. The plant clings to structures with its aerial roots, making it a perfect choice for vertical growth. Its fruit can be eaten raw or used in jams, and its dense foliage provides great coverage, contributing to the food forest's aesthetics and productivity.

8. FUNGAL LAYER:

Wood Ear Mushroom (*Auricularia auricula-judae*): Wood ear mushrooms are edible and have a rubbery texture, often used in Asian cuisines. These fungi grow on decaying wood, particularly in moist, shaded environments. Wood ear mushrooms are valuable in the food forest for their ability to decompose wood and return nutrients to the soil, helping to maintain soil health.

Lion's Mane Mushroom (*Hericium erinaceus*): Lion's mane is an edible mushroom that has been lauded for its potential cognitive and nerve-regenerating benefits. It grows on decaying hardwood and forms white, shaggy clusters resembling a lion's mane. This mushroom contributes to the food forest by breaking down organic matter and improving nutrient cycling while providing a unique, medicinal food source.

Inspired by the energy of tropical lagoons, this layout arranges layers of edible species in a refreshing, dynamic design. It is meant to produce an ongoing bounty of flavors while reinforcing natural support mechanisms.

1. CANOPY LAYER:

Bunya Pine (*Araucaria bidwillii*): The Bunya Pine is an ancient tree that can grow up to 150 feet tall, though it typically stays smaller in cultivated settings. Its large cones produce edible seeds, similar to pine nuts, that can be roasted or boiled. Bunya Pines provide an impressive canopy and contribute to the diversity and resilience of the food forest ecosystem.

Mountain Soursop (*Annona montana*): Mountain Soursop is a tropical tree that grows up to 50 feet tall and produces spiny, green fruits with a tart, juicy pulp. The fruits are edible and often used in juices, desserts, or eaten fresh. This tree adds diversity to the canopy layer and its dense foliage helps protect smaller plants from intense sun.

2. UNDERSTORY LAYER:

Nutmeg (*Myristica fragrans*): Nutmeg is a tropical tree that thrives in the shaded understory of USDA Zone 10. The edible parts are the seeds, from which nutmeg spice is derived, and the arils, which produce mace. This slow-growing tree requires protection from direct sun when young, making it a great understory choice.

Langsat (*Lansium parasiticum*): Langsat is a tropical fruit tree native to Southeast Asia that thrives in the partial shade of larger canopy trees. Its small, translucent fruits are sweet and tangy, making it a rare delicacy. This slow-growing tree adds diversity to the food forest and is relatively pest-resistant.

3. SHRUB LAYER:

Jujube (*Ziziphus jujuba*): This drought-tolerant shrub produces small, sweet, and chewy fruits known as Chinese dates. It can thrive in a variety of soils and is highly resilient, offering a nutritious fruit source that is rich in vitamins and antioxidants.

Acerola Cherry (*Malpighia emarginata*): This tropical shrub produces bright red cherries rich in vitamin C. It grows well in USDA Zone 10, providing a prolific harvest of fruit. The cherries are eaten fresh, juiced, or made into preserves, making them an excellent addition to any food forest.

4. HERBACEOUS LAYER:

Lemon Balm (*Melissa officinalis*): Lemon balm is a fragrant herb with edible leaves that are used in teas, salads, and as a garnish. It has calming properties and attracts pollinators like bees, making it an excellent plant for promoting biodiversity. This perennial herb grows well in partial shade and can thrive in hot climates like Zone 10.

Betel Leaf (*Piper betle*): Betel leaf is an herbaceous climbing plant with edible leaves commonly used in Asian cuisine for their digestive benefits. The plant thrives in tropical environments and can grow in both sun and shade. It provides vertical growth in the herbaceous layer, maximizing space usage in the food forest while adding cultural and culinary value.

5. ROOT LAYER:

Mashua (*Tropaeolum tuberosum*): Mashua produces edible, spicy-tasting tubers that are high in antioxidants and vitamin C. Native to the Andes, it grows well in warm climates like Zone 10 and has a climbing habit, allowing it to thrive in vertical spaces. The tubers can be used in stews or eaten raw and are a valuable crop in the food forest's root layer.

Scorzonera (*Scorzonera hispanica*): Scorzonera, also known as black salsify, has edible roots that are similar in taste to salsify but with a more earthy, sweet flavor. This hardy root crop grows in a variety of soils and is particularly suited for Zone 10's warm climate. The deep taproot helps improve soil structure while providing a unique root vegetable for harvesting.

6. GROUNDCOVER LAYER:

Lamb's Lettuce (*Valerianella locusta*): Lamb's lettuce, also known as corn salad, is a fast-growing, low-maintenance groundcover that produces tender, edible leaves. It is a cool-season crop that thrives in the subtropical conditions of Zone 10 and can be used in salads. Its spreading nature helps suppress weeds and contributes to healthy soil.

Mallow (*Malva sylvestris*): Mallow is a versatile groundcover with edible leaves, flowers, and seeds. The young leaves can be used in salads, and the flowers can be used for making herbal teas. Mallow grows well in Zone 10's heat, helping to reduce soil erosion and attract beneficial insects to the food forest.

7. VERTICAL LAYER:

Karaka (*Corynocarpus laevigatus*): Karaka is a climbing or sprawling vine that produces edible fruit, which can be eaten when cooked or processed. While native to New Zealand, it can grow well in the warm climate of Zone 10. The leaves and fruit have been traditionally used in indigenous cuisines, making it an intriguing and rare addition to a food forest.

Moringa (*Moringa oleifera*): Moringa, also known as the drumstick tree, is a fast-growing vertical plant that produces edible leaves, pods, and flowers. Known for its rich nutrient profile, Moringa's leaves are packed with vitamins, minerals, and antioxidants. The plant thrives in hot, dry conditions and can be trained to grow vertically along structures, offering valuable food and health benefits.

8. FUNGAL LAYER:

Maitake Mushroom (*Grifola frondosa*): Maitake mushrooms, also known as hen-of-the-woods, are edible fungi that grow in clusters at the base of trees. These mushrooms have a rich, earthy flavor and are highly valued for their immune-boosting properties. Maitake contributes to the food forest by decomposing organic material and forming mycorrhizal relationships with trees, benefiting both the plants and the soil.

Porcini Mushroom (*Boletus edulis*): Porcini mushrooms are prized for their meaty texture and rich, nutty flavor, commonly used in Italian cuisine. They form mycorrhizal relationships with trees, particularly pines, and help in nutrient cycling. In a food forest, porcini mushrooms contribute to soil fertility by breaking down organic material and supporting the health of surrounding plants.

Zone 10 Plan Seven: Paradise Plant Plan

Evoking the idyllic charm of tropical paradises, this plan layers carefully chosen plants to create a lush food forest. It focuses on producing sumptuous yields and nurturing the ecosystem with enduring support.

1. CANOPY LAYER:

African Breadnut (*Brosimum alicastrum*): This fast-growing tropical tree can reach heights of 80 feet and is known for its edible seeds, which can be ground into flour or roasted. The tree is highly drought-tolerant and enriches the soil with its nitrogen-fixing properties. African Breadnut trees provide valuable canopy cover while producing a sustainable food source.

Ilama (*Annona diversifolia*): Ilama is a rare tropical tree that can reach up to 40 feet tall and produces large, heart-shaped fruits with sweet, pink or white flesh. The fruits are highly prized and often eaten fresh or used in beverages. The tree's unique canopy structure adds biodiversity to the food forest, while its fallen leaves contribute to soil health.

2. UNDERSTORY LAYER:

Cinnamon Myrtle (*Backhousia myrtifolia*): Cinnamon Myrtle is an Australian native that can grow in USDA Zone 10 understory conditions. Its leaves are edible and have a cinnamon-like flavor, often used as a spice or tea. This shrub adds aromatic diversity to the food forest and has potential medicinal uses.

Muntingia (*Muntingia calabura*): Muntingia, also known as Jamaican cherry, produces small, sweet red berries. It thrives in the understory layer of tropical environments and can grow in a wide range of soils. The fruit is highly attractive to birds and humans alike, making it a great plant for both wildlife and food production.

3. SHRUB LAYER:

Roselle (*Hibiscus sabdariffa*): Roselle is a fast-growing shrub that thrives in warm climates, producing edible calyces used to make refreshing beverages or jams. It's also known for its health benefits, including being high in vitamin C and antioxidants.

Mountain Pepper (*Tasmannia lanceolata*): This rare shrub, native to Australia, produces small, peppery berries and leaves used as a spice. It thrives in temperate to tropical climates, offering unique culinary uses and adding biodiversity to the food forest.

4. HERBACEOUS LAYER:

Wood Sorrel (*Oxalis acetosella*): Wood sorrel has edible leaves, flowers, and seed pods with a tart, lemony flavor, often used in salads and garnishes. This herbaceous perennial thrives in shady areas, making it ideal for the lower levels of a food forest. It also attracts pollinators and provides ground cover, helping to suppress weeds.

Marshmallow (*Althaea officinalis*): Marshmallow is a herbaceous plant with edible roots, leaves, and flowers. Historically, the roots were used to make the original marshmallow confections. The plant thrives in moist soils and adds medicinal and culinary diversity to the food forest. Its deep roots also improve soil structure and moisture retention.

5. ROOT LAYER:

Kohlrabi (*Brassica oleracea var. gongylodes*): Kohlrabi, though typically grown for its edible stem, also has edible roots that add mild, sweet flavor to dishes. It thrives in cooler temperatures but can be grown in the subtropical climate of Zone 10 with careful timing. It also brings biodiversity by attracting pollinators and is an excellent choice for food forest health.

Zingiber (*Zingiber zerumbet*): Known as shampoo ginger, Zingiber produces edible rhizomes that have a slightly spicy flavor and are used in traditional remedies. The plant grows well in the moist understory of the food forest and provides medicinal properties as well as an interesting, edible root to harvest. Its lush foliage adds beauty and texture to the root layer.

6. GROUNDCOVER LAYER:

Oregano (*Origanum vulgare*): Oregano is a hardy, low-growing herb that produces flavorful leaves used in many culinary dishes. This edible plant is perfect for growing as a groundcover, as it spreads quickly and helps suppress weeds. Oregano thrives in full sun and dry, well-drained soils, providing both a flavorful herb and an effective ground cover.

Borage (*Borago officinalis*): Borage is a low-growing plant with edible leaves and beautiful blue flowers. The leaves can be used in salads or as a garnish, and the flowers attract pollinators to the food forest. It spreads rapidly and serves as an excellent groundcover, while its deep roots help improve soil health.

7. VERTICAL LAYER:

Kangkong (*Ipomoea aquatica*): Kangkong, also known as water spinach, is an aquatic plant that thrives in warm, wet conditions. The tender leaves and stems are edible and are commonly used in Southeast Asian cuisine. This climbing plant provides a quick-growing, vertical ground cover while producing nutritious and delicious greens.

Coffee (*Coffea arabica*): Coffee plants can be grown as vertical shrubs or small trees that produce edible coffee beans. The plant thrives in partial shade and warm, subtropical conditions like those found in Zone 10. While coffee beans require careful processing, they are highly valued and contribute both to the food forest's productivity and to the livelihood of the grower.

8. FUNGAL LAYER:

Polypore Mushroom (*Fomes fomentarius*)**:** Known for its medicinal properties, particularly in traditional cultures, polypore mushrooms are used to treat various ailments. They grow on decaying wood and contribute to the fungal layer by decomposing wood and recycling nutrients into the soil. This mushroom plays an essential role in forest ecosystems by promoting healthy soil and supporting plant growth.

Hen of the Woods (*Grifola frondosa*)**:** Also known as maitake, hen of the woods is an edible mushroom that grows in clusters at the base of trees. It is known for its earthy flavor and has been traditionally used for its health benefits, particularly for its immune-boosting properties. This mushroom plays an important role in the food forest by breaking down organic material and improving the fertility of the soil.

Celebrating the rich biodiversity of equatorial regions, this design creates a canopy of layered, food-producing plants. It aims to honor cultural flavors while building a system that supports itself through thoughtful ecological pairings.

1. CANOPY LAYER:

Spanish Lime (*Melicoccus bijugatus*): Also known as Genip or Quenepa, this tree grows up to 85 feet tall and produces small, round fruits with a sweet, tangy flavor. The edible pulp is rich in vitamins and commonly eaten fresh or used in drinks. The Spanish Lime tree provides dense shade and helps regulate soil moisture, making it a valuable addition to a food forest canopy.

Peanut Butter Fruit Tree (*Bunchosia glandulifera*): This small to medium-sized tree, reaching up to 20-30 feet, produces small, red fruits with a sweet, peanut butter-like flavor. The fruit is edible and can be eaten fresh or used in desserts. Its dense canopy provides good shade, and the tree is well-suited for tropical food forests in USDA Zone 10.

2. UNDERSTORY LAYER:

Alligator Apple (*Annona glabra*): Alligator apple is a tropical fruit tree that grows well in partially shaded understory conditions. It produces edible, apple-like fruits that are popular in tropical cuisine. This tree can also help improve soil moisture retention, making it a valuable addition to a food forest ecosystem.

Cinnamon (*Cinnamomum verum*): A small tree known for its aromatic bark, used as a spice and beneficial for pollinators.

3. SHRUB LAYER:

Aronia Berry (*Aronia melanocarpa*): Also known as

chokeberry, this shrub produces dark berries that are highly nutritious, rich in antioxidants, and used in juices and jams. It grows well in USDA Zone 10 and is tolerant of a wide range of soils, offering food and wildlife support.

Coffeeberry (*Frangula californica*): Coffeeberry is an adaptable shrub that grows in a range of climates, producing small, dark berries eaten by birds and wildlife. While not a major human food source, it adds ecological diversity to the food forest and improves soil health.

4. HERBACEOUS LAYER:

Salsify (*Tragopogon porrifolius*): Salsify is an herbaceous biennial with edible roots and young shoots. The roots have a flavor similar to oysters and are often used in soups and stews. It grows well in sunny locations and helps break up compacted soils, improving aeration and drainage in the food forest.

Sorrel (*Rumex acetosa*): Sorrel is an herb with edible leaves that have a sharp, tangy flavor, often used in soups, sauces, and salads. The plant is rich in vitamin C and thrives in full sun or partial shade, making it a versatile addition to the food forest. Sorrel also attracts pollinators and helps suppress weeds with its dense foliage.

5. ROOT LAYER:

Cassava (*Manihot esculenta*): A root crop that grows well in tropical zones,providing starchy edible roots and enriching the soil.

Cumin (*Cuminum cyminum*): While primarily known for its seeds, cumin also produces edible roots. The plant is often used in cooking for its aromatic, nutty flavor. It grows well in sunny areas of the food forest and can be used to help diversify the root layer with an herbaceous crop that attracts pollinators.

6. GROUNDCOVER LAYER:

Chickweed (*Stellaria media*): An edible ground cover that suppresses weeds, provides food for wildlife, and helps improve soil health.

Creeping Fennel (*Foeniculum vulgare var. dulce*): Creeping fennel is a flavorful herb that can be used in cooking, with edible leaves, seeds, and stems. It is a low-growing perennial that spreads across the ground, making it an excellent groundcover for Zone 10's hot conditions. It also attracts beneficial insects, helping to support the ecosystem of the food forest.

7. VERTICAL LAYER:

Akebia Vine (*Akebia quinata*): The Akebia vine, also known as chocolate vine, produces edible fruit that has a sweet, vanilla-like flavor. This fast-growing climber adds both vertical height and a source of edible fruit. It is an ideal addition to the food forest, as its vines can be trained along trellises, and it also provides shelter and food for beneficial insects.

Cucumber (*Cucumis sativus*): Cucumber vines grow rapidly and produce refreshing, crisp fruits that are commonly eaten raw or used in salads. The plant is a great space-saver, providing vertical coverage in the food forest, and is easy to grow in the warm climate of Zone 10. Its vigorous growth helps shade the ground and protect more delicate plants from excessive sun.

8. FUNGAL LAYER:

Beech Mushroom (*Hypsizygus tessellatus*): Beech mushrooms are edible fungi that grow in clusters on beech trees and are valued for their tender texture and delicate flavor. They can be used in a variety of dishes and add value to the food forest by breaking down organic material. These mushrooms help decompose wood and improve soil health, creating a balanced and fertile ecosystem.

Oyster Mushroom (*Pleurotus ostreatus*): A fast-growing mushroom species that helps decompose organic matter, fostering a healthy soil ecosystem.

Zone 10 Plan Nine: Fiery Foliage Formula

This formula harnesses the dynamic energy of equatorial climates to layer plants with vibrant, fiery foliage. It is designed to yield a robust harvest and construct a resilient, interdependent ecosystem.

1. CANOPY LAYER:

Chocolate Sapote (*Diospyros digyna*): Also known as Black Sapote, this tropical tree can grow up to 40 feet tall and produces fruits with dark, chocolate-like pulp when ripe. The fruit is edible and often used in desserts and smoothies. Chocolate Sapote trees offer valuable canopy cover and help protect other plants from excessive heat, while also contributing edible yields.

Imbu (*Spondias tuberosa*): Imbu, also known as Umbu, is a drought-tolerant tree that can grow up to 30 feet tall and produces green, round fruits with a tangy, refreshing taste. The fruits are edible and commonly used in juices, jellies, and sauces. Imbu trees provide shade and help regulate water retention in the soil, making them a resilient canopy tree for hot climates.

2. UNDERSTORY LAYER:

Carob (*Ceratonia siliqua*): A drought-tolerant tree that produces edible, sweet pods used as a chocolate substitute.

Nutmeg Bush (*Tetradenia riparia*): The nutmeg bush is an aromatic shrub that grows well in USDA Zone 10, producing fragrant leaves with a nutmeg-like scent. While the leaves aren't commonly consumed, the plant is known for its essential oils and can be used in herbal remedies. It brings biodiversity and aromatic qualities to the food forest.

3. SHRUB LAYER:

Yellow Mombin (*Spondias mombin*): This tropical

shrub produces yellow, oval fruits with a tart flavor. The fruits are eaten fresh, used in beverages, or made into jellies. It thrives in USDA Zone 10 and is drought-resistant, making it an excellent addition to the food forest.

Mexican Weeping Bamboo (*Otatea acuminata*): This bamboo shrub provides dense foliage that can be used as a privacy screen while offering edible shoots. It thrives in tropical and subtropical climates and adds a unique, airy texture to the landscape, supporting biodiversity.

4. HERBACEOUS LAYER:

Peppermint (*Mentha piperita*): Peppermint is a fast-growing herb with edible leaves commonly used in teas, desserts, and as a garnish. It thrives in moist, shady environments and can help improve soil health by providing ground cover. Peppermint's strong scent helps repel pests, making it a valuable plant for natural pest control in the food forest.

Comfrey (*Symphytum officinale*): Comfrey is a deep-rooted herb with edible young leaves and flowers. It is often used in herbal medicine for its healing properties and as a nutrient-rich mulch in the garden. Comfrey helps improve soil fertility by mining nutrients from deep in the soil, making it a valuable addition to the food forest's nutrient cycle.

5. ROOT LAYER:

Sunchoke (*Helianthus tuberosus*): Sunchoke, also known as Jerusalem artichoke, produces edible tubers that are rich in inulin, a prebiotic fiber that aids digestion. The plant is hardy and easy to grow, thriving in Zone 10's warm conditions. Sunchokes help improve soil aeration with their deep roots while providing a valuable food source.

Sweet Potato (*Ipomoea batatas*): A versatile root crop that produces nutritious tubers, supports soil health, and attracts beneficial insects.

6. GROUNDCOVER LAYER:

Purslane (*Portulaca oleracea*): Purslane is an edible groundcover that produces thick, fleshy leaves and stems that are high in omega-3 fatty acids. It thrives in hot, sunny conditions and is a great addition to the food forest, adding both nutrition and ground coverage. Purslane is also drought-tolerant, making it an excellent choice for low-water areas.

Ajwain (*Trachyspermum ammi*): Ajwain is a low-growing herb known for its aromatic, edible seeds that are commonly used in Indian cuisine. It thrives in full sun and well-drained soils, providing not only a flavorful herb but also excellent ground coverage. Ajwain's dense foliage also helps suppress weeds and improves the soil structure.

7. VERTICAL LAYER:

Winged Bean (*Psophocarpus tetragonolobus*): The winged bean is a unique climbing vine that produces edible pods, leaves, flowers, and tubers. All parts of the plant are edible and highly nutritious, making it a valuable resource in a food forest. Its ability to climb and provide vertical coverage makes it an excellent choice for maximizing space and increasing food production.

Gourds (*Cucurbita spp.*): Gourds, which include varieties like bottle gourds and ornamental squashes, are excellent vertical plants that can grow up trellises or arbors. The young fruits can be eaten or used as vegetables in cooking. These plants are fast-growing and provide vertical coverage in the food forest, offering food while also being useful for making tools, containers, and crafts.

8. FUNGAL LAYER:

Fairy Ring Mushroom (*Marasmius oreades*): Fairy ring mushrooms are known for their magical appearance, often growing in rings in the forest.

While they are edible, they are more commonly appreciated for their aesthetic qualities. These mushrooms decompose organic material, helping to enrich the soil and support the surrounding plant life in a food forest.

King Oyster Mushroom (*Pleurotus eryngii*): A large, flavorful mushroom that helps decompose organic matter and contributes to soil fertility.

Zone 10 Plan Ten: Exotic Eco

Bringing together an ensemble of exotic species in a harmonious layered system, this plan is all about balance. It provides a rich array of edibles while ensuring that every component plays a role in long-term sustainability.

1. CANOPY LAYER:

Caimito (*Chrysophyllum cainito*): Also known as Star Apple, this tree can grow up to 50 feet tall and produces round, purple or green fruits with sweet, juicy flesh. The fruit is edible and often eaten fresh or used in desserts. Caimito trees provide a thick canopy and are highly ornamental, contributing to both the aesthetic and ecological health of the food forest.

Papaya (*Carica papaya*): A fast-growing tree producing sweet, nutrient-rich fruit, perfect for tropical climates and providing habitat for pollinators.

2. UNDERSTORY LAYER:

Dwarf Banana (*Musa acuminata*): Dwarf banana trees are ideal for the understory, growing well in USDA Zone 10. They produce edible bananas, and their large leaves create a lush, tropical look while providing shelter for other understory plants. These trees are fast-growing and can yield fruit within a year or two of planting.

Cardamom (*Elettaria cardamomum*): Cardamom is a tropical herb that grows in the shaded understory, producing edible pods that are used as a spice. It requires a warm, humid environment and thrives in USDA Zone 10. Cardamom adds valuable spices to the food forest while improving biodiversity and supporting soil health.

3. SHRUB LAYER:

Peruvian Peppertree (*Schinus molle*): Known for its

pink peppercorn-like berries, this hardy shrub thrives in warm climates like USDA Zone 10. The fruits are used as a spice, and the tree itself offers shade and shelter to smaller plants in a food forest.

Bay Laurel (*Laurus nobilis*): This Mediterranean shrub produces aromatic leaves used in cooking to flavor soups, stews, and sauces. It grows well in USDA Zone 10, tolerating drought and poor soils, and serves both culinary and ornamental purposes.

4. HERBACEOUS LAYER:

Dandelion (*Taraxacum officinale*): Dandelion is an herbaceous perennial with edible leaves, flowers, and roots. The leaves are rich in vitamins A, C, and K, and are often used in salads. Dandelions thrive in poor soils and help improve soil health by loosening compacted soil and attracting pollinators.

Epazote (*Dysphania ambrosioides*): Epazote is an herb with edible leaves commonly used in Mexican cuisine for its unique flavor and digestive benefits. It grows well in hot, sunny environments and is drought-tolerant, making it ideal for Zone 10. Epazote also helps repel pests, contributing to the natural balance of the food forest ecosystem.

5. ROOT LAYER:

Elephant Garlic (*Allium ampeloprasum*): Elephant garlic is a hardy root crop with large, mild-flavored bulbs that resemble garlic but have a sweeter taste. The roots grow deep into the soil, helping with soil aeration and structure. The plant thrives in warm climates and provides a unique, edible bulb for harvest while attracting beneficial insects.

Yacon (*Smallanthus sonchifolius*): Yacon produces edible tubers that are sweet, crunchy, and high in prebiotics. Known for its health benefits, particularly for digestive health, Yacon thrives in moist, fertile soils. It is an excellent addition to the root layer, adding both food and medicinal value to the food forest.

6. GROUNDCOVER LAYER:

Radish Greens (*Raphanus sativus*): The leaves of the radish plant are edible and can be used in salads or as a garnish. Radishes grow quickly and spread across the ground, making them an excellent groundcover. This plant also helps break up compacted soil and can be used to add nutrient-rich organic matter to the soil when it decomposes.

Yarrow (*Achillea millefolium*): Yarrow is an edible herb with feathery leaves and small white or pink flowers. The leaves can be used in teas or salads, and the plant also provides medicinal benefits. Yarrow is drought-tolerant and thrives in Zone 10's sunny conditions, spreading quickly to cover the ground while attracting pollinators and supporting soil health.

7. VERTICAL LAYER:

Passionflower (*Passiflora caerulea*): Passionflower is a climbing vine known for its beautiful flowers and edible fruit, known as passion fruit. The fruit can be eaten raw or used in juices, offering a tropical flavor. This vine thrives in the warm, subtropical climate of Zone 10, attracting pollinators with its stunning blooms and adding vertical height and productivity to the food forest.

Sweet Peas (*Lathyrus odoratus*): Sweet peas are fragrant climbing plants that produce edible flowers, which can be used as a garnish or in salads. These vines grow quickly and can be trained to climb, providing vertical coverage while offering a touch of beauty and fragrance to the food forest. They also attract beneficial pollinators, supporting the ecosystem.

8. FUNGAL LAYER:

Mycorrhizal Fungi (*Glomus spp.*): Mycorrhizal

fungi form symbiotic relationships with plant roots, helping them absorb nutrients more efficiently. These fungi are vital for the health of a food forest, as they improve plant growth and contribute to soil fertility. By enhancing nutrient uptake, mycorrhizal fungi support the growth of edible plants and other beneficial organisms in the ecosystem.

Birch Polypore (*Fomitopsis betulina*): This parasitic fungus grows on birch trees and has long been used in traditional medicine. While not typically consumed for culinary purposes, it has health-promoting properties and contributes to the forest by breaking down birch wood. Birch polypore aids in nutrient cycling and improves the soil structure, supporting the overall health of the food forest.

ZONE 11

Zone 11 represents one of the warmest growing zones in the USDA Plant Hardiness Zone Map, and it is predominantly found in tropical and subtropical regions.

This zone is characterized by its consistently warm temperatures and long growing seasons, making it ideal for tropical plants that require warmth year-round.

Locations within Zone 11

Here is a list of states and locations that fall within Zone 11:

American Samoa (U.S. Territory): Entire island group, including Tutuila and surrounding islands.

California: Some isolated areas of coastal Southern California, such as parts of Los Angeles and San Diego's coastal regions (microclimates may fall in this zone).

Florida: Southernmost parts of Florida, including: Florida Keys, Key West and parts of the southern tip of the mainland near Everglades National Park.

Guam (U.S. Territory): Entire island of Guam, located in the western Pacific Ocean.

Hawaii: Most lowland areas of the Hawaiian Islands, particularly coastal regions such as: Honolulu (Oahu), Hilo (Big Island), Kailua-Kona (Big Island), Maui's coastal regions.

Northern Mariana Islands (U.S. Territory): Includes islands like Saipan, Tinian, and Rota.

Puerto Rico (U.S. Territory): Coastal and lowland regions across much of Puerto Rico, including: San Juan, Ponce. Mayagüez, Arecibo.

U.S. Virgin Islands: Entirety of the U.S. Virgin Islands, including: St. Thomas, St. John, St. Croix.

Zone 11

Climate Characteristics of Zone 11

Average Minimum Temperature: Zone 11 experiences temperatures that do not dip below 40°F (4°C) during the winter months. The average minimum temperature typically ranges from 40°F to 50°F (4°C to 10°C), which allows for year-round growing of frost-sensitive plants.

Growing Season: The growing season in Zone 11 is virtually year-round. There are no hard freezes in this zone, so many plants can continue growing during the winter months, with some species even flowering or producing fruit during colder periods.

Humidity: Zone 11 often experiences high humidity, particularly in coastal areas and tropical climates. Humidity levels can influence plant growth, particularly in terms of water needs and susceptibility to fungal diseases.

Rainfall: Rainfall in Zone 11 is generally abundant, and it may either come in steady, frequent showers or in seasonal monsoons, depending on the region. Many tropical and subtropical plants thrive in the ample moisture.

Zone 11 Soil Conditions

Soils in Zone 11 vary but are often rich and fertile, particularly in coastal and tropical areas. Due to the warm, humid conditions:

Well-draining soils are essential for many plants, particularly in areas with high rainfall.

Acidic to neutral pH soils are common in tropical zones, with many plants favoring this soil range.

Organic matter is abundant in tropical soils, as decomposing vegetation contributes nutrients to the soil, making it fertile for plant growth.

However, gardeners may need to amend soils with compost or organic materials to improve drainage in some areas, especially if the soil tends to be clay-heavy or compacted.

Growing Potential in Zone 11

Zone 11 supports a vast array of plant life, including both tropical and subtropical species. The warm, frost-free conditions make this zone perfect for growing many types of tropical fruits, herbs, and flowering plants that would struggle in cooler climates. Some of the key types of vegetation that thrive in Zone 11 include:

Tropical Fruits: Mango, papaya, coconut, banana, guava, pineapple, and avocados are commonly grown in Zone 11. The consistent warmth allows these plants to grow year-round, producing fruit without interruption.

Subtropical Trees: Trees like citrus, loquat, and olive also grow well in this zone.

Tropical Flowers and Shrubs: Bougainvillea, hibiscus, and plumeria are just a few examples of the colorful, vibrant flowers that thrive in Zone 11's climate.

Perennials: Zone 11 is also ideal for growing perennials such as ginger, turmeric, and yams. These plants require long warm seasons to grow and mature.

Potential Challenges of Zone 11

While Zone 11's warmth offers numerous growing opportunities, there are a few challenges to be aware of:

High Humidity: The consistently high humidity levels can encourage the growth of mold, mildew, and fungal diseases. It is important to select plants that are resistant to these conditions and to provide adequate air circulation in the garden.

Saltwater Exposure: Coastal areas in Zone 11 may face issues with saltwater from the ocean, particularly during storms. Some plants may struggle in these saline environments, so it is important to select salt-tolerant species.

Pests: Due to the warm and humid conditions, pest populations can be higher, including insects like aphids, whiteflies, and caterpillars, as well as larger pests such as rodents or iguanas. Integrated pest management strategies will be important for keeping these under control.

Agricultural Uses of Zone 11

Zone 11 is often used for cultivating crops that cannot grow in colder regions. Some specialty crops that flourish in Zone 11 include:

Coffee: Particularly in areas of Hawaii and Puerto Rico, coffee thrives in this zone's tropical climate.

Cacao: The chocolate-producing plant requires constant warmth and humidity, which is offered in Zone 11.

Spices: Many spices such as pepper, vanilla, and cardamom grow well in the warmth of Zone 11's tropical conditions.

Zone 11 also sees a variety of ornamental plants being cultivated, including palms, flowering shrubs, and other exotic species.

Permaculture Considerations

For permaculture enthusiasts, Zone 11 offers a variety of opportunities to design food forests, gardens, and sustainable landscapes.

Water Management: Given the potential for high rainfall and humidity, it's important to incorporate systems for water management, such as rainwater harvesting, swales, or proper drainage.

Diversity: In Zone 11, having a mix of crops that provide both edible and medicinal benefits can help with sustainability. Many tropical plants offer multiple uses, such as fruit, flowers, medicinal leaves, and edible roots.

Companion Planting: The year-round growing season means that companion planting strategies can be continuously in play to help with pest control, nutrient cycling, and maximizing space.

Zone 11's consistently warm temperatures and frost-free climate allow gardeners to cultivate a wide variety of tropical and subtropical plants, making it one of the most productive growing zones for both food and ornamental plants. Whether you're growing tropical fruits, spices, or flowering plants, the conditions of Zone 11 provide endless possibilities for gardening enthusiasts and permaculturists.

However, careful attention must be paid to humidity levels, soil conditions, and pest management to ensure the success of plantings in this zone.

Zone 11 Plan One: Balmy Bloom Blueprint

ZONE 11 PLAN ONE: BALMY BLOOM BLUEPRINT

Crafted for the warmest stretches, this blueprint layers species that burst into bloom under balmy skies. It focuses on generating delicious harvests while ensuring the ecosystem remains vibrant and self-sustaining.

1. CANOPY LAYER:

Allspice (*Pimenta dioica*): The allspice tree produces aromatic leaves and berries, both of which are used as flavorings. The leaves can be used fresh or dried in cooking, while the berries are harvested and dried to produce the spice. This tree can grow up to 30 feet tall and thrives in warm climates with rich, well-drained soil, making it a valuable addition to the canopy layer.

Atemoya (*Annona squamosa* × *Annona cherimola*): A hybrid between cherimoya and sugar apple, atemoya is known for its sweet, custard-like fruit that is enjoyed fresh or in desserts. The tree can grow 15-20 feet tall and requires full sun and protection from strong winds. It's an excellent fruiting tree for subtropical and tropical climates, adding both beauty and bounty to the food forest.

2. UNDERSTORY LAYER:

Chilean Guava (*Ugni molinae*): This low-growing shrub produces aromatic, sweet berries that are perfect for fresh eating or making jams. Known for its evergreen foliage, it can thrive in shady areas, making it ideal for understory planting. Chilean Guava is also drought-tolerant and a great addition to a food forest for providing year-round edible fruit.

Indian Gooseberry (*Phyllanthus emblica*): Often called Amla, this small tree produces tangy, vitamin C-rich fruit that is used in both culinary and medicinal applications. It can handle full sun but grows well in partial shade, making it suitable for the understory layer. It is a hardy, low-maintenance plant that contributes to the overall health of a food forest.

3. SHRUB LAYER:

Cranberry Hibiscus (*Hibiscus acetosella*): This striking shrub features dark maroon, almost purple foliage and edible leaves that can be used in salads or as garnishes. It thrives in partial shade and is a hardy shrub, adding both visual appeal and edible biomass to your food forest. The plant also attracts beneficial insects, enhancing biodiversity.

Chaya (*Cnidoscolus aconitifolius*): Also known as Tree Spinach, Chaya is a nutrient-packed shrub with edible, large green leaves that are rich in vitamins. The leaves need to be cooked before consumption to neutralize toxins, but once prepared, they are an excellent source of protein and fiber. Chaya thrives in full sun to partial shade and is extremely drought-tolerant.

4. HERBACEOUS LAYER:

Culantro (*Eryngium foetidum*): Often confused with cilantro, Culantro has a stronger flavor and is used in many Latin American and Caribbean dishes. The leaves are edible and packed with flavor, perfect for seasoning. It thrives in partial shade and is known for its resilience in hot, humid climates, making it ideal for food forests.

Moringa (*Moringa oleifera*): Also known as the "drumstick tree," Moringa is a highly nutritious herbaceous plant. The leaves are edible and rich in vitamins, minerals, and antioxidants, while the seeds and pods can also be eaten. Moringa is drought-tolerant and grows quickly, providing multiple food sources in the understory layer.

5. ROOT LAYER:

Ulluco (*Ullucus tuberosus*): Ulluco is a tuber-producing plant native to the Andes, offering colorful, edible roots that are rich in starch and nutrients. The roots are used in stews, soups, and as a side dish. Ulluco thrives in well-drained, fertile soils and adds a unique texture to the food forest with its vibrant tubers.

Yacon (*Smallanthus sonchifolius*): Known for its sweet, edible tubers, Yacon is a root crop that thrives in tropical climates. The tubers are low in calories and high in inulin, which is beneficial for gut health. Yacon plants grow best in well-drained, slightly acidic soil and are an excellent addition to the root layer, offering both culinary and health benefits.

6. GROUNDCOVER LAYER:

Groundnut (*Apios americana*): Groundnut is a perennial vine that produces edible tubers which have a nutty, sweet flavor. The tubers are high in protein and can be eaten boiled, roasted, or mashed. It thrives in moist, well-drained soil and adds nitrogen-fixing benefits to the soil while also serving as a food source for the forest.

Sweet Potato (*Ipomoea batatas*): Known for its edible tubers, Sweet Potato is a groundcover that spreads rapidly and thrives in warm conditions. The tubers are rich in nutrients, including beta-carotene, and can be used in a variety of dishes. This plant also enriches the soil by reducing erosion and helping maintain moisture levels.

7. VERTICAL LAYER:

Moringa (*Moringa oleifera*): Known as the "drumstick tree," Moringa is a fast-growing tree that can be trained vertically. The leaves, pods, and flowers are all edible and rich in vitamins, antioxidants, and minerals. Moringa thrives in well-drained soils and full sun, adding a nutritional powerhouse to the food forest while attracting beneficial insects.

Fuzzy Kiwi (*Actinidia arguta*): A smaller, fuzz-free variety of kiwi, the Fuzzy Kiwi produces edible fruits that are rich in vitamin C. It grows on a vine, making it ideal for vertical planting, and requires a sunny spot with well-drained soil. The vine is fast-growing and helps provide shade to lower layers of the food forest while offering sweet, tangy fruits.

8. FUNGAL LAYER:

Maitake Mushroom (*Grifola frondosa*): Maitake, also known as "Hen of the Woods," is an edible mushroom known for its delicate flavor and medicinal properties. It grows at the base of hardwood trees and thrives in a forested environment, making it a perfect addition to the fungal layer. Maitake mushrooms are rich in nutrients and have been used in traditional medicine to boost the immune system.

Porcini Mushroom (*Boletus edulis*): Known for its rich, nutty flavor, Porcini mushrooms are highly prized in gourmet cooking. These mushrooms grow near conifers and hardwoods and can be foraged in the wild or cultivated in shaded, moist conditions. Porcini mushrooms add a savory depth to dishes and have health benefits due to their high levels of antioxidants.

ZONE 11 PLAN TWO: SUSTAINING SUNRISE

This plan leverages the promise of each new day by arranging early-blooming and steady-yield plants in layers. Its purpose is to capture the energy of the sunrise to provide both food and enduring ecological support.

1. CANOPY LAYER:

Breadfruit (*Artocarpus altilis*): Breadfruit trees are large, reaching up to 85 feet in tropical climates, and their large, starchy fruit can be cooked like potatoes. It is a great source of carbohydrates and adds a tropical feel to the food forest. Breadfruit trees require a warm climate, and the trees provide excellent shade and a dense canopy for other layers.

Cacao (*Theobroma cacao*): The cacao tree, famous for producing the seeds that become chocolate, thrives in USDA Hardiness Zone 11. Cacao is grown for its edible seeds, which can be harvested and processed into chocolate. The tree itself has beautiful large leaves and adds both visual and functional value to the canopy layer.

2. UNDERSTORY LAYER:

Nance (*Byrsonima crassifolia*): Nance trees yield small, yellow-orange fruits that are tangy and edible raw or in preserves. This plant grows as a dense shrub or small tree and thrives in slightly shaded environments. The Nance tree is drought-resistant and attracts pollinators, making it valuable in a diverse food forest.

Tropical Almond (*Terminalia catappa*): This tree produces edible nuts, and its leaves are known for their use in traditional medicine. Although a larger tree, its spreading canopy allows it to thrive in partial shade and contribute to nutrient cycling in the soil. Its attractive, broad leaves also serve as mulch material, benefitting the surrounding plants.

3. SHRUB LAYER:

Cranberry Hibiscus (*Hibiscus acetosella*): This striking shrub features dark maroon, almost purple foliage and edible leaves that can be used in salads or as garnishes. It thrives in partial shade and is a hardy shrub, adding both visual appeal and edible biomass to your food forest. The plant also attracts beneficial insects, enhancing biodiversity.

Chaya (*Cnidoscolus aconitifolius*): Also known as Tree Spinach, Chaya is a nutrient-packed shrub with edible, large green leaves that are rich in vitamins. The leaves need to be cooked before consumption to neutralize toxins, but once prepared, they are an excellent source of protein and fiber. Chaya thrives in full sun to partial shade and is extremely drought-tolerant.

4. HERBACEOUS LAYER:

African Blue Basil (*Ocimum kilimandscharicum* × *Ocimum basilicum*): A hybrid of basil with a more robust and pungent flavor, African Blue Basil is used in both culinary and medicinal applications. The leaves are edible and offer a rich, savory taste. It grows well in partial shade and attracts pollinators like bees, which benefits the overall ecosystem of the food forest.

Kangkong (*Ipomoea aquatica*): Known as water spinach, Kangkong is a fast-growing herbaceous plant with edible leaves and stems. It thrives in moist, tropical conditions and provides a nutritious addition to stir-fries, soups, and salads. This plant is a great source of vitamins A and C and grows easily in the understory layer.

5. ROOT LAYER:

Sunchoke (*Helianthus tuberosus*): Also known as Jerusalem artichoke, Sunchoke produces edible, nutty-flavored tubers that are rich in fiber and antioxidants. These tubers are great for roasting or adding to soups and stews. The plant is hardy and easy to grow, contributing both food and structural support to the root layer of the food forest.

Chufa (*Cyperus esculentus*): Chufa, also known as tiger nuts or earth almonds, produces small edible tubers that are sweet and rich in healthy fats. The tubers can be eaten raw, roasted, or used to make a milk-like drink. This plant grows well in moist, well-drained soil and is an excellent addition to the root layer.

6. GROUNDCOVER LAYER:

Creeping Raspberry (*Rubus pentalobus*): A low-growing perennial that produces edible, sweet-tasting raspberries. The plant spreads quickly, making it an excellent choice for groundcover, and it helps with soil erosion control. The leaves can also be used for tea, offering both food and medicinal benefits to the food forest.

New Zealand Spinach (*Tetragonia tetragonioides*): This groundcover plant produces edible, spinach-like leaves that are rich in vitamins and minerals. Unlike true spinach, it tolerates heat and drought better, making it an ideal choice for tropical zones like Zone 11. It helps retain soil moisture and provides nutritious greens for the food forest.

7. VERTICAL LAYER:

Gourds (*Lagenaria spp.*): Gourds such as bottle gourd (Lagenaria siceraria) grow quickly on vertical supports and are used for both culinary and crafting purposes. The young gourds are edible, and their mature forms are often used to create containers or utensils. Gourds help to maximize space in the food forest and provide a unique harvest.

Cucamelon (*Melothria scabra*): Often called "Mexican sour gherkin," cucamelon is a small, cucumber-like fruit with a tangy flavor. The plant is a vigorous climber, ideal for vertical growing, and produces tiny fruits that can be eaten raw or pickled. Cucamelon is drought-tolerant and adds a fun, productive element to the food forest.

8. FUNGAL LAYER:

Chicken of the Woods (*Laetiporus sulphureus*): A strikingly bright yellow-orange mushroom, Chicken of the Woods is an edible fungus with a meaty texture, often used as a vegetarian alternative to chicken in cooking. It grows on decaying wood, particularly oak trees, and can be easily incorporated into the food forest's fungal layer. Known for its medicinal qualities, it also provides an intriguing and flavorful food source.

King Oyster Mushroom (*Pleurotus eryngii*): King Oyster mushrooms have thick, meaty stems and a mild, earthy flavor that makes them versatile in many dishes. They thrive in moist, shaded environments and can be grown on logs or wood chips. These mushrooms contribute to the food forest by improving soil health through their mycorrhizal relationships and offering a valuable food source.

Zone 11 Plan Three: Zesty Zenith

Zone 11 Plan Three: Zesty Zenith

Bold and full of life, this design plan pairs vibrant, zesty species across layers to reach the zenith of flavor. It's engineered to produce a dynamic harvest while maintaining a balanced, sustainable ecosystem.

1. CANOPY LAYER:

Calamondin (*Citrus × microcarpa*): Often used as a souring agent in cooking, the calamondin is a small citrus tree that produces small, tangy fruits. The tree has a bushy growth habit and is well-suited for container growth as well as food forests. The fruit is used in Filipino and Southeast Asian cuisines and can be used in marmalades, sauces, and even drinks.

Cashew (*Anacardium occidentale*): The cashew tree, best known for its edible seeds, produces nuts that are both nutritious and delicious. It grows as a small to medium tree and thrives in sandy, well-drained soil. Besides its valuable nuts, the tree produces a fruit, the cashew apple, which is used in jams and beverages.

2. UNDERSTORY LAYER:

Jabuticaba (*Plinia cauliflora*): A unique fruiting tree native to Brazil, Jabuticaba produces dark purple, grape-like fruits that grow directly on the trunk. The fruits are edible and rich in antioxidants, making them perfect for a healthy diet. The tree is slow-growing but thrives in partial shade, ideal for the understory of a food forest.

Hibiscus (*Hibiscus rosa-sinensis*): While commonly known for its ornamental use, the leaves and flowers of Hibiscus are edible and often used to make teas. This evergreen shrub grows well in the understory, tolerating shade and adding color and beauty to the space. It's also easy to grow and care for.

3. SHRUB LAYER:

Nance (*Byrsonima crassifolia*): Nance shrubs

produce small, aromatic yellow to orange fruits that are both sweet and sour, perfect for fresh eating or making preserves. This hardy plant thrives in tropical climates and partial shade, contributing to the understory with its edible fruits and flowers that attract pollinators.

Tamarillo (*Solanum betaceum*): Also known as the tree tomato, Tamarillo produces egg-shaped fruits with tangy to sweet flavors, depending on ripeness. This plant thrives in partial shade and is well-suited to the understory, providing a unique edible fruit that adds diversity and flavor to the food forest. It's also a relatively low-maintenance shrub.

4. HERBACEOUS LAYER:

Lemon Balm (*Melissa officinalis*): Lemon Balm is an aromatic herb with lemon-scented leaves that are used in teas, salads, and as a flavoring agent. The leaves are edible and have calming properties. This herb thrives in partial shade, and its growth attracts pollinators, making it a great addition to your food forest's herbaceous layer.

Purslane (*Portulaca oleracea*): A succulent herb that thrives in hot, dry conditions, Purslane is a highly nutritious plant with edible leaves and stems. It is rich in omega-3 fatty acids, vitamins, and minerals. This low-maintenance plant is perfect for ground coverage and grows well in partial shade.

5. ROOT LAYER:

Kangkung (*Ipomoea aquatica*): Although better known for its edible leaves, Kangkung's roots are also edible and packed with nutrients. The plant thrives in moist, tropical conditions and provides a good source of vitamins A and C. Its growth in the root layer also helps in erosion control and soil enrichment in the food forest.

Chinese Yam (*Dioscorea polystachya*): A climbing yam species with edible tubers, Chinese Yam thrives in the tropical climate of Zone 11. The tubers are used in soups, stews, and stir-fries and have medicinal properties. It grows quickly, producing both food and ground cover for the root layer.

6. GROUNDCOVER LAYER:

Oregano (*Origanum vulgare var. vulgare*): A low-growing herb that spreads easily and is known for its aromatic leaves, which are used in cooking and for medicinal purposes. The leaves are edible and can be used fresh or dried for seasoning. Oregano thrives in well-drained, sunny spots and helps improve soil health by acting as a groundcover.

Wild Strawberry (*Fragaria vesca*): Wild Strawberry produces small, sweet berries that are great for snacking or adding to dishes. The plant spreads by runners and thrives in a variety of soil types. It is perfect for adding a touch of sweetness to the ground layer of the food forest while providing ground coverage.

7. VERTICAL LAYER:

Chilean Guava (*Ugni molinae*): Also known as the "Tazzito" or "Strawberry Myrtle," this shrub produces edible, aromatic berries that are used in jams, jellies, and desserts. The plant grows well as a small climbing vine, making it suitable for vertical growing. It thrives in slightly acidic soil and provides a delicious fruiting option for the vertical layer.

Jujube (*Ziziphus jujuba*): This small tree grows well as a vertical feature, producing sweet, apple-like fruits. The fruits, seeds, and leaves are edible and have been used for their medicinal properties for centuries. Jujube is a hardy plant that requires minimal maintenance and provides nutritious, flavorful fruits.

8. FUNGAL LAYER:

Wine Cap Mushroom (*Stropharia rugosoannulata*):

Also known as "Garden Giant," the Wine Cap Mushroom is a large, edible fungus that grows well in wood chip beds. This mushroom is known for its meaty texture and ability to improve soil health by breaking down organic material. It grows in shaded areas and can be easily incorporated into the fungal layer of the food forest to increase biodiversity and yield.

Black Trumpet Mushroom (*Craterellus cornucopioides*): Often called the "Horn of Plenty," Black Trumpet mushrooms have a rich, smoky flavor and are considered a delicacy. These mushrooms grow in the forest floor and can form beneficial relationships with tree roots. They are nutrient-dense and provide a unique addition to the food forest ecosystem while enhancing soil health.

Zone 11 Plan Four: Gilded Grove

This plan crafts a "gilded" grove of layered edibles designed for warm, sunlit climates. Its focus is on creating a visually and nutritionally rich food forest that stands as a model of sustainability.

1. CANOPY LAYER:

Cinnamon (*Cinnamomum verum*): Known for its aromatic bark, cinnamon trees are a great addition to a food forest canopy. The bark can be harvested and dried to make the spice we know and love. Growing up to 30 feet tall, cinnamon trees do well in warm, humid climates with well-drained soil.

Coffee (*Coffea arabica*): Coffee plants are small trees that thrive in the understory but can also grow as part of the canopy when allowed to mature. The seeds, or coffee beans, are harvested from the plant's berries and processed into coffee. Coffee trees require shaded, moist environments and benefit from nutrient-rich soil.

2. UNDERSTORY LAYER:

Soursop (*Annona muricata*): Known for its creamy, sweet fruit, Soursop trees thrive in tropical climates and can tolerate some shade. The fruit is eaten fresh or used in smoothies, and the tree contributes to the biodiversity of the food forest. Its broad leaves can be used in traditional medicine for various ailments.

Starfruit (*Averrhoa carambola*): This small tree produces the distinctively shaped, juicy star-shaped fruit that is sweet or tangy depending on its ripeness. Starfruit trees are ideal for the understory, growing in partial shade and bringing unique flavor to your food forest. The tree is also attractive with its ornamental foliage.

3. SHRUB LAYER:

Brambleberry (*Rubus spp.*): A thorny shrub that

produces sweet, small blackberries or raspberries, brambleberries are ideal for the food forest's shrub layer. The fruit is edible and used in jams, pies, or eaten fresh. This shrub prefers partially shaded environments and can also be used as a natural hedge or barrier.

Moringa (*Moringa oleifera*): Known for its highly nutritious leaves, Moringa is a fast-growing shrub that thrives in full sun but also tolerates partial shade. The leaves, flowers, and pods are edible and offer a rich source of vitamins, minerals, and antioxidants. It is drought-tolerant and contributes significantly to soil improvement by fixing nitrogen.

4. HERBACEOUS LAYER:

Giant Lobelia (*Lobelia gigantea*): A tall, perennial herb with striking blue flowers, Giant Lobelia has edible leaves that can be used as a vegetable when cooked. It thrives in partial shade and moist, well-drained soil. The plant adds beauty to the herbaceous layer while providing an edible resource for the food forest.

Sambung Nyawa (*Gynura procumbens*): Known for its medicinal properties, Sambung Nyawa is an herbaceous plant with edible leaves that are used to treat a variety of ailments. The plant grows well in partial shade and can be grown as a ground cover or low shrub. Its leaves are rich in antioxidants and vitamins.

5. ROOT LAYER:

Radish (*Raphanus sativus*): A quick-growing root vegetable, Radish has edible roots that can be eaten raw in salads or pickled. Radishes come in many varieties, including red, white, and black, and are rich in antioxidants. They thrive in well-drained soil and help to break up compacted soil, promoting overall health in the food forest.

Salsify (*Tragopogon porrifolius*): Known as the

"oyster plant" due to the flavor of its roots, Salsify produces edible roots that are delicious when roasted, boiled, or fried. The plant is easy to grow and thrives in well-drained soil. Salsify is a hardy perennial and contributes both edible roots and pollinator-friendly flowers to the food forest.

6. GROUNDCOVER LAYER:

Lemon Balm (*Melissa officinalis*): A fragrant herb that produces edible leaves used for teas, cooking, and medicinal purposes. Lemon Balm is fast-growing and acts as an effective groundcover, suppressing weeds and attracting beneficial insects. It thrives in well-drained, moderately moist soils and provides aromatic flavor to dishes and teas.

Thyme (*Thymus vulgaris*): A versatile herb that forms a dense groundcover, Thyme produces small edible leaves used for seasoning a wide variety of dishes. It thrives in dry, well-drained soils and tolerates drought well. Thyme also attracts pollinators to the food forest and improves biodiversity.

7. VERTICAL LAYER:

Chayote (Sechium edule): Chayote, also known as vegetable pear, is a climbing plant that produces edible fruits that are often used in savory dishes. The young fruits and shoots are edible, and the plant thrives in vertical spaces, offering both culinary and aesthetic benefits. It is also a fast-growing plant that can fill vertical niches quickly.

Clematis (Clematis spp.): Some varieties of Clematis produce edible flowers that can be used for garnishing salads or making teas. It is a versatile vine that can climb vertical supports and brings color to the food forest with its striking flowers. It thrives in a sunny location and adds ornamental value alongside its edible blooms.

8. FUNGAL LAYER:

Hedgehog Mushroom (*Hydnum repandum*): Known for its distinctive spiky appearance, the Hedgehog mushroom has a mild, nutty flavor and is highly prized in culinary use. It grows in the forest floor under pine and hardwood trees, and its mycorrhizal relationship benefits tree roots. The mushroom is a valuable addition to the fungal layer, offering both food and ecosystem support.

Mushroom Coral (*Ramaria spp.*): This group of edible mushrooms has a unique coral-like structure and is known for its delicate flavor and texture. It thrives in shaded, moist forest environments and helps break down organic matter, improving soil health. Although some species are poisonous, the edible varieties add a fascinating aesthetic and valuable nutritional resource to the fungal layer.

Zone 11 Plan Five: Mosaic Manifesto

Zone 11 Plan Five: Mosaic Manifesto

A manifesto for balance and abundance, this plan arranges a mosaic of mellow, warm-climate species into layered harmony. It is designed to produce a steady yield of food while reinforcing natural system resilience.

1. CANOPY LAYER:

Durian (*Durio spp.*): Known as the "king of fruits," durian trees produce large, spiky fruit with a strong smell but a rich, custardy taste. The fruit is prized in Southeast Asia, and the tree can grow up to 100 feet tall. Durian trees thrive in tropical climates, and the large, green leaves and towering height provide both a canopy and a unique addition to the food forest.

Feijoa (*Acca sellowiana*): Known as the pineapple guava, feijoa produces sweet, aromatic fruits that are enjoyed fresh or in jams. The tree has an attractive, evergreen foliage and a dense canopy, reaching 10 to 15 feet in height. It grows well in subtropical regions and thrives in well-drained soil with plenty of sunlight.

2. UNDERSTORY LAYER:

Bitter Melon (*Momordica charantia*): This vine produces edible, bitter fruit often used in Asian cooking and herbal medicine. While it requires a vertical structure to climb, it grows well in partial shade and brings a nutritious, low-maintenance edible crop to the food forest.

Tamarillo (*Solanum betaceum*): Known as the tree tomato, Tamarillo trees produce egg-shaped fruits that vary in flavor from tangy to sweet. These trees grow well in shaded or partially shaded areas and add a unique flavor to the food forest while contributing to the understory layer.

3. SHRUB LAYER:

Aloe Vera (*Aloe barbadensis miller*): This succulent shrub has many uses, with its thick, fleshy leaves providing an edible gel used in health and beauty products. While it's often associated with medicinal uses, Aloe Vera can also contribute to the food forest by offering shelter for small creatures and providing an edible resource in the form of gel from its leaves.

White Sapote (*Casimiroa edulis*): A subtropical shrub that produces sweet, creamy, and edible fruit, White Sapote thrives in full to partial sun. The tree can grow in various soil types and is low-maintenance, making it a valuable shrub for the food forest. Its fruit is often eaten fresh and is used in smoothies and desserts.

4. HERBACEOUS LAYER:

Bitter Melon (*Momordica charantia*): Bitter Melon is an herbaceous vine that produces edible fruit with a distinctive bitter taste, often used in Asian cuisines and for medicinal purposes. The leaves and fruit are edible, and the plant thrives in warm climates with partial shade, offering both nutritional and medicinal benefits.

Sweet Potato Greens (*Ipomoea batatas*): The edible leaves of Sweet Potato are used in a variety of dishes, providing a rich source of vitamins A and C. The plant grows well in partial shade, and its vine spreads rapidly, making it useful for ground cover. Sweet Potato greens are a great way to add both nutrition and ground stabilization to the food forest.

5. ROOT LAYER:

Sweet Cicely (*Myrrhis odorata*): While the leaves of Sweet Cicely are often used in cooking, its edible roots can also be harvested for their sweet, anise-like flavor. This perennial herb grows well in partial shade and well-drained soil, making it a great addition to the root layer of the food forest.

Zingiber zerumbet (*Zingiber zerumbet*): Known

as Shampoo Ginger, this plant produces edible rhizomes that have a unique, soapy substance. While the rhizomes are often used for medicinal purposes, they can also be used in cooking for their subtle flavor. The plant is perfect for the tropical conditions of Zone 11 and contributes to the biodiversity of the food forest.

6. GROUNDCOVER LAYER:

Chamomile (*Matricaria chamomilla*): Chamomile is a low-growing herb with edible flowers that are used in teas for their calming and medicinal properties. It grows well in the groundcover layer, offering both culinary and medicinal benefits. Chamomile attracts pollinators and contributes to the overall health and wellness of the food forest.

Creeping Jenny (*Lysimachia nummularia*): This groundcover plant forms a dense carpet of round, green leaves and produces small yellow flowers. While not traditionally consumed in large quantities, the leaves are occasionally used in herbal teas. It helps control soil erosion and thrives in moist, shaded areas of the food forest.

7. VERTICAL LAYER:

Chili Pepper Tree (*Capsicum annuum var. glabriusculum*): This is a perennial chili pepper plant that can grow vertically with proper support. The small, potent peppers are edible and can be used for cooking, adding heat and flavor. The plant can be grown in containers or directly in the ground and thrives in full sun.

Goji Berry (*Lycium barbarum*): Goji berries grow on a shrub that can be trained to climb or spread vertically. The bright red berries are packed with antioxidants and can be eaten fresh, dried, or used in cooking. Goji berries thrive in well-drained soils and are a great addition to a food forest due to their health benefits and versatility.

8. FUNGAL LAYER:

Earthstar Mushroom (*Geastrum spp.*): Earthstar mushrooms have a distinctive, star-shaped appearance and are primarily found in woodlands. While not commonly consumed, they are interesting additions to the fungal layer due to their unique structure and ability to enrich soil through decomposition processes. Their mycorrhizal relationship with trees can benefit the overall health of the food forest.

Angel's Wings (*Pleurocybella porrigens*): This edible mushroom has a delicate, mild flavor and grows in clusters on decaying wood. It thrives in humid, shaded areas, adding a gourmet touch to the food forest while contributing to soil health through its decomposition activities. Its appearance resembles angel wings, making it a visually appealing and functional mushroom for the fungal layer.

Zone 11 Plan Six: Radiant Rain Regimen

Zone 11 Plan Six: Radiant Rain Regimen

Embracing the refreshing promise of morning dew and soft rains, this regimen layers plants to capture radiant energy. It balances immediate culinary rewards with a structure that supports long-term ecosystem health.

1. CANOPY LAYER:

Gingerbread Plum (*Diospyros lotus*): This small tree produces sweet, round fruit that resembles a plum in appearance but with a flavor often likened to gingerbread. It grows to about 15 feet tall and prefers well-drained soil. Its edible fruit is harvested in late summer and can be eaten fresh or dried.

Indian Jujube (*Ziziphus mauritiana*): The Indian jujube, also known as the ber tree, produces small, sweet, and tangy fruits that are high in vitamin C. The tree is hardy and drought-tolerant, growing well in tropical and subtropical climates. It's a valuable fruiting tree for the food forest and can thrive in poor soil conditions.

2. UNDERSTORY LAYER:

Cranberry Hibiscus (*Hibiscus acetosella*): This striking shrub has deep maroon foliage and edible leaves that can be used in salads or as a garnish. The plant is relatively low maintenance and can tolerate some shade, adding both visual appeal and edible biomass to the understory.

Chaya (*Cnidoscolus aconitifolius*): Known as the "Tree Spinach," Chaya is a hardy shrub with highly nutritious leaves that can be cooked and eaten. It thrives in partial shade and is known for its drought tolerance, making it an excellent understory plant for low-maintenance food forests.

3. SHRUB LAYER:

Chilean Wine Palm (*Jubaea chilensis*): A slow-growing shrub that produces edible, sweet seeds known as "chilean wine nuts." This plant thrives in subtropical climates and tolerates some shade. The tree has also been traditionally used for its sap, which can be processed into a sweet syrup.

Blueberry (*Vaccinium spp.*): Known for its antioxidant-rich berries, Blueberry shrubs are excellent for the shrub layer. These shrubs prefer acidic soils and partial shade, making them perfect for the understory. The fruit is edible and nutritious, providing a valuable addition to your food forest while supporting wildlife.

4. HERBACEOUS LAYER:

Chaya (*Cnidoscolus aconitifolius*): Chaya is a fast-growing, perennial herb with large, nutritious leaves that are used in soups, stews, and stir-fries. The leaves are edible when cooked, but should not be eaten raw. It thrives in partial shade and is drought-tolerant, making it a valuable plant for low-maintenance food forests.

Dandelion (*Taraxacum officinale*): Dandelion is a well-known herb with edible leaves, flowers, and roots. The leaves are rich in vitamins A, C, and K and can be eaten in salads or cooked. Dandelion is highly beneficial for the soil, improving soil health through its deep taproot, and grows well in a variety of conditions, including partial shade.

5. ROOT LAYER:

Vigna subterranea (*Vigna subterranea*): Also known as Bambara groundnut, this plant produces edible roots and tubers that are rich in protein. The tubers are often used as a staple food in Africa, and the plant grows well in warm, well-drained soils. Its ability to fix nitrogen benefits the soil health of the food forest.

Oca (*Oxalis tuberosa*): Known for its brightly colored, edible tubers, Oca is a root vegetable native to the Andes. The tubers come in a variety of colors and can be eaten raw, boiled, or roasted. Oca thrives in well-drained soils and provides both a nutritious food source and ground coverage in the root layer.

6. GROUNDCOVER LAYER:

Mitsuba (*Cryptotaenia japonica*): Mitsuba is a Japanese herb with edible leaves and stems that are often used in soups, salads, and teas. The plant grows in clumps and tolerates partial shade, making it an ideal groundcover for forested areas. Mitsuba adds both flavor and texture to dishes while contributing to the ecosystem by attracting beneficial insects.

Lamb's Quarters (*Chenopodium album*): Known for its edible young leaves and seeds, Lamb's Quarters is a nutritious, fast-growing groundcover plant. The leaves can be eaten raw in salads or cooked like spinach, while the seeds can be ground into flour. It thrives in disturbed soils and helps stabilize the ground layer.

7. VERTICAL LAYER:

Fennel (*Foeniculum vulgare*): While primarily a herb, fennel can grow vertically, especially if supported. Its edible fronds and seeds are used in cooking and for medicinal purposes. Fennel attracts pollinators and beneficial insects while also adding a unique flavor to the vertical layer of the food forest.

Tromboncino Squash (*Cucurbita moschata*): This Italian squash grows as a vertical vine and produces long, curved fruits that are edible at various stages of growth. The tender young fruits can be eaten like zucchini, while the mature squashes can be used as winter storage. Tromboncino is a prolific grower and an excellent addition to the vertical layer of a food forest.

8. FUNGAL LAYER:

Chicken of the Woods (*Laetiporus cincinnatus*): A variant of the more common Chicken of the Woods, this mushroom is similarly edible with a rich, savory flavor. It grows on decaying hardwoods and provides a bold addition to the fungal layer, both as a nutrient source and for its role in decomposition. The mushroom can help restore forest soil by breaking down organic matter.

Beefsteak Mushroom (*Fistulina hepatica*)**:** This large, reddish-brown mushroom grows on oak trees and has a meaty, beef-like texture. It is edible when young and can be used in a variety of savory dishes. Beefsteak mushrooms are often found in the forest floor and provide a good source of protein while contributing to the breakdown of wood and organic material.

Zone 11 Plan Seven: Sunrise Sustainability

Built around the renewing power of the dawn, this scheme layers a variety of species that flourish in warm conditions. Its design ensures that each level contributes to a harvest that is both abundant and environmentally sound.

1. CANOPY LAYER:

Lemon Myrtle (*Backhousia citriodora*): This tree is known for its lemon-scented leaves, which are often used in cooking for their tangy flavor. Lemon myrtle trees can grow up to 30 feet tall and are a great addition to the canopy, adding fragrance and flavor. Its leaves are harvested and dried to make teas or used as a spice.

Longan (*Dimocarpus longan*): A small tropical tree, longan produces small, round, translucent fruit that tastes similar to lychee. The tree grows up to 30 feet tall and thrives in well-drained, sandy soils. Its fruit is consumed fresh or in desserts and is a refreshing addition to tropical food forests.

2. UNDERSTORY LAYER:

Feijoa (*Acca sellowiana*): This shrub produces fragrant, green fruit that is tangy-sweet and is used in a variety of culinary dishes, from smoothies to jams. It thrives in partially shaded conditions and attracts pollinators to the food forest. Feijoa is an evergreen that also provides dense foliage.

Guavaberry (*Myrciaria floribunda*): A rare fruiting tree from the Brazilian rainforest, Guavaberry produces small, edible berries that are sweet and aromatic. It grows well in partial shade and adds an exotic, flavorful element to the food forest while supporting pollinators and beneficial insects.

3. SHRUB LAYER:

Tropical Almond (*Terminalia catappa*): This

tropical tree produces edible, nut-like seeds that are rich in fats and proteins. The plant thrives in well-drained soil and can handle partial shade. Tropical Almonds add valuable nutrients to the food forest ecosystem, and their broad, evergreen leaves contribute to nutrient cycling.

Indian Gooseberry (*Phyllanthus emblica*): Known for its tangy and vitamin C-rich fruit, Indian Gooseberry (Amla) is a hardy shrub that thrives in tropical and subtropical climates. It grows well in partial shade and can be used in herbal medicine and cooking. The fruit has numerous health benefits and is a great edible addition to the food forest.

4. HERBACEOUS LAYER:

Red Amaranth (*Amaranthus cruentus*): A striking herb with vibrant red leaves, Red Amaranth produces edible leaves and seeds that are high in protein and other nutrients. It grows well in full sun to partial shade and is a versatile addition to the herbaceous layer, offering edible greens and grains for the food forest.

Watercress (*Nasturtium officinale*): A highly nutritious aquatic herb, Watercress has edible leaves that are rich in vitamins and antioxidants. It thrives in moist environments, growing in partial shade along streams, ponds, or damp areas of the food forest. This plant adds a peppery flavor to salads and sandwiches.

5. ROOT LAYER:

Bitterroot (*Lewisia rediviva*): Bitterroot produces edible, starchy tubers that were historically used by Native American tribes. The roots are rich in carbohydrates and can be used as a survival food. This perennial plant grows well in dry, well-drained soils, making it ideal for the root layer in the food forest.

Kohlrabi (*Brassica oleracea var. gongylodes*):

Known for its edible, bulbous stem, Kohlrabi also produces edible leaves that are often used in salads or cooked like collards. The plant grows well in cool, moist conditions and can be harvested year-round, contributing both food and organic matter to the soil in the root layer

6. GROUNDCOVER LAYER:

Sweet Woodruff (*Galium odoratum*): This perennial groundcover plant produces fragrant, edible leaves that can be used to flavor beverages and desserts. Sweet Woodruff grows best in shaded, moist soils and forms a dense carpet, which helps to suppress weeds. Its ability to thrive in low-light conditions makes it valuable for shaded parts of the food forest.

Chickweed (*Stellaria media*): Chickweed is an edible groundcover plant that produces tender, mild-tasting leaves and flowers that can be used in salads and cooked dishes. The plant thrives in moist, nutrient-rich soils and is a fast grower. Chickweed helps prevent soil erosion and provides ground cover while offering nutrition.

7. VERTICAL LAYER:

Hops (*Humulus lupulus*): Known for its use in beer production, hops grow as a vigorous vine and produce edible flowers (known as hop cones). These cones are used for brewing, but the young shoots and leaves are also edible and can be used in cooking. Hops help cover vertical spaces quickly, providing both a harvest and ornamental appeal.

Passionfruit (*Passiflora edulis*): A climbing vine that produces delicious, tangy fruits. The edible pulp of the fruit is a great source of vitamins, and the plant thrives in vertical spaces, making it an ideal option for small gardens. Passionfruit also attracts pollinators and adds an exotic touch to the food forest.

8. FUNGAL LAYER:

Shaggy Mane Mushroom (*Coprinus comatus*): This mushroom has a distinctive appearance with shaggy white caps, and when mature, its black ink can be used in cooking. It is an edible species that grows in nutrient-rich soil, typically under trees. The Shaggy Mane adds both culinary diversity and soil health benefits as it breaks down organic matter.

Enokitake (*Flammulina filiformis*): This mushroom is known for its long, thin white stems and delicate flavor, often used in Asian cuisine. It thrives in cooler temperatures and moist environments, making it a suitable addition to shaded areas of the fungal layer. Enokitake mushrooms are high in antioxidants and have numerous health benefits.

Zone 11 Plan Eight: Warmth Weave

Zone 11 Plan Eight: Warmth Weave

I designed this plan to intricately weave warm-climate plants into a layered design that supports both food production and ecological resilience. It's all about interlacing complementary species to build a robust, sustainable system.

1. CANOPY LAYER:

Macadamia (*Macadamia integrifolia*): The macadamia tree produces rich, buttery nuts that are highly sought after in culinary uses. It grows to about 40 feet in height and has a dense canopy. The tree prefers well-drained soils and full sun, making it an ideal food forest addition for nut production.

Mulberry (*Morus alba*): Mulberry trees are fast-growing, with some varieties producing edible berries within the first few years of growth. The trees can grow up to 30 feet tall and thrive in a variety of soils. Their fruit is rich in vitamins and can be eaten fresh or used in jams and pies.

2. UNDERSTORY LAYER:

Malabar Plum (*Syzygium cumini*): Malabar Plum produces small, purple-black fruits that are sweet when ripe and used in traditional herbal medicine. This tree grows well in partial shade, making it an excellent option for the understory. Its flowers also attract pollinators, adding biodiversity to the food forest.

Kaffir Lime (*Citrus hystrix*): Known for its aromatic leaves and tangy fruit, Kaffir Lime is an evergreen shrub that thrives in partial shade. Its leaves are commonly used in Southeast Asian cooking, and the fruit has a unique flavor that enhances culinary dishes. It also provides fragrant foliage, adding a sensory element to the food forest.

3. SHRUB LAYER:

Clove (*Syzygium aromaticum*): This evergreen shrub produces aromatic flowers that are dried to make cloves, a spice used in both culinary and medicinal applications. Clove shrubs prefer warm climates and can tolerate partial shade, adding both flavor and medicinal benefits to the food forest.

Winter Cherry (*Withania somnifera*): Also known as Ashwagandha, this shrub is used for its medicinal properties, especially as an adaptogen. Its roots and berries are edible, with the roots used in health supplements. It grows well in partial shade and is a valuable addition to a food forest for medicinal purposes.

4. HERBACEOUS LAYER:

Chickweed (*Stellaria media*): Chickweed is a low-growing herb that produces edible leaves and stems. It is rich in vitamins and minerals and can be used in salads or cooked like spinach. This herb grows in partial shade and adds an early spring green to the food forest.

Fennel (*Foeniculum vulgare*): Fennel is an aromatic herb with edible feathery fronds and seeds. The seeds are used in cooking and for medicinal purposes, while the leaves can be added to salads. This herb thrives in well-drained soil and can be grown in partial shade, attracting beneficial insects like ladybugs.

5. ROOT LAYER:

Chickpea (*Cicer arietinum*): A leguminous plant that produces edible seeds (chickpeas), which are commonly used in cooking for their high protein content. Chickpeas also improve soil fertility by fixing nitrogen. The plant grows best in well-drained soils and provides both food and soil enrichment benefits.

Shallots (*Allium cepa var. aggregatum*): Shallots are edible bulbs that are used in cooking for their mild, sweet flavor. They grow well in well-drained soils and require minimal care. Shallots are a great addition to the root layer, offering an edible root and contributing to soil health through their bulb production.

6. GROUNDCOVER LAYER:

Cress (*Lepidium sativum*): Known for its peppery-flavored leaves, Cress is a fast-growing groundcover that produces edible leaves that can be used in salads or sandwiches. It thrives in moist, nutrient-rich soils and grows quickly, helping to stabilize the soil while providing fresh greens for the food forest.

Dandelion (*Taraxacum officinale*): A well-known edible plant, Dandelion offers nutrient-rich leaves, roots, and flowers that can be used in teas, salads, or as a cooked green. It is a hardy perennial that thrives in a variety of soils and grows quickly. Dandelion helps aerate the soil with its taproot and attracts beneficial insects to the food forest.

7. VERTICAL LAYER:

Pineapple Guava (*Feijoa sellowiana*): This small tree or shrub can be trained to grow vertically and produces edible, aromatic fruits. The fruits have a unique flavor and are rich in vitamin C. The leaves, flowers, and fruit are all edible, and the plant thrives in full sun with minimal maintenance.

Sweet Potato Vine (*Ipomoea batatas*): While known for its ground-covering roots, Sweet Potato Vine can also be grown vertically, with its edible leaves and tubers being the main benefits. The leaves can be eaten raw in salads or cooked like spinach. The vines can be trained up vertical supports to save space in the food forest while providing a nutritional harvest.

8. FUNGAL LAYER:

Pink Oyster Mushroom (*Pleurotus djamor*): The Pink Oyster Mushroom stands out with its vibrant color and delicate, mild taste. It can be grown on logs or in wood chips and adds aesthetic value along with its edible fruiting bodies. The mushroom helps to break down organic matter, enriching the soil and improving the health of the surrounding environment.

Shiitake (*Lentinula edodes*): Although it was previously excluded, Shiitake is worth mentioning as it is an iconic edible mushroom known for its rich umami flavor and health benefits. It grows on decaying hardwoods and can be cultivated in shaded areas of the fungal layer. Its medicinal properties, including immune-boosting effects, make it a valuable resource for the food forest.

Zone 11 Plan Nine: Lush Life Layout

Designed for the ultimate warm-climate experience, this layout arranges vibrant, life-giving plants in a structured, layered pattern. Its focus is on delivering a lush bounty of edibles while reinforcing the ecosystem's long-term health.

1. CANOPY LAYER:

Olive (*Olea europaea*): Olive trees are hardy and drought-tolerant, producing small, nutrient-rich fruit that can be harvested for oil or eaten whole. These trees thrive in full sun and well-drained soil. Their silvery leaves and attractive appearance make them an excellent addition to the food forest canopy.

Papaya (*Carica papaya*): Known for its sweet and juicy fruit, papaya trees grow quickly and can reach heights of 10 to 15 feet. The fruit is harvested when ripe and can be eaten fresh, dried, or in smoothies. Papayas thrive in warm, tropical climates and require well-drained soil to flourish.

2. UNDERSTORY LAYER:

Surinam Cherry (*Eugenia uniflora*): This small evergreen tree produces small, red to purple fruits that are sweet-tart and often used in jams. It grows well in the understory, tolerating shade and providing year-round fruit production in the food forest. Its compact size makes it ideal for smaller spaces.

Longan (*Dimocarpus longan*): This tropical fruit tree produces small, translucent fruit similar to lychee but with a sweeter flavor. The tree prefers some shade, making it ideal for the understory, and adds a rich, edible crop to the food forest. It also attracts pollinators with its fragrant flowers.

3. SHRUB LAYER:

Rhubarb (*Rheum rhabarbarum*): Rhubarb is a

perennial shrub that produces tangy, edible stalks used in pies, jams, and sauces. While the leaves are toxic, the stalks are rich in fiber and vitamins. Rhubarb grows best in partial shade and is a hardy, low-maintenance plant for the food forest.

Strawberry Tree (*Arbutus unedo*): Known for its unique, strawberry-like fruit, the Strawberry Tree is an evergreen shrub that thrives in Mediterranean climates. The fruits are edible, though a bit tart, and can be used in jams or eaten fresh. It also provides visual interest with its attractive, evergreen foliage and clusters of flowers.

4. HERBACEOUS LAYER:

Malabar Spinach (*Basella alba*): A tropical vine with thick, succulent, edible leaves, Malabar Spinach is a heat-loving plant that grows well in partial shade. The leaves are used in soups, stews, and salads, providing a nutritious source of vitamins. The plant is also a vigorous climber, offering ground cover and vertical growing options in the food forest.

Culinary Ginger (*Zingiber officinale*): A common herbaceous plant grown for its edible rhizomes, Ginger thrives in warm climates and partial shade. The rhizomes are used in cooking and for medicinal purposes. Ginger is a valuable addition to the herbaceous layer, providing both culinary and health benefits.

5. ROOT LAYER:

Malanga (*Xanthosoma sagittifolium*):A root crop similar to yam, malanga produces edible tubers rich in starch. It grows well in tropical food forests and helps with soil stabilization.

Crosne (*Stachys affinis*): Known as Chinese artichoke or Crosne, this plant produces edible tubers that have a nutty, crunchy texture. The tubers are often eaten raw in salads or cooked, and they are a good source of inulin. Crosne thrives in moist, well-drained soils and adds both food and diversity to the root layer.

6. GROUNDCOVER LAYER:

Chili Pepper (*Capsicum annuum*):A resilient ground cover that thrives in the tropical heat, chili peppers are not only edible but also help to deter pests while adding vibrant color to the forest.

Borage (*Borago officinalis*): Borage is a low-growing herb with edible leaves and flowers that have a cucumber-like flavor. It thrives in well-drained soils and is a great attractor of pollinators. Borage also contributes to soil health and helps with nitrogen fixation in the food forest.

7. VERTICAL LAYER:

Kiwiberry (*Actinidia arguta*): A smaller, fuzz-free variety of kiwi that grows on vines and produces edible, sweet fruits. These fruits are rich in vitamin C and can be eaten raw or used in cooking. Kiwiberries are perfect for vertical gardening due to their climbing nature and compact size.

Cucuzzi (*Cucurbita pepo*): A fast-growing climbing squash that produces edible, tender fruits known for their mild, cucumber-like flavor. Cucuzzi grows quickly on vertical supports and is easy to harvest. It thrives in warm, sunny climates and adds both an edible and aesthetic element to the vertical layer.

8. FUNGAL LAYER:

Pine Mushroom (*Tricholoma matsutake*): A rare and highly prized mushroom, the Pine Mushroom is known for its rich, earthy flavor. It grows in pine forests and forms a mycorrhizal relationship with pine trees, benefiting both the mushrooms and the trees. As a gourmet delicacy, it provides a rare food source while enhancing forest biodiversity.

Wood Ear Mushroom (*Auricularia auricula-judae*): This edible mushroom is known for its gelatinous texture and is commonly used in Asian dishes. It grows on decaying wood, typically from trees like elderberry, and is a valuable decomposer that contributes to soil fertility. Wood Ear mushrooms can thrive in the shaded, humid environment of the fungal layer.

Zone 11 Plan Ten: Sunset Grove

This plan is designed for a subtropical climate with a focus on providing high yields and habitat for pollinators, while enhancing soil health and biodiversity.

With a variety of rare and underutilized plants, it provides high-yield edible crops, medicinal plants, and pollinator-friendly species. The design utilizes plants that are resilient to the subtropical climate and includes mushrooms like **Shiitake** and **Oyster** to boost soil health and enhance plant growth. This food forest maximizes vertical, ground, and root layers for a lush and productive environment.

1. CANOPY LAYER:

Peach Palm (*Bactris gasipaes*): This tropical tree produces hearts of palm as well as small, edible fruits. The tree can grow up to 30 feet tall and thrives in humid environments with well-drained soil. The hearts of palm are harvested for culinary uses, while the fruit is enjoyed fresh or in desserts.

Pineapple (*Ananas comosus*): While often grown as a groundcover or herbaceous layer, pineapple can also be grown in the lower canopy of food forests. It is a tropical plant that produces large, sweet, spiky fruit. Pineapple plants grow in clumps and can be an ornamental addition to your food forest.

2. UNDERSTORY LAYER:

Brahmi (*Bacopa monnieri*): This creeping herb is renowned for its medicinal properties, particularly in improving cognitive function. Brahmi thrives in partial shade and wet conditions, making it a great ground cover or understory plant in damp areas of the food forest. Its leaves can be used fresh or dried in herbal remedies.

Pineapple Guava (*Feijoa sellowiana*): This unique shrub produces small, tangy fruits that taste like a combination of pineapple and guava. It thrives in partial shade, making it a perfect addition to the food forest's understory. The fragrant flowers also attract pollinators, contributing to the ecosystem's biodiversity.

3. SHRUB LAYER:

Dwarf Banana (*Musa spp.*): A small, ornamental shrub that produces edible bananas in subtropical climates, Dwarf Banana plants do well in partial shade. The fruit is sweet and versatile, and the plant's large, tropical leaves add beauty to the food forest while providing a habitat for wildlife.

Figs (*Ficus carica*): Figs are low-maintenance shrubs that produce sweet, soft fruit when grown in sunny or partly shaded areas. The fruits are edible and can be eaten fresh or dried. This shrub also contributes to biodiversity by attracting birds and pollinators to the food forest.

4. HERBACEOUS LAYER:

Taro (*Colocasia esculenta*): Taro is a root vegetable with edible corms and leaves, though the leaves must be cooked before consumption. It grows well in moist, shaded environments and adds a unique texture and flavor to dishes. Taro is a hardy, low-maintenance plant that enhances the diversity of the food forest.

Celosia (*Celosia argentea*): This herbaceous plant produces colorful, edible flowers that can be used in salads and garnishes. The leaves are also edible and can be cooked as a green. Celosia thrives in tropical climates, offering both beauty and nutritional value to the food forest.

5. ROOT LAYER:

Japanese Sweet Potato (*Ipomoea batatas*): A unique variety of sweet potato with purple skin and white flesh, the tubers are edible and highly nutritious. Japanese sweet potatoes thrive in warm, well-drained soil and can be grown in partial shade, offering a tasty and nutrient-rich food source to the root layer of the food forest.

Jicama (*Pachyrhizus erosus*): A root vegetable known for its crunchy, mild, and sweet flavor, Jicama is eaten raw in salads or cooked in stir-fries. The tubers are rich in fiber and vitamin C. This plant thrives in well-drained soils and is an excellent addition to the root layer for both its nutritional value and its ability to prevent soil erosion.

6. GROUNDCOVER LAYER:

Fennel (*Foeniculum vulgare*): Fennel is a perennial herb with edible leaves, stems, and seeds. The seeds are often used as a spice, while the leaves and stems can be eaten raw or cooked. Fennel's feathery foliage makes it a beautiful groundcover plant, and it thrives in well-drained soils and full sun, adding both culinary value and aesthetics to the food forest.

Mallow (*Malva sylvestris*): Mallow is a hardy, edible groundcover plant that produces soft, mucilaginous leaves that can be used in salads or soups. It tolerates a variety of soil types and provides ground cover while offering medicinal properties. The flowers can also be used in teas, making Mallow a versatile addition to the food forest.

7. VERTICAL LAYER:

Black Raspberry (*Rubus occidentalis*): This fruiting vine produces edible, sweet black raspberries that can be eaten fresh or used in preserves. It grows best in vertical spaces, where it can be trained along trellises or fences. Black raspberries are a great addition to a vertical food forest due to their productivity and ease of growth.

Chilean Hazelnut (*Garrya integrifolia*): This shrub produces edible nuts that grow in clusters. While the plant is typically grown as a shrub, with support, it can be trained to grow vertically. The edible seeds are rich in oils and proteins, and the plant provides a source of nuts for the food forest while also being aesthetically pleasing.

8. FUNGAL LAYER:

Honey Mushroom (*Armillaria mellea*): A parasitic fungus that feeds on tree roots, the Honey Mushroom is edible and has a mild, sweet flavor. It grows in clusters at the base of trees, making it easy to harvest in the forest. While it may harm some trees, it contributes to the decomposition process and improves soil health in the food forest.

Sulphur Shelf Mushroom (*Laetiporus sulphureus*): Known for its bright orange and yellow color, this mushroom has a distinctive flavor and is edible when young. It grows on hardwood trees, especially oaks, and adds a flavorful source of protein to the food forest. Its vibrant color and unique taste make it a valuable and functional addition to the fungal layer.

ZONES 12 & 13

In the context of the USDA Plant Hardiness Zones, **Zones 12 & 13** represent some of the warmest and most tropical environments on Earth. The USDA has classified zones up to 13 for specific tropical and subtropical regions in places like Hawaii and certain other islands, as well as a few select desert or coastal areas in the U.S.

Since **Zones 12 & 13** are categorized as part of the broader tropical and subtropical zones, they primarily include tropical islands and certain protected inland areas near the equator, where temperatures remain consistently warm and frost is essentially non-existent. These zones feature extremely mild to hot climates with high humidity, year-round growing conditions, and long seasons. In essence, these zones offer an ideal environment for growing tropical, sub-tropical, and desert plants that require intense warmth, sunlight, and moisture.

Locations within Zones 12 & 13

Here is a breakdown of U.S. locations that fall within USDA Zones 12 & 13:

American Samoa: The island of Tutuila and surrounding islands in the American Samoa territory fall under Zone 12 and 13.

Florida: The southernmost part of the Florida Keys (rarely reaches Zone 12, but certain microclimates in the Keys might come close).

Guam: The entire island of Guam, located in the western Pacific Ocean, is generally classified in Zone 12 and 13.

Hawaii: Some very low-lying coastal areas in

Hawaii, especially in the warmest parts of: The Big Island (Hilo, Kona, some coastal regions), Maui (southern coast), Oahu (around Honolulu in specific microclimates)

Northern Mariana Islands: Islands such as Saipan, Tinian, and Rota.

Puerto Rico: Coastal and lowland areas of Puerto Rico, particularly: San Juan, Ponce Fajardo, Vieques (a smaller island off Puerto Rico)

U.S. Virgin Islands: All islands in the U.S. Virgin Islands are likely Zone 12 and 13: St. Thomas, St. John, St. Croix,

Climate Characteristics

Temperature: The defining feature of Zone 12 and 13 is the consistently warm temperatures throughout the year. In these zones, the minimum average temperatures do not drop below 50°F (10°C), meaning frost is non-existent. The average temperature is typically in the mid-60s to low 80s (°F) year-round, with warmer areas possibly reaching into the 90s during peak summer months.

Growing Season: The growing season in Zones 12 & 13 is effectively perpetual due to the lack of cold temperatures. In tropical regions, there are no seasons in the traditional sense, and plants can be grown year-round. In desert-adjacent regions, the growing season may be slightly reduced due to dry periods but is still extended compared to cooler climates.

Precipitation: Rainfall is highly variable depending on the specific area. Tropical regions (like those in Hawaii, parts of the Caribbean, and

Southeast Asia) may experience high rainfall, including seasonal monsoons, while desert-adjacent regions (such as parts of the U.S. southwest or islands near the equator) could experience drier conditions punctuated by intense, localized rains. Rainwater is crucial for sustaining the dense vegetation typical of these zones, especially in tropical rainforests.

Humidity: Humidity levels in Zones 12 & 13 are consistently high, particularly in coastal and lowland tropical regions. This high moisture level supports the dense, lush vegetation that thrives in these areas. In desert-adjacent regions, humidity may fluctuate, but it tends to be relatively low compared to the tropical zones.

Sunlight: Zones 12 & 13 are subject to abundant sunlight, with these regions often located near the equator. Daylight hours are long year-round, leading to consistent photosynthesis. This consistency of light is ideal for tropical crops, flowering plants, and fruit trees.

Soil Types: Soil conditions in these zones vary. In tropical areas, the soil tends to be acidic, rich in organic matter, and highly fertile, though it can sometimes be prone to leaching in areas with high rainfall. In desert-adjacent zones, the soil may be sandy or loamy, with the need for amendments to enhance fertility, water retention, and soil structure. Well-draining soils are essential in both tropical and desert conditions to avoid root rot, but the tropical soil is more likely to be moisture-retentive than desert soils.

Ecological Features and Biodiversity

Zones 12 & 13 support an immense variety of plant and animal life, much of which is adapted to extreme heat, high humidity, or dry conditions depending on the location. These areas may feature:

Tropical Rainforests: Dense, multi-layered forests filled with towering trees, shrubs, vines, and undergrowth plants. These areas can include plants like coconut palms, bananas, breadfruit, and other tropical fruits, as well as valuable timber species.

Coastal Ecosystems: Salt-tolerant plants like mangroves, seagrape, and coastal palms thrive in regions with saltwater exposure. These ecosystems support biodiversity and offer numerous benefits to local climates, including erosion control and habitat for marine species.

Desert and Semi-arid Regions: In the desert areas within Zones 12 & 13, xerophytic plants like cacti, yucca, agave, and other succulents dominate. These areas may have specialized plant life adapted to long dry seasons with short bursts of intense rainfall. These plants often have deep roots, thick skin, and waxy coatings to retain moisture and prevent excessive water loss.

Wildlife: Due to the constant warmth and plentiful vegetation, wildlife in Zones 12 & 13 is abundant. Many tropical birds, such as parrots, toucans, and hornbills, as well as insects, lizards, and amphibians, make their home in these zones. Desert areas may host reptiles, small mammals, and insects adapted to the heat.

Pollinators: Pollination is a critical aspect of plant reproduction in Zones 12 & 13, where bees, butterflies, bats, and even hummingbirds contribute to the healthy growth of tropical fruit trees and flowering plants. The high diversity of plants supports a broad spectrum of pollinators that are critical for food production.

Challenges in Zones 12 & 13

Water Management: In tropical zones, heavy rains can lead to waterlogging and erosion if not managed well. Conversely, in desert-adjacent zones, maintaining consistent water access during dry periods is crucial for plant survival. Water collection and irrigation systems may be necessary in these dry regions to maintain healthy ecosystems.

Extreme Heat and Sun: The consistent intense sunlight can cause sunburn on plants that are not properly shaded. Heat stress can also damage sensitive crops, making it necessary to provide some degree of shading, whether by other plants or shade structures.

Soil Management: While tropical soils tend to be fertile, they are also vulnerable to leaching, which can strip essential nutrients if not replenished regularly. In arid desert areas, soils may require significant amendments to enhance fertility and water retention.

Pests and Diseases: Warm climates support a wide range of pests and diseases that can cause damage to plants. Regular monitoring for pests such as aphids, mites, and weevils, as well as fungal diseases like rust and mildew, is essential to maintaining healthy crops. Effective pest control may include the use of organic methods, such as companion planting, natural predators, and beneficial insects.

Biodiversity Loss: As tropical forests and desert ecosystems are increasingly threatened by human development, maintaining a balanced, biodiverse food forest system can be a challenge. Sustainable farming practices that integrate native plants and local species can help to preserve the ecological balance and ensure the health of the environment.

Zones 12 & 13 represent some of the most ecologically rich and warm climates in the world. From the lush, dense vegetation of tropical rainforests to the dry, resilient plant life of desert regions, these zones offer a wide range of opportunities for food forests. The main challenges lie in managing water resources, protecting plants from the extreme sun, and maintaining soil fertility. However, with appropriate planning and care, these zones can support incredibly productive, year-round food systems. The plants that thrive here can provide abundant food and materials, and the surrounding biodiversity offers numerous ecological benefits.

Zones 12 & 13 Plan One: Equinox Empire

Zones 12 & 13 Plan One: Equinox Empire

Designed for the most tropical and challenging climates, this plan builds an empire of edible layers that work in unison. It's engineered to maximize food production and maintain ecological stability in lush, warm environments.

1. CANOPY LAYER:

Ceylon Cinnamon (*Cinnamomum verum*)**:** This tropical evergreen tree is known for its aromatic bark, which can be harvested as cinnamon. The inner bark is edible and used for its sweet, spicy flavor in culinary applications. It is a fast-growing tree that provides shade, enhances soil health, and attracts beneficial insects. Cinnamon's essential oils are also used for medicinal purposes.

Breadfruit (*Artocarpus altilis*)**:** This large tree produces large, starchy fruits that are highly nutritious and versatile in cooking. The fruit can be eaten when ripe or unripe, often compared to potatoes when cooked. It thrives in tropical climates, making it an excellent choice for zones 12 and 13. Breadfruit trees also produce large leaves that provide habitat and cover for wildlife.

2. UNDERSTORY LAYER:

Cinnamon Myrtle (*Backhousia citriodora*)**:** A native Australian shrub with aromatic leaves, this plant is great for culinary uses, as the leaves and bark are both edible. The leaves are often used as a spice for cooking, especially in savory dishes. It thrives in shady, well-drained soil and can provide aromatic benefits to the understory layer.

Kaffir Lime (*Citrus hystrix*)**:** This small tree offers both edible fruit and aromatic leaves, making it valuable for culinary applications. The leaves are commonly used in Southeast Asian cuisine, while the fruits are used for their zest and juice. It thrives in partially shaded areas and provides a tropical citrus aroma to the understory.

3. SHRUB LAYER:

Brazilian Pepper Tree (*Schinus terebinthifolius*): The Brazilian pepper tree produces small, red berries that are edible and used to flavor dishes in many parts of the world. The leaves also have medicinal uses, particularly for anti-inflammatory properties. It thrives in tropical and subtropical climates and can act as an attractive, low-maintenance shrub in your food forest.

Moringa (*Moringa oleifera*): Known as the "drumstick tree," Moringa produces edible leaves, pods, and flowers. The leaves are highly nutritious, packed with vitamins and minerals, and the pods can be cooked like green beans. It grows quickly and can provide a steady supply of food, especially in the understory layer, with both food and medicinal value.

4. HERBACEOUS LAYER:

Chaya (*Cnidoscolus aconitifolius*): Chaya, also known as the "tree spinach," is a highly nutritious perennial herb that produces edible leaves rich in vitamins and minerals. The leaves need to be cooked to neutralize toxins, but when prepared properly, they are a great source of protein and fiber. Chaya thrives in tropical climates and can serve as a versatile vegetable in the food forest.

Malabar Spinach (*Basella alba*): This fast-growing, vining herb produces succulent, edible leaves and stems that are commonly used in salads or stir-fries. Known for its resistance to heat, it thrives in subtropical climates and can tolerate partial shade, making it an excellent addition to the understory and herbaceous layers.

5. ROOT LAYER:

Chayote (*Sechium edule*): Chayote, also known as vegetable pear, is a unique root vegetable that produces edible tubers, as well as edible shoots and fruit. The tubers are mild and can be cooked in a variety of ways, while the shoots and fruit offer additional food sources. Chayote is a vining plant that thrives in tropical climates, providing valuable, nutritious roots and additional edible parts for the food forest.

Skirret *(Sium sisarum)*: Skirret is a perennial plant that produces clusters of sweet, edible tubers. It grows well in temperate and subtropical climates and is known for its crisp, slightly sweet flavor. The tubers are high in starch and can be used in soups, stews, or as a root vegetable similar to parsnips or carrots. Skirret also helps enrich the soil by improving its structure over time.

6. GROUNDCOVER LAYER:

Bitter Cress (*Barbarea vulgaris*): Bitter Cress is an edible herb whose leaves and flowers are used in salads or as garnish. This plant grows well in shaded areas, making it a great groundcover under taller plants. It is rich in vitamins and antioxidants, adding both nutritional and ecological value to the food forest by attracting beneficial insects. Its fast-growing nature helps with soil stabilization.

Chickweed (*Stellaria media*): Chickweed is a nutritious edible plant with tender, mild-tasting leaves and stems that can be used in salads, soups, or teas. It thrives in partial shade and can spread quickly, providing excellent groundcover. Known for its ability to improve soil quality, chickweed can also help with weed suppression while enriching the ecosystem.

7. VERTICAL LAYER:

Chilean Guava (*Ugni molinae*): Chilean Guava produces small, aromatic red or purple fruits that are edible and used in jams, desserts, and sauces. This plant is well-suited for vertical growth and thrives in temperate climates, creating an attractive, flavorful addition to the food forest. It also provides a natural habitat for pollinators and attracts beneficial insects.

Bittermelon (*Momordica charantia*): Bittermelon is a tropical vine known for its edible fruits and leaves. The fruit has a distinct bitter taste and is commonly used in stir-fries and teas for its medicinal benefits. It grows well in vertical spaces and provides shade for other plants while contributing to the diversity of edible crops in the food forest.

8. FUNGAL LAYER:

Chicken of the Woods (*Laetiporus sulphureus*): Chicken of the Woods is a unique and edible mushroom that grows in clusters on trees, particularly oak and other hardwoods. Its tender texture resembles chicken, making it a favorite for culinary use. This fungus can be an important part of the fungal layer in the food forest, contributing to the decomposition process and nutrient cycling while providing a delicious and nutritious food source.

Shaggy Mane Mushroom (*Coprinus comatus*): Shaggy Mane is an edible mushroom known for its unique appearance, resembling a white, shaggy cone. The mushroom has a delicate flavor and is typically eaten fresh. It plays a role in breaking down organic matter and can contribute to the health of the soil in the food forest by enhancing microbial diversity.

Zones 12 & 13 Plan Two: Dusky Delights

Zones 12 & 13 Plan Two: Dusky Delights

As the day winds down, this design harnesses the subtle energy of dusk to create a layered food forest. It pairs species that thrive in warm, humid conditions, delivering both nutritional delight and lasting system support.

1. CANOPY LAYER:

Morinda (Noni) (*Morinda citrifolia*): This tropical tree produces small, aromatic fruits that are known for their health benefits. The fruit, leaves, and root are edible, though the fruit has a strong, acquired taste. Noni is valued in traditional medicine for its wide array of health-promoting properties and can be used to create teas, juices, and topical treatments.

Jackfruit (*Artocarpus heterophyllus*): Known for its enormous, sweet fruit, jackfruit is an incredible addition to any food forest. The fruit can be eaten fresh or cooked, with the unripe fruit often used as a meat substitute due to its fibrous texture. This tree grows quickly, providing substantial shade and attracting pollinators with its flowers.

2. UNDERSTORY LAYER:

Ylang Ylang (*Cananga odorata*): Known for its sweet and fragrant flowers, Ylang Ylang's blossoms can be used to make essential oils. The plant also produces edible fruit, although it's not commonly consumed in many regions. It's a slow-growing tree that benefits the understory with its fragrance and light fruit production.

Tacca Palm (*Tacca palmata*): This plant produces edible tubers similar to yam or taro, which are starchy and can be used in various dishes. The plant has large, attractive leaves and can grow well in shaded or semi-shaded conditions, providing both food and beauty to the food forest.

3. SHRUB LAYER:

Chilean Guava (*Ugni molinae*): This evergreen shrub produces small, aromatic berries that are edible and flavorful, often used in jams and desserts. The leaves are also aromatic and can be used to make herbal teas. It thrives in temperate climates and can be grown in shaded areas, contributing both food and fragrance to the food forest.

Honeyberry (*Lonicera caerulea*): A close relative of the honeysuckle, this shrub produces early-season edible berries that are similar to blueberries but with a more tart flavor. The plant is cold-hardy, making it ideal for subtropical climates, and offers early fruit for the food forest, contributing to biodiversity.

4. HERBACEOUS LAYER:

Moringa (*Moringa oleifera*): Moringa, often referred to as the "drumstick tree," offers edible, nutrient-dense leaves, flowers, and pods. The leaves are highly valued for their rich vitamin content and are commonly used in soups, salads, and smoothies. Moringa grows quickly in tropical and subtropical climates and provides a diverse, health-boosting resource for the food forest.

Ceylon Spinach (*Basella rubra*): This plant produces red-stemmed, edible spinach-like leaves that are delicious and full of antioxidants. It thrives in tropical climates and is commonly used in a variety of culinary dishes. The plant is heat-tolerant and offers a nutritious alternative to traditional spinach in the food forest.

5. ROOT LAYER:

Chinese Artichoke (*Stachys affinis*): Also known as crosne, Chinese artichoke is a unique tuber-producing plant with edible, knobby tubers that have a nutty, crunchy texture. The plant grows well in well-drained soil and provides a reliable food source.

Its small tubers can be roasted, boiled, or eaten raw in salads, offering a nutritious alternative to potatoes.

Oca (*Oxalis tuberosa*): Oca, also known as the New Zealand yam, produces small, edible tubers that are used similarly to potatoes. The tubers come in a variety of colors and are rich in vitamin C and starch. Oca thrives in temperate climates and can grow well in the root layer of the food forest, offering a colorful and nutritious root crop.

6. GROUNDCOVER LAYER:

Sweet Woodruff (*Galium odoratum*): Sweet Woodruff is a fragrant, ground-hugging herb with edible leaves that can be used to flavor beverages or desserts. This perennial thrives in shaded areas and brings a sweet aroma to the food forest. It acts as an effective groundcover and helps retain soil moisture, which benefits surrounding plants.

Lamb's Ear (*Stachys byzantina*): Known for its soft, silvery leaves, Lamb's Ear can be used as groundcover while adding beauty to the landscape. The leaves are not only ornamental but can also be used to make a mild herbal tea. This plant thrives in well-drained soil and provides excellent erosion control in the food forest.

7. VERTICAL LAYER:

Aunt Molly's Ground Cherry (*Physalis pruinosa*): A relative of the tomatillo, Aunt Molly's Ground Cherry produces sweet, edible, golden berries encased in papery husks. It is perfect for vertical growth, climbing with the support of nearby structures or other plants. This plant thrives in warm climates, providing a delicious, nutritious fruit while also attracting pollinators.

Winged Sumac (*Rhus copallina*): Winged Sumac is a deciduous shrub with vibrant red berries that are edible when prepared properly and often used to make tangy beverages or flavorings. This plant is ideal for vertical planting and can be used to add ecological value by improving soil quality. Its dense foliage also provides shelter for beneficial insects.

8. FUNGAL LAYER:

Maitake Mushroom (*Grifola frondosa*): Maitake, also known as Hen of the Woods, is a prized edible mushroom that grows at the base of trees, particularly oaks. Its ruffled, grayish-brown clusters are used in a variety of dishes, known for their meaty texture and rich umami flavor. Maitake enhances the food forest by promoting soil health through decomposition and contributing to the symbiotic relationships with tree roots.

Chaga Mushroom (*Inonotus obliquus*): Chaga is a medicinal mushroom that grows on birch trees and is prized for its high antioxidant content. While it's not typically eaten in large amounts, its bark can be harvested to make medicinal teas, known for boosting immunity and promoting overall health. Chaga adds value to the fungal layer by assisting in the decomposition of wood and enriching the ecosystem.

Zones 12 & 13 Plan Three: Paradise Pulse

This plan taps into the vibrant pulse of tropical life, arranging a series of layered species for peak productivity. It's focused on turning challenging conditions into a rhythmic, sustainable food harvest.

1. CANOPY LAYER:

Breadfruit (*Artocarpus altilis*): This is another variety of breadfruit with a similar large, starchy, and edible fruit. It offers a nutritious, high-calorie food source and thrives in tropical environments, growing quickly and providing large leaves for shade. The tree's low-maintenance nature and adaptability to various soil conditions make it an excellent choice for canopy growth.

Guaruba Palm (*Syagrus coronata*): This palm tree produces edible fruits known as "guaruba nuts," which are rich in protein and fats. The tree also provides shelter for smaller plants and animals and helps enhance soil quality through its leaf litter. Guaruba palm thrives in humid tropical climates and adds a unique vertical element to the food forest.

2. UNDERSTORY LAYER:

Black Mulberry (*Morus nigra*): This tree produces edible fruit that is both sweet and nutritious. It grows well in shaded understory areas and can provide a continuous supply of fresh berries. The leaves are also used for silkworm farming, adding additional value to the food forest ecosystem.

Ceylon Spinach (*Basella alba*): A tropical vine that thrives in partial shade, Ceylon spinach is a fast-growing edible that can be used in salads, soups, and stir-fries. The leaves and young shoots are edible and high in vitamins A and C. It's a perfect addition to the understory layer due to its vigorous growth.

3. SHRUB LAYER:

Mulberry (*Morus alba*): The mulberry tree or shrub produces sweet, edible fruit that can be eaten fresh, dried, or made into jams. Mulberry leaves are also used as food for silkworms, making the plant valuable for both food production and potential silk farming. It grows quickly and can be grown in semi-shaded environments.

Naseberry (*Manilkara zapota*): This tropical shrub produces sweet, grainy-textured fruit that is edible and delicious. The fruit can be eaten fresh or used in desserts. Naseberry grows well in tropical climates and adds a unique, exotic fruit to the food forest's shrub layer.

4. HERBACEOUS LAYER:

Sweet Potato Vine (*Ipomoea batatas*): While often grown for its tubers, the sweet potato vine also produces edible, tender leaves and young shoots. The leaves are high in vitamins A and C and can be eaten cooked or raw. It thrives in tropical and subtropical zones, making it a low-maintenance ground cover in the food forest.

Amaranth (*Amaranthus spp.*): This versatile herb produces edible leaves, seeds, and flowers. The leaves are rich in protein, vitamins, and minerals, while the seeds can be used as a grain. Amaranth is easy to grow and can thrive in both sun and partial shade, providing nutritious, fast-growing food for the food forest.

5. ROOT LAYER:

Jerusalem Artichoke (*Helianthus tuberosus*): Jerusalem artichoke, or sunchoke, is a tuber-producing plant that grows well in a variety of soils and climates. The edible tubers have a nutty flavor and can be roasted, boiled, or made into soups. Jerusalem artichoke is also a great source of inulin, which is beneficial for digestive health and can be used as a prebiotic.

Cranberry Hibiscus (*Hibiscus acetosella*): Although not a traditional root crop, cranberry hibiscus offers edible leaves and has root structures that can be utilized for medicinal purposes. The plant thrives in tropical climates and adds vibrant color and flavor to the root layer while its roots hold medicinal value. The leaves, which have a tangy, sour flavor, are often used in salads and cooked dishes.

6. GROUNDCOVER LAYER:

Green Bean Groundcover (*Phaseolus vulgaris*): While typically grown vertically, certain varieties of Green Bean can also function as a groundcover. The leaves, young pods, and beans are edible and provide valuable protein. It thrives in sun-dappled areas and adds nitrogen to the soil, benefiting nearby plants.

Creeping Jenny (*Lysimachia nummularia*): Creeping Jenny is a low-growing plant with edible leaves that can be used in salads. This hardy groundcover works well in moist, shaded areas and adds beauty with its vibrant green foliage. It's also known for its ability to improve soil health by preventing erosion and fostering biodiversity.

7. VERTICAL LAYER

Yard Long Bean (*Vigna unguiculata subsp. sesquipedalis*): Yard Long Bean produces edible, tender beans that are used in a variety of Asian dishes. The plant grows quickly and reaches impressive heights, providing a vertical option for a productive ground layer. Its nitrogen-fixing properties benefit the soil and help maintain a balanced ecosystem in the food forest.

Cardoon (*Cynara cardunculus*): Cardoon is a thistle-like plant whose stalks are edible and used in Mediterranean cooking. Growing in vertical layers, it provides an attractive structural element to the food forest while producing a delicious, nutritious crop. The leaves and flowers are also edible and bring biodiversity to the environment.

8. FUNGAL LAYER

Turkey Tail Mushroom (*Trametes versicolor*): Turkey Tail is an edible and medicinal mushroom commonly found on fallen logs. It is known for its colorful, fan-shaped caps and is used in teas and tinctures for its immune-boosting properties. This mushroom contributes to the decomposition of organic matter in the food forest, promoting a healthy fungal network.

Blewit Mushroom (*Clitocybe nuda*): The Blewit Mushroom is an edible fungus found in deciduous forests, especially in rich, undisturbed soil. It is known for its lavender-blue color and its rich, savory taste when cooked. As a saprophyte, it plays an important role in breaking down organic matter and enriching the soil's microbial diversity.

Zones 12 & 13 Plan Four: Exuberant Equator

Built for equatorial exuberance, this edifice layers robust, exotic species that deliver high yields. It's intended to flourish in intense warmth while supporting a diverse and sustainable ecosystem.

1. CANOPY LAYER

Papaya (*Carica papaya*): This fast-growing tree produces delicious, tropical fruit, which can be eaten fresh or used in smoothies and desserts. Both the fruit and leaves are edible, with the leaves often used as a tenderizer for meat. Papaya trees are great for providing quick shade and support for climbing plants.

Soursop (*Annona muricata*): Known for its tangy, sweet fruit, soursop is a tropical tree that bears edible fruit and leaves. The fruit is rich in vitamins, antioxidants, and fiber, and the leaves are used in traditional medicine. This tree provides substantial shade and adds a unique flavor to your food forest.

2. UNDERSTORY LAYER

Pineapple Sage (*Salvia elegans*): This aromatic herb produces edible leaves and bright red flowers that attract pollinators. The leaves have a pineapple-like flavor and can be used in teas, salads, and desserts. It grows well in partial shade, making it a great choice for the understory layer.

Paprika Pepper (*Capsicum annuum*): A variety of pepper that is often grown for its dried, powdered fruit, used as a spice in cooking. The peppers can be harvested when ripe and are great for seasoning or adding to dishes for heat. It thrives in semi-shaded environments with well-drained soil.

3. SHRUB LAYER

Jujube (*Ziziphus jujuba*): Also known as the Chinese date, the jujube produces small, sweet fruit that can be eaten fresh, dried, or used in cooking. It's a hardy shrub that thrives in subtropical climates and offers a productive, edible resource for the food forest throughout the year.

Chokeberry (*Aronia melanocarpa*): Known for its dark purple to black berries, chokeberry is a hardy shrub that produces edible fruit high in antioxidants. The berries can be used in juices, jams, and baked goods. It grows well in a variety of soil types and adds both nutrition and color to the food forest.

4. HERBACEOUS LAYER

Chickweed (*Stellaria media*): Chickweed is a fast-growing, low-spreading herb with edible leaves, flowers, and stems. It is often used in salads, soups, or as a garnish, and is known for its high vitamin C content. Chickweed grows well in moist, shaded areas, making it a valuable herb for the understory layer.

Purslane (*Portulaca oleracea*): Purslane is a hardy herb that produces succulent, tangy-tasting leaves rich in omega-3 fatty acids, antioxidants, and vitamins. It thrives in hot climates and can be eaten raw in salads or cooked in stews and soups. Purslane is also a drought-tolerant plant, making it perfect for the food forest in dry conditions.

5. ROOT LAYER

Salsify (*Tragopogon porrifolius*): Salsify, also known as oyster plant, produces long, edible roots with a mild, oyster-like flavor. It grows best in well-drained soil and provides a nutrient-rich root crop. Salsify is a hardy plant that can withstand cooler temperatures and enrich the soil by fixing nitrogen, making it a valuable addition to the root layer.

Elephant Foot Yam (*Amorphophallus paeoniifolius*): Elephant foot yam produces large, starchy tubers that are edible when cooked. It thrives in tropical climates and is known for its unique appearance. The tubers are rich in carbohydrates and fiber, making them an excellent energy source. It is a great addition to the root layer due to its fast-growing nature and high yield.

6. GROUNDCOVER LAYER

Yarrow (*Achillea millefolium*): Though better known for its medicinal uses, Yarrow's edible young leaves can be used in salads or as a herb. It thrives in full sun or partial shade and provides attractive blooms while serving as an excellent groundcover. It also helps attract beneficial insects and adds to the food forest's ecological balance.

Creeping Thyme (*Thymus serpyllum*): Creeping Thyme is a low-growing herb that produces edible leaves and flowers. It's perfect for sunny spots and adds a lovely fragrance to the groundcover layer. Thyme's culinary uses are numerous, and it also attracts pollinators, helping the overall health of the food forest.

7. VERTICAL LAYER

Cucamelon (*Melothria scabra*): Cucamelon, also known as Mexican sour gherkin, is a small, cucumber-like fruit with a tart flavor. It is perfect for vertical growth on trellises, making it an ideal addition to the vertical layer in food forests. The plant is productive, and its tiny fruits provide a unique, tangy snack while attracting beneficial insects.

Goji Berry (*Lycium barbarum*): Goji Berry is a climbing shrub that produces small, edible red berries, known for their high antioxidant content. It is commonly used in smoothies, teas, or dried as a snack. This plant thrives in vertical spaces and provides both medicinal and nutritional benefits, enhancing the food forest's overall productivity and biodiversity.

8. FUNGAL LAYER

Pig's Ear Fungus (*Gomphus clavatus*): Pig's Ear is an edible fungus with a distinct ear-like shape, found growing in coniferous forests. The mushroom has a mild flavor and is typically sautéed in dishes. It contributes to the fungal layer of the food forest by decomposing organic material, particularly pine needles, and helping maintain soil fertility.

Hen of the Woods (*Grifola frondosa*): Hen of the Woods, or Maitake, is a nutritious, edible mushroom that grows at the base of trees. The mushroom has an earthy flavor and is used in soups, stir-fries, and teas. Its role in the fungal layer extends to nutrient cycling and supporting the surrounding plant community through the decomposition process.

Zones 12 & 13 Plan Five: Fiesta Flora Formula

Celebrating the festive spirit of the tropics, this design plan arranges plant layers like the components of a joyful celebration. Its goal is to produce a lively harvest and create an ecosystem that dances with sustainability.

1. CANOPY LAYER:

Cacao (*Theobroma cacao*): The source of chocolate, cacao trees produce pods that contain edible seeds, known as cacao beans. These beans can be harvested and processed into cocoa, offering a valuable food product. The trees also have large leaves that provide significant shade and help maintain moisture in the forest ecosystem.

Pili Nut Tree (*Canarium ovatum*): This tropical tree produces edible nuts that are highly nutritious and rich in healthy fats. The nuts can be eaten raw, roasted, or used in various dishes. Pili nut trees are also excellent for attracting wildlife and contribute to soil health by decomposing leaf litter.

2. UNDERSTORY LAYER:

Ginger Lily (*Hedychium coronarium*): The rhizomes of this plant are edible and can be used to make ginger-flavored tea or in cooking. It thrives in shaded, moist areas and produces beautiful, fragrant white flowers, adding aesthetic value to the understory layer while providing edible rootstock.

Curry Tree (*Murraya koenigii*): Native to India, the curry tree's leaves are widely used in South Asian cuisine, offering a unique flavor. The tree grows well in subtropical to tropical environments, providing a valuable edible resource for flavoring foods in the understory.

3. SHRUB LAYER:

Fuyu Persimmon (*Diospyros kaki*): This deciduous shrub produces sweet, edible fruit that is commonly eaten fresh or dried. The fruit is known for its firm texture and mild sweetness. Fuyu persimmon thrives in subtropical climates and provides an attractive, flavorful addition to the food forest.

Cranberry (*Vaccinium macrocarpon*): This evergreen shrub produces tart red berries that are often used in sauces, juices, and baked goods. The berries are rich in antioxidants and have medicinal uses as well. Cranberries grow well in acidic, boggy soils and can thrive in the food forest as a low-growing shrub.

4. HERBACEOUS LAYER:

Egyptian Walking Onion (*Allium × proliferum*): This unique herb produces small, bulb-like clusters at the top of its stalks, as well as edible leaves and bulbs that are milder than regular onions. The plant can grow in a variety of conditions, making it a resilient addition to the herbaceous layer and offering an easy-to-harvest perennial food source.

Oca (*Oxalis tuberosa*): Oca, or the "New Zealand yam," is a tuber-producing plant that thrives in tropical and subtropical regions. The edible tubers are rich in vitamin C and are used similarly to potatoes in cooking. Oca grows well in the understory and provides an excellent root crop for the food forest.

5. ROOT LAYER:

Yam Bean (*Pachyrhizus spp.*): Yam bean, also known as jícama, produces large, edible roots that are crunchy and slightly sweet. The root is a good source of fiber and vitamin C and can be eaten raw in salads or cooked in a variety of dishes. It is a hardy plant that grows well in tropical and subtropical climates, making it an excellent choice for the root layer.

Edible Canna (*Canna indica*): Known for its vibrant flowers, the edible canna produces starchy, tuberous rhizomes that are rich in carbohydrates and can be cooked and eaten much like potatoes. It thrives in warm climates and moist soils, making it a good choice for the root layer in tropical food forests. The rhizomes are high in energy and provide a nutrient-rich root crop for the food forest.

6. GROUNDCOVER LAYER:

Oregano (*Origanum vulgare*): Oregano is a hardy herb with edible leaves that are commonly used in cooking. It thrives as groundcover in well-drained soils and full sun or partial shade. It attracts beneficial insects and can also serve as a medicinal herb, promoting the biodiversity and sustainability of the food forest.

Mullein (*Verbascum thapsus*): Mullein is a towering herb with edible leaves that can be used for teas or in medicinal preparations. It works well as a groundcover by stabilizing the soil and attracting pollinators with its bright flowers. Its deep taproot also helps with soil aeration and water retention.

7. VERTICAL LAYER:

Kiwiberry (*Actinidia arguta*): Kiwiberry is a miniature version of the kiwi fruit, with smooth skin and a sweet-tart flavor. This vine grows vertically, producing clusters of small, edible fruits that can be eaten fresh or added to recipes. It is well-suited for the vertical layer and can tolerate cooler temperatures, making it a great choice for food forests in zones 12 & 13.

Scarlet Runner Bean (*Phaseolus coccineus*): Scarlet Runner Bean is a climbing plant that produces vibrant red flowers and edible beans. It is a vigorous grower, ideal for vertical gardens or trellises. The beans are delicious when cooked and provide a source of protein, while the plant's ability to fix nitrogen in the soil benefits surrounding plants.

8. FUNGAL LAYER:

Wood Ear Mushroom (*Auricularia auricula-*

judae): Known for its gelatinous texture, the Wood Ear Mushroom is an edible fungus commonly used in Asian cooking, particularly in soups and stir-fries. It grows on decaying wood, aiding in the decomposition process. This fungus is beneficial in the fungal layer of the food forest, as it breaks down organic matter and enhances the health of the soil.

Pine Mushroom (*Tricholoma magnivelare*): The Pine Mushroom, also known as Matsutake, is a highly prized edible mushroom, particularly in Asian cuisines. It grows in association with pine trees and has a distinct, spicy aroma and flavor. The mushroom enhances the fungal layer by breaking down organic material and providing important nutrients for the soil ecosystem.

Zones 12 & 13 Plan Six: Celestial Canopy

Zones 12 & 13 Plan Six: Celestial Canopy

With a nod to the grandeur of the tropical skies, this conception layers species into a celestial canopy of edibles. It's designed to capture both the beauty and the bounty of a warm, humid climate in one sustainable system.

1. CANOPY LAYER:

Moringa (*Moringa oleifera*): Often referred to as the "drumstick tree," moringa offers a wide range of edible parts, including leaves, flowers, pods, and seeds. The leaves are particularly nutrient-dense, often used in soups, salads, or smoothies. Moringa trees are fast-growing, drought-tolerant, and help fix nitrogen in the soil.

Lychee (*Litchi chinensis*): Lychee trees produce small, sweet, and aromatic fruit that is rich in vitamin C. Both the fruit and flowers are edible, and the tree also offers fragrant blossoms that attract pollinators. Lychee trees thrive in tropical and subtropical climates, offering shade while enhancing the biodiversity of the food forest.

2. UNDERSTORY LAYER:

Strawberry Guava (*Psidium cattleianum*): This small tree or shrub produces sweet and tangy edible fruit that is often used in jellies, juices, and sauces. It's a low-maintenance plant that thrives in shaded understory areas and provides both a food and habitat source for wildlife.

Moringa (*Moringa oleifera*): Known as the "drumstick tree," Moringa produces edible pods, leaves, and flowers. The leaves are highly nutritious, containing essential vitamins and minerals. It can thrive in the understory layer as long as there is sufficient light and water.

3. SHRUB LAYER:

Goji Berry (*Lycium barbarum*): Known for its high nutritional value, goji berries are edible and packed with vitamins and antioxidants. The plant produces small red berries that are commonly used in dried form or as a fruit in smoothies and teas. It grows as a hardy shrub and can offer both food and medicinal benefits.

Blackberry (*Rubus fruticosus*): This thorny shrub produces dark purple-black berries that are sweet and perfect for pies, jams, and eating fresh. It thrives in full sun and can spread quickly, providing both edible fruit and an attractive ground cover for the food forest.

4. HERBACEOUS LAYER:

Lemongrass (*Cymbopogon citratus*): Known for its citrusy aroma, lemongrass produces edible leaves that can be used in teas, soups, and curries. It grows well in tropical and subtropical climates and can be a valuable herb for flavoring dishes in the food forest. Lemongrass also acts as a natural insect repellent, providing additional benefits.

Taro (*Colocasia esculenta*): Taro is a staple in many tropical food systems, providing edible tubers and leaves. The tubers are used in a variety of dishes, while the young leaves can be cooked and used in stews. It thrives in moist, shaded environments, making it an ideal addition to the understory layer.

5. ROOT LAYER:

Dahlia (*Dahlia pinnata*): Dahlia tubers are edible and can be used in a variety of culinary dishes. The tubers are starchy and have a slightly sweet taste, resembling sweet potatoes. Dahlia thrives in temperate climates and can be grown as a low-maintenance root crop in the food forest. The plant also adds beautiful flowers to the landscape.

Spicebush (*Lindera benzoin*): Spicebush, while primarily known for its aromatic leaves and berries, also produces edible roots with a spicy flavor. The roots have been traditionally used for their medicinal properties and can be used in cooking to add a unique flavor. Spicebush grows well in shaded areas and adds both food and medicine to the food forest.

6. GROUNDCOVER LAYER:

Chives (*Allium schoenoprasum*): Chives provide an edible, onion-like flavor to dishes with their tender leaves. This plant thrives as groundcover, particularly in well-drained, slightly alkaline soils, and can withstand both sun and partial shade. Its vibrant purple flowers attract pollinators, adding to the ecological diversity of the food forest.

Daylilies (*Hemerocallis spp.*): Though primarily used for their ornamental flowers, Daylilies also produce edible blooms and tuber-like roots. These plants thrive in full sun or partial shade and can help fill in gaps in the food forest's groundcover layer. Their resilience and beauty contribute to the aesthetic and ecological health of the food forest.

7. VERTICAL LAYER:

Climbing Spinach (*Tetragonia tetragonioides*): Also known as New Zealand spinach, Climbing Spinach is a leafy green that thrives in vertical spaces. Its edible leaves can be used in salads or cooked like regular spinach. This plant is drought-tolerant, grows quickly, and adds valuable greens to the food forest while helping to control soil erosion.

Chayote Vine (*Sechium edule*): While it's technically a vine, Chayote is commonly grown as a vertical crop for its edible fruit. The fruits can be eaten raw or cooked, and the young shoots and leaves are also edible. Chayote is a hardy, fast-growing vine that is perfect for adding vertical interest while producing a nutritious, edible crop.

8. FUNGAL LAYER:

Mycena Mushroom (*Mycena spp.*): Various species of Mycena mushrooms are edible and thrive in decaying wood and leaf litter. Known for their delicate structure, they add to the fungal diversity of the food forest. These mushrooms are particularly useful in breaking down organic materials and adding nutrients back into the soil, promoting a healthy environment for plants.

Lions Mane Mushroom (*Hericium erinaceus*): Lions Mane is an edible mushroom that grows on hardwood trees, prized for its unique, fluffy white appearance and medicinal properties. It has a mild, sweet flavor and is used in both culinary dishes and as a supplement for cognitive health. This mushroom helps decompose dead wood and provides an important source of food in the fungal layer.

Zones 12 & 13 Plan Seven: Cosmic Orchard

Zones 12 & 13 Plan Seven: Cosmic Orchard

Venturing beyond the ordinary, this plan creates a "cosmic" orchard where layered plantings align for mutual benefit. It pairs complementary species to yield delicious produce and build a self-sustaining ecosystem in tropical conditions.

1. CANOPY LAYER:

Cherimoya (*Annona cherimola*): Cherimoya produces sweet, creamy fruit that is often referred to as the "ice cream fruit." The fruit is highly prized for its unique texture and flavor. Cherimoya trees are small to medium-sized, making them ideal for planting in the canopy layer while providing habitat for wildlife.

Cinnamon Myrtle (*Backhousia myrtifolia*): This small evergreen tree produces aromatic leaves and flowers, which are used in culinary dishes and to make essential oils. Its edible leaves offer a cinnamon-like flavor, adding a unique spice to recipes. The tree thrives in tropical climates and can serve as both a food source and a fragrant addition to the forest.

2. UNDERSTORY LAYER:

Chaya (*Cnidoscolus aconitifolius*): Known as tree spinach, Chaya produces large, nutritious leaves that are edible when cooked. It's a highly productive shrub that grows in shaded areas and can provide substantial amounts of food throughout the year.

Betel Leaf (*Piper betle*): This climbing vine's leaves are used in traditional cooking, particularly in Southeast Asia, and are often chewed for their stimulant properties. It thrives in shady conditions and is an excellent addition to the understory with its high productivity.

3. SHRUB LAYER:

Wattleseed (*Acacia spp.*): Native to Australia, wattleseed shrubs produce edible seeds that can be ground into flour for baking. These seeds have a nutty flavor and are used in a variety of native dishes. The shrub is well-suited for arid or semi-arid climates, providing a unique, sustainable source of food.

Cape Gooseberry (*Physalis peruviana*): This small shrub produces small, golden fruit encased in a papery husk, which are edible and sweet. They are commonly used in jams, desserts, and as a snack. The plant thrives in subtropical climates and offers unique, vitamin-rich fruit for the food forest.

4. HERBACEOUS LAYER:

Nasturtium (*Tropaeolum majus*): This herb produces edible flowers, leaves, and seeds that have a peppery flavor, often used in salads or as garnishes. The plant is easy to grow and can tolerate a range of soil types. Nasturtium also attracts pollinators and provides color and flavor to the food forest.

Shiso (*Perilla frutescens*): Known for its aromatic leaves, shiso is a herb commonly used in Japanese cuisine. The leaves can be eaten fresh or used in pickles and sushi. It thrives in tropical climates and adds a unique, fragrant herb to the food forest.

5. ROOT LAYER:

Yacon (*Smallanthus sonchifolius*): A root vegetable that produces sweet, juicy tubers. Yacon improves soil health with its extensive root system and attracts beneficial microorganisms.

Chickpea (*Cicer arietinum*): Chickpeas are widely known for their edible seeds, but they also produce edible roots. The roots are high in protein and fiber and help enrich the soil by fixing nitrogen. Chickpeas are well-suited for the root layer as they provide both nutrition and soil fertility benefits to the food forest.

6. GROUNDCOVER LAYER:

Sunshine Mimosa (*Mimosa strigillosa*): A nitrogen-fixing ground cover with pink, fluffy flowers. It attracts pollinators and enhances soil fertility, creating a resilient base layer for the forest.

Lamb's Lettuce (*Valerianella locusta*): Known for its mild, nutty flavor, Lamb's Lettuce can be harvested for its edible leaves. It grows well as a groundcover in partial shade or full sun and is an excellent choice for nutrient-dense greens in the food forest. It is quick to establish and helps to keep the soil covered and healthy.

7. VERTICAL LAYER:

Fuzzy Kiwi (*Actinidia deliciosa*): Fuzzy Kiwi, or regular kiwi, is a climbing vine known for its edible, sweet fruits. The vine grows well vertically and produces a large quantity of kiwis in a relatively small space. This plant thrives in zones 12 & 13 and provides not only a nutritious fruit but also structural beauty to the food forest.

Tromboncino Squash (*Cucurbita moschata*): Tromboncino Squash is a unique, long-necked squash that can grow vertically with the right support. Its tender, edible fruit is a versatile ingredient in a variety of dishes. As a vigorous grower, this plant produces an abundance of food while enriching the food forest by improving soil health and attracting pollinators.

8. FUNGAL LAYER:

Wood Blewit (*Clitocybe nuda*): Wood Blewit is another edible mushroom, with a distinctive lavender-blue color, growing on wood and organic matter. Its mild flavor and versatility in cooking make it a popular choice in culinary dishes. As it decomposes wood, it helps maintain soil health, promoting the growth of plants in the surrounding areas.

Candy Cap Mushroom *(Lactarius rubidus)*: Candy Cap is an edible mushroom known for its sweet, maple-like scent, and it's used in desserts and savory dishes. It grows in symbiosis with trees, especially oaks, and can be found on the forest floor. This mushroom's role in the fungal layer extends to nutrient cycling and enhancing the flavors of the food forest ecosystem.

Zones 12 & 13 Plan Eight: Tropical Triumph Tapestry

Zones 12 & 13 Plan Eight: Tropical Triumph Tapestry

This tapestry weaves together vibrant, tropical species into a layered design that triumphs over challenging conditions. Its focus is on celebrating food production and natural support in an exotic, sustainable manner.

1. CANOPY LAYER:

Tamarillo (*Solanum betaceum*): Known as the "tree tomato," tamarillo produces edible fruits that are tangy and rich in vitamins. The fruit can be eaten raw or cooked, often used in sauces or salsas. Tamarillo trees grow in tropical to subtropical climates, providing shelter and enhancing the diversity of a food forest.

Tamarind (*Tamarindus indica*): A medium-sized tree with sweet-sour edible pods. Tamarind trees are drought-tolerant, fix nitrogen in the soil, and attract pollinators with their fragrant flowers.

2. UNDERSTORY LAYER:

Soursop (*Annona muricata*): A small tropical tree that produces large, sweet-sour fruits. It attracts beneficial pollinators such as bees and butterflies, and its fallen leaves improve soil structure.

Ice Cream Bean (*Inga edulis*): A nitrogen-fixing tree that produces long pods with sweet, edible pulp. Its ability to improve soil fertility makes it an excellent companion for nearby plants.

3. SHRUB LAYER:

Lemon Verbena (*Aloysia citriodora*): While primarily known for its aromatic leaves, which are used in teas, lemon verbena also provides small edible flowers. It thrives in subtropical regions and adds both flavor and fragrance to the understory while attracting pollinators.

Snow Pea Bush (*Pisum sativum*): This low-growing

shrub produces edible pods that are consumed in many cuisines. The pods can be eaten fresh or cooked in stir-fries and soups. It thrives in cool, subtropical climates and provides a nitrogen-fixing legume to the food forest.

4. HERBACEOUS LAYER:

Elephant Foot Yam (*Amorphophallus paeoniifolius*): This edible tuber plant produces large, starchy tubers that are used in a variety of traditional dishes. The plant thrives in warm, subtropical climates and adds a unique, nutritious root crop to the food forest. The young shoots and leaves are also edible when properly prepared.

Sweet Basil (*Ocimum basilicum*): Sweet basil produces aromatic, edible leaves that are used extensively in cooking, particularly in Italian and Mediterranean dishes. It thrives in warm, sunny climates and can be grown as an annual herb. Sweet basil adds both flavor and fragrance to the herbaceous layer in the food forest.

5. ROOT LAYER:

Cassava (*Manihot esculenta*): A hardy root crop producing starchy tubers. Cassava thrives in tropical climates and improves soil structure by loosening compacted earth.

Tobacco Root (*Nicotiana tabacum*): While typically associated with its leaves for smoking, tobacco plants also produce edible roots that are sometimes used in traditional medicine and cooking. The roots can be boiled and eaten for their nutritional value. The plant thrives in hot, sunny environments and adds a unique, multi-use crop to the food forest.

6. GROUNDCOVER LAYER:

Taro Leaf (*Colocasia esculenta*): While typically grown for its tuber, Taro leaves can also be eaten when cooked. Taro thrives in wet, partially shaded areas, making it a perfect addition to the groundcover layer in food forests located near water. It's not only edible but also helps to prevent soil erosion in damp environments.

Japanese Knotweed (*Fallopia japonica*): This plant's edible young shoots are commonly harvested and eaten like asparagus. It is a hardy perennial that spreads rapidly as groundcover, especially in moist, nutrient-rich soils. While invasive in some regions, it can be a useful addition to food forests when controlled, offering nutritious edible shoots and a quick-growing ground cover.

7. VERTICAL LAYER:

Dragon Fruit (*Hylocereus undatus*): A climbing cactus producing large, sweet fruits. It thrives in dry conditions and provides habitat for pollinators such as bats and bees.

Cucumber (*Cucumis sativus*): While typically a horizontal grower, cucumber vines can thrive in vertical planting systems. The fruits are widely known for their refreshing, edible taste and are a staple in many cuisines. This plant contributes to the food forest's productivity by providing a quick-growing and widely enjoyed vegetable.

8. FUNGAL LAYER:

Craterellus Tuberatus (*Craterellus tuberatus*): Craterellus Tuberatus, also known as the Yellow Foot or Trumpet Chanterelle, is an edible mushroom that grows in mossy, damp forest floors. The mushroom is prized for its earthy flavor and is often used in gourmet cooking. Its role in the food forest is to help break down organic matter, contributing to soil health and supporting other plant life.

Fuzzy Foot Mushroom (*Gymnopilus underwoodii*): The Fuzzy Foot Mushroom is an edible species that grows on decaying wood in humid, temperate forests. Known for its yellow to orange colors, it adds a unique flavor to dishes and aids in the decomposition of dead wood. This mushroom helps improve soil conditions by contributing to organic matter breakdown.

Zones 12 & 13 Plan Nine: Zenith Zest Zone

At the zenith of tropical abundance, this plan layers zesty species to create a zone of flavor and resilience. It is crafted to harness the region's intense energy for a robust harvest and lasting ecological balance.

1. CANOPY LAYER:

Longan (*Dimocarpus longan*): A relative of lychee, longan trees produce sweet, translucent fruit that is often eaten fresh or dried. The fruit is rich in antioxidants and vitamin C. Longan trees also provide a fast-growing canopy layer, contributing to the overall health of the food forest ecosystem.

Cocoa Palm (*Elaeis guineensis*): This palm tree produces edible fruits known as coconuts, which provide both water and flesh that can be eaten or used for making oil. Coconut palms are essential for a tropical food forest, adding a vertical element that supports climbing plants and provides shelter for wildlife.

2. UNDERSTORY LAYER:

Clove (*Syzygium aromaticum*): An aromatic spice tree that produces edible flower buds. Clove trees are excellent companions in tropical forests, promoting beneficial insects and adding fragrance to the food forest.

Indian Curry Leaf Tree (*Murraya koenigii*): A small tree that produces aromatic leaves used in Indian cuisine to flavor curries. The leaves are highly valued for their unique taste and nutritional benefits, especially in medicinal practices. This tree grows well in shaded understory areas.

3. SHRUB LAYER:

Barbados Cherry (*Malpighia emarginata*): This shrub produces small, tart cherries that are edible and high in vitamin C. The fruit can be eaten fresh or used in drinks and jellies. It grows well in tropical and subtropical climates and offers a nutritious fruit for the food forest.

Roselle (*Hibiscus sabdariffa*): A tropical shrub known for its tart, cranberry-like calyces used in teas and jellies. Roselle flowers attract pollinators, and its leaves and stems add nutrients to the soil as they decompose.

4. HERBACEOUS LAYER:

Kale (*Brassica oleracea)*: Kale is a hardy leafy green that is rich in vitamins A, C, and K. The leaves are edible and can be used in salads, soups, or as a garnish. It grows well in cooler climates, making it an excellent addition to the herbaceous layer for year-round food production.

Ginger (*Zingiber officinale*): A medicinal and culinary herb with vibrant yellow flowers that attract pollinators. Ginger improves soil aeration with its sprawling root system.

5. ROOT LAYER:

Arrowroot (*Maranta arundinacea*): Arrowroot is a starchy root vegetable that is prized for its edible tubers, which are often ground into flour. It grows best in tropical and subtropical climates and provides a gluten-free flour option for baking and cooking. Arrowroot is an excellent addition to the root layer for its culinary and medicinal uses.

Burdock (*Arctium lappa*): Burdock produces large, edible roots that are commonly used in traditional medicine and culinary dishes, such as soups and stews. The root has a mild, earthy flavor and is high in fiber and nutrients. It thrives in temperate and subtropical climates, making it a hardy and nutritious addition to the root layer.

6. GROUNDCOVER LAYER:

Water Spinach *(Ipomoea aquatica*): Water Spinach

is a tropical, edible leafy green that thrives in moist, swampy environments, making it ideal for damp areas of the food forest. It is rich in nutrients and can be harvested for its tender leaves. This plant is a fast grower, acting as an excellent groundcover in wetter spots.

Wild Mustard *(Sinapis arvensis)*: Wild Mustard produces edible leaves and seeds, which have a sharp, spicy flavor. It grows quickly as a groundcover in disturbed soils and provides a fast-growing green that attracts pollinators. Additionally, its seeds can be harvested for making mustard, contributing both culinary and ecological benefits.

7. VERTICAL LAYER:

Cape Gooseberry (*Physalis peruviana*): Cape Gooseberry, also known as Goldenberry, is a small fruit-bearing plant that produces edible, tangy berries encased in a papery husk. This vine grows well in vertical spaces and provides a nutritious addition to the food forest. The fruit can be eaten raw, added to jams, or used in sauces and desserts.

Moringa Vine (*Moringa oleifera*): Known for its nutritional value, Moringa is a fast-growing vine that provides edible leaves, flowers, and pods. Its leaves are rich in vitamins and can be used in salads or smoothies, while the pods are also edible when cooked. This plant's rapid growth makes it an excellent option for vertical growing and adding edible diversity to the food forest.

8. FUNGAL LAYER:

Ink Cap Mushroom (*Coprinopsis atramentaria*): The Ink Cap Mushroom is known for its black ink-like secretion when the mushroom is mature. It is edible when fresh and is often cooked in soups or used as a garnish. It contributes to the fungal layer by decomposing organic matter and providing a food source for other organisms in the food forest.

Honey Fungus (*Armillaria mellea*): Honey Fungus is an edible, parasitic fungus that typically grows on tree roots. While it can be destructive to trees, it also plays an essential role in breaking down decaying organic matter, enriching the food forest's soil. The honey-colored caps are edible when cooked and have a sweet, nutty flavor.

Zones 12 & 13 Plan Ten: Luminous Life

I created this food forest plan to focus on a laboratory of layered life, this design plan experiments with bold tropical pairings for maximum yield. It's built to provide a luminous display of edible abundance while pioneering sustainable techniques in the warmest zones.

1. CANOPY LAYER:

Tahitian Lime (*Citrus latifolia*): This citrus tree produces tangy, flavorful limes that are prized in culinary dishes and beverages. Both the fruit and zest are edible, with the fruit offering a rich source of vitamin C. Tahitian lime trees also contribute to soil health and provide shade to the understory layer.

Indian Jujube (*Ziziphus mauritiana*): Known for its small, sweet fruit, the Indian jujube tree is an excellent addition to a food forest. The fruit can be eaten fresh, dried, or used in jams and jellies. The tree thrives in tropical climates and offers shade while also providing a habitat for various forms of wildlife.

2. UNDERSTORY LAYER:

Mango Ginger (*Curcuma amada*): The rhizomes of this plant are edible and have a ginger-like taste, often used as a spice in cooking. It grows well in moist, shaded conditions, adding an edible root crop to the understory layer.

Fennel (*Foeniculum vulgare*): This herb is known for its aromatic fronds and edible bulb. It's used widely in cooking for its mild licorice flavor and can thrive in partial shade, adding both beauty and food value to the understory layer.

3. SHRUB LAYER:

Pigeon Pea (*Cajanus cajan*): This leguminous shrub produces edible, protein-rich pods and seeds. It is often used as a staple food in many tropical regions. The plant also fixes nitrogen, enriching the soil, and is a great companion plant for the food forest.

Fijian Long Bean (*Vigna unguiculata*): This climbing shrub produces long, edible beans that are commonly used in stir-fries and soups. It grows well in tropical and subtropical climates, producing large quantities of food for the food forest, and also provides nitrogen fixation benefits.

4. HERBACEOUS LAYER:

Chard (*Beta vulgaris*): Chard is a leafy green with edible stems and leaves, commonly used in Mediterranean dishes. It is rich in vitamins A, C, and K, and grows well in both sun and partial shade. Chard is a hardy herbaceous plant that thrives in tropical and subtropical climates, making it ideal for the food forest.

Malanga (*Xanthosoma sagittifolium*): Malanga is a tuber plant related to taro, producing edible, starchy roots that are often used in soups and stews. The plant thrives in warm, humid climates and can provide a valuable, easy-to-grow root crop for the food forest. Malanga is also a nutritious source of carbohydrates and dietary fiber.

5. ROOT LAYER:

Beetroot (*Beta vulgaris*): Beetroots produce edible, sweet roots that can be used in a variety of dishes, from salads to juices. The roots are high in vitamins and minerals, particularly iron. Beetroot grows well in temperate climates and can be used both as a root crop and for its greens, making it a versatile addition to the root layer.

Yam (*Dioscorea spp.*): Not to be confused with sweet potatoes, yams produce edible, starchy tubers that can be cooked in a variety of ways. The roots are rich in carbohydrates and can serve as a main source of energy in the food forest. Yams grow well in tropical and subtropical climates and can be a key food source in the root layer.

6. GROUND COVER LAYER:

Lemon Balm (*Melissa officinalis*): Lemon Balm is a fragrant herb with edible leaves used for teas, salads, and flavoring. It thrives as a groundcover in partial shade and well-drained soil. This plant attracts pollinators, making it a beneficial addition to the food forest's groundcover layer, while its culinary and medicinal uses add value to the ecosystem.

White Clover (*Trifolium repens*): White Clover's leaves, flowers, and young seed pods are edible, and it is widely used for groundcover due to its ability to fix nitrogen in the soil. This plant thrives in full sun and spreads quickly, helping to maintain soil fertility and prevent erosion. Its flowers attract pollinators, making it an excellent ecological asset to the food forest.

7. VERTICAL LAYER:

Malabar Tamarind (*Garcinia cambogia*): Malabar Tamarind is a tropical vine whose sour fruit is used in cooking and has potential medicinal benefits. The vine grows quickly, producing small fruits that are used in sauces or as a souring agent. This plant brings culinary value and unique flavor to the food forest's vertical layer.

Jasmine Vine (*Jasminum spp.*): While typically known for its fragrant flowers, certain varieties of Jasmine also produce edible parts. The flowers can be infused to create aromatic teas or used as flavoring in desserts. The plant's ability to grow vertically while attracting pollinators makes it an excellent addition to the food forest.

8. FUNGAL LAYER:

Earthstar Fungus (*Geastrum spp.*): Earthstar mushrooms are interesting fungi that release spores from a star-shaped structure. They can be used medicinally or in some cases, for culinary purposes. As part of the fungal layer, they contribute to soil health and enhance the decomposition process of organic matter.

Bolete Mushroom (*Boletus spp.*): Boletes are a group of edible mushrooms with spongy undersides instead of gills. Many species, such as the King Bolete, are highly prized for their flavor and meaty texture. These mushrooms decompose wood and organic material in the food forest, contributing to nutrient cycling and providing a diverse food source for foragers.

ABOUT THE AUTHOR

Audra Haldane is an agricultural engineer, author and passionate advocate for self-sufficiency whose expertise spans from the sprawling landscapes of Appalachia to the misty hills of Scotland, where she operates her own farms.

Known for her accessible, hands-on approach to homesteading, Audra's books—including The Homestead Blueprint: Introductory Self-Sufficiency With Homesteading, Advanced Homesteading Skills and Techniques, and Sustainable Homestead Design and Permaculture Principles—have become essential reads for those looking to cultivate independence, resilience, and sustainability.

When she's not tending to her farms or sharing practical homesteading wisdom, Audra indulges her love for travel and reading, constantly seeking inspiration from new places and stories. Readers from urban high-rises to remote countryside homes discover the joys of sustainable living, making her a beloved voice in the modern homesteading movement.

THANK YOU

Thank you for reading *The Food Forest Solution*, Volume Two!

I hope you found as much pleasure in this book as I did researching and designing the food forest plans presented here. If you enjoyed this book, please consider spreading your good word in the form of leaving a review and sharing on your socials. It's my goal to empower my readers in creating a food-secure environment and getting the word out about this book, and others related to homesteading and permaculture really help.

Thank you again, dear friends. Until next time!

Audra xx

INDEX

Symbols

B

D

F

P

Plants Featured In The Designs In This Book:

R

U

Z

www.ingramcontent.com/pod-product-compliance
Lightning Source LLC
LaVergne TN
LVHW052014200926
848481LV00005B/1488